Nanotechnology in Water Treatment Applications

Edited by

T. Eugene Cloete
Michele de Kwaadsteniet
Marelize Botes

Faculty of Science, Stellenbosch University, South Africa

and

J. Manuel López-Romero

Faculty of Science, University of Malaga, Spain

Caister Academic Press

Caister Academic Press
Norfolk, UK

www.caister.com

British Library Cataloguing-in-Publication Data
A catalogue record for this book is available from the British Library

ISBN: 978-1-904455-66-0

Cover image adapted from Figure 1.4

Printed and bound in Great Britain by Cromwell Press Group, Trowbridge, Wiltshire.

Contents

Contributors

Marelize Botes
Department of Microbiology
Stellenbosch University
South Africa

mbr@sun.ac.za

Mehmet Cakmakci
Istanbul Technical University
Civil Engineering Faculty
Environmental Engineering Department
Maslak
Istanbul
Turkey

cakmakci@itu.edu.tr

Thomas Eugene Cloete
Faculty of Science
Stellenbosch University
South Africa

eugenecloete@sun.ac.za

Elena Garrido-Pérez
Dep. Química Orgánica
Facultad de Ciencias
Universidad de Málaga
29071 Málaga
Spain

elengarrido@gmail.com

Jesús Hierrezuelo-León
Dep. Química Orgánica
Facultad de Ciencias
Universidad de Málaga
29071 Málaga
Spain

jhierrezuelo@uma.es

Ismail Koyuncu
Istanbul Technical University
Civil Engineering Faculty
Environmental Engineering Department
Maslak
Istanbul
Turkey

koyuncu@itu.edu.tr

Michele de Kwaadsteniet
Department of Microbiology
Stellenbosch University
South Africa

mdekwaad@sun.ac.za

J. Manuel López-Romero
Dep. Química Orgánica
Facultad de Ciencias
Universidad de Málaga
Málaga
Spain

jmromero@uma.es

Melanie Richards
Department of Microbiology and Plant Pathology
Faculty of Natural and Agricultural Sciences
University of Pretoria
South Africa

melanie.richards1@gmail.com

Eugene Smit
Electrospinning and Nanofiber Yarns Group
Department of Chemistry and Polymer Science
Stellenbosch University
South Africa

asmit@sun.ac.za

Jacques Theron
Department of Microbiology and Plant Pathology
Faculty of Natural and Agricultural Sciences
University of Pretoria
Pretoria
South Africa

jacques.theron@postino.up.ac.za

Joseph Adrian Walker
Department of Microbiology and Plant Pathology
Faculty of Natural and Agricultural Sciences
University of Pretoria
Pretoria
South Africa

joseph.walker@up.ac.za

Preface

The world is facing formidable challenges in meeting the rising demands of potable water as the available supplies of freshwater are decreasing due to extended droughts, population growth, more stringent health-based regulations, and competing demands from a variety of users. Moreover, increasing pollution of groundwater and surface water from a wide variety of industrial, municipal, and agricultural sources has seriously tainted water quality in these sources, effectively reducing the supply of freshwater for human. Although the nature of pollution problems may vary, they are typically due to inadequate sanitation, algal blooms fertilized by the phosphorus and nitrogen contained in human and animal wastes, detergents and fertilizers, pesticides, chemicals, heavy metals, salinity caused by widespread and inefficient irrigation, and high sediment loads resulting from upstream soil erosion.

Given the importance of potable water to people in both the developed and developing countries, and taking into account concerns regarding the viability of current practices of meeting the increasing demands of all water users, there is a clear need for the development of innovative new technologies and materials whereby challenges associated with the provision of safe potable water can be addressed. Although new approaches are continually being examined, these need to be lower in overall cost, durable and more effective than current options for the removal of contaminants from water, either *in situ* or in water purification systems. Nanotechnology has been identified as a technology that could play an important role in resolving or ameliorating many of the problems involving water purification and quality.

Nanotechnology, the engineering and art of manipulating matter at the nanoscale (1–100 nm), offers the potential of novel nanomaterials for treatment of surface water, groundwater, and wastewater contaminated by toxic metal ions, organic and inorganic solutes, and microorganisms. Owing to their unique activity towards recalcitrant contaminants and application flexibility, many nanomaterials are under active research and development.

The potential impact areas for nanotechnology are divided into three categories, i.e. treatment and remediation, sensing and detection, and pollution prevention. Within the category of treatment and remediation, nanotechnology has the potential to contribute to long-term water quality, availability, and viability of water resources, such as through the use of advanced filtration materials that enable greater water reuse, recycling, and desalinization. Within the category of sensing and detection, of particular interest is the development of new and enhanced sensors to detect biological and chemical contaminants at very low concentration levels in the environment, including water. Moreover, nanotechnology has the potential to facilitate the development of continuous monitoring devices capable of yielding real-time measurements at a lower cost and improved specificity.

While much attention has been focused on the development and potential benefits of nanomaterials in water treatment processes, concerns have also been raised regarding their potential human and environmental toxicity. Indeed, studies have indicated that the selfsame properties of nanomaterials that make them attractive (e.g. size, shape, structure, reactivity) may also cause them to be toxic. It is difficult, however, to assess the effect of nanomaterials on health and the environment because the methods and tools for such a task have not been well developed yet. In addition, common frameworks for risk research, risk assessment, and risk management are lacking at present. It is vital that these processes be developed and investigated to ensure that nanomaterials are as safe as possible, while reaching their full potential. Notwithstanding these information gaps, it is certain that new nanomaterials, particularly in water and wastewater treatment, will play key roles in ensuring sufficient and good quality water to meet the ever-increasing demand for potable water.

Accordingly, literature about current research on different nanomaterials (nanostructured catalytic membranes, nanosorbents, nanocatalysts, and bioactive nanoparticles) and their application in water treatment, purification and disinfection is reviewed in this book. Moreover, knowledge regarding toxicological effects of engineered nanomaterials on humans and the environment is presented.

T. Eugene Cloete

Nanotechnology and Water Treatment: Applications and Emerging Opportunities

1

Jacques Theron, Joseph A. Walker and T. Eugene Cloete

Abstract
Nanotechnology, the engineering and art of manipulating matter at the nanoscale (1–100 nm), offers the potential of novel nanomaterials for treatment of surface water, groundwater and wastewater contaminated by toxic metal ions, organic and inorganic solutes, and microorganisms. Due to their unique activity towards recalcitrant contaminants and application flexibility, many nanomaterials are under active research and development. Accordingly, literature about current research on different nanomaterials (nanostructured catalytic membranes, nanosorbents, nanocatalysts and bioactive nanoparticles) and their application in water treatment, purification and disinfection is reviewed. Moreover, knowledge regarding toxicological effects of engineered nanomaterials on humans and the environment is presented.

Introduction

There are about 1 billion people in the world, mostly in developing countries, who have no access to potable water, and a further 2.6 billion people lack access to adequate sanitation (WHO, 2004). In addition to the well-documented economic, social and environmental impacts of poor water supply and sanitation (Mara, 2003; Moore *et al.*, 2003; Johnson *et al.*, 2007; Montgomery and Elimelech, 2007), the health and welfare of people, especially of vulnerable groups such as children, the elderly and poor, are closely connected to the availability of adequate, safe and affordable water supplies (Theron and Cloete, 2002; Ashbolt, 2004; Eshelby, 2007). The world is facing formidable challenges in meeting the rising demands of potable water as the available supplies of freshwater are decreasing due to extended droughts, population growth, more stringent health-based regulations, and competing demands from a variety of users (Vörösmarty *et al.*, 2000; Lee and Schwab, 2005; Savage and Diallo, 2005; Moe and Rheitigans, 2006). Moreover, increasing pollution of groundwater and surface water from a wide variety of industrial, municipal and agricultural sources has seriously tainted water quality in these sources, effectively reducing the supply of freshwater for human use (Kemper, 2004; Foley *et al.*, 2005; Coetser *et al.*, 2007). Although the nature of pollution problems may vary, they are typically due to inadequate sanitation, algal blooms fertilized by the phosphorus and nitrogen contained in human and animal wastes, detergents and fertilizers, pesticides, chemicals, heavy metals, salinity caused by widespread and inefficient irrigation, and high sediment loads resulting from upstream soil erosion (Ritter *et al.*, 2002; Fawell and Nieuwenhuijsen, 2003; Mara, 2003; Rodriguez-Mozaz, *et al.*, 2004; Falconer and Humpage, 2005; Foley *et al.*, 2005).

Given the importance of potable water to people in both the developed and developing countries, and taking into account concerns regarding the viability of current practices of meeting the increasing demands of all water users (Weber, 2002), there is a clear need for the development of innovative new technologies and materials whereby challenges associated with the provision of safe potable water can be addressed. Although new approaches are continually being examined, these need to be lower in overall cost, durable and more effective than current options for the removal of contaminants from water, either *in situ* or in water purification systems. Nanotechnology has been identified as a technology that could play an important role in resolving or ameliorating many of the problems involving water purification and quality (Savage and Diallo, 2005; Bottero *et al.*, 2006). Nanotechnology encompasses the creation and utilization of materials, devices and systems at the level of atoms and molecules, cutting across such disciplines as chemistry, physics, biology, engineering and materials science (Masciangioli and Zhang, 2003). Nanomaterials and nanostructures have nanoscale dimensions that range from 1 to 100 nm, and often exhibit novel and significantly changed physical, chemical and biological properties. These are as a result of their structure, larger surface area per unit of volume and quantum effects that occur at the nanoscale. The potential impact areas for nanotechnology are divided into three categories, i.e. treatment and remediation, sensing and detection, and pollution prevention (Masciangioli and Zhang, 2003; Rickerby and Morrison, 2007; Vaseashta *et al.*, 2007). Within the category of treatment and remediation, nanotechnology has the potential to contribute to long-term water quality, availability and viability of water resources, such as through the use of advanced filtration materials that enable greater water re-use, recycling and desalinization. Within the category of sensing and detection, of particular interest is the development of new and enhanced sensors to detect biological and chemical contaminants at very low concentration levels in the environment, including water. Moreover, nanotechnology has the potential to facilitate the development of continuous monitoring devices capable of yielding real-time measurements at a lower cost and improved specificity (Riu *et al.*, 2006; Vaseashta *et al.*, 2007).

This review will highlight the uses of nanotechnology in areas relevant to water purification, including separation and reactive media for water filtration, as well as nanomaterials and nanoparticles for use in water bioremediation and disinfection. The toxic potential of engineered nanomaterials for human health and the environment is also discussed.

Nanomaterials and water filtration

Membrane processes are playing an important role for water purification, since conventional water treatment techniques such as sedimentation, flocculation, coagulation and activated carbon are not able to remove organic pollutants to meet laid down specifications (Viessman and Hammer, 1998; Strathmann, 2001). The types of membrane processes that are commonly used in water purification include microfiltration (MF), ultrafiltration (UF) and nanofiltration (NF) for water and wastewater treatment (Peltier *et al.*, 2003; Van der Bruggen and Vandercasteele, 2003; Ahmad *et al.*, 2004; Qin *et al.*, 2007), and nanofiltration and reverse osmosis (RO) for desalination and water reclamation (Van der Bruggen and Vandercasteele, 2002; Mohsen *et al.*, 2003; Walha *et al.*, 2007). Because membrane processes are considered as key components of advanced water purification and desalination technologies, there is a continuous search for new materials and water purification technologies. In this regard, nanomaterials (e.g. carbon nanotubes, nanoparticles and dendrimers) are contributing to the development of more efficient and cost-effective water filtration processes (Table 1.1).

Nanostructured membranes

Carbon nanotubes (CNTs) have been attracting interest since their discovery (Iijima, 1991). Although the use of carbon nanotube structures in separation technology has been proposed (Miller *et al.*, 2001), the building of macroscopic structures that have controlled geometric shapes, density and dimensions for specific filtration applications remained a challenge. Srivastava *et al.* (2004) reported the fabrication of carbon nanotube filters that consist entirely of hollow cylinders with radially aligned carbon nanotube walls. The cylindrical membranes were shown to remove bacterial pathogens (*Escherichia coli* and *Staphylococcus aureus*) and poliovirus Sabin 1 from contaminated water. These carbon nanotube filters were also re-usable and could be cleaned by ultrasonication or autoclaving. It is anticipated that the properties of carbon nanotube filters, which combine the

Table 1.1 Examples of nanostructured and nanoreactive membranes for use in water filtration

Membrane	Pollutant	Reference
Nanostructured membranes		
Carbon nanotubes	Bacteria and viruses	Srivastava *et al.* (2004)
Nanocapillary array membranes (NCA)	NT	Chatterjee *et al.* (2005)
Nanoreactive membranes		
Alumina membrane formed from A-alumoxane	Synthetic dyes	DeFriend *et al.* (2003)
Alumina membranes functionalized with poly(styrene sulphonate) or poly(allylamine hydrochloride)	Divalent cations	Stanton *et al.* (2003)
Silica and cellulose-based membranes functionalized with amino acid homopolymers	Metal ions	Bhattacharayya *et al.* (1998), Ritchie *et al.* (1999, 2001)
Polycarbonate track-etched membranes functionalized with amino acid homopolymers	Metal ions	Hollman and Bhattacharayya (2004)
Pt/Fe laden cellulose acetate film	Trichloroethylene (TCE)	Meyer *et al.* (2004)
Zero-valent Fe laden cellulose acetate membrane	TCE	Wu *et al.* (2005)
Ni/Fe or Pd/Fe laden polyacrylic acid/polyether sulphone composite membranes	TCE	Xu *et al.* (2005a)
Ni/Fe laden cellulose acetate membrane	TCE	Wu and Ritchie (2006)
Alumina or polymeric membranes with gold nanoparticles	4-nitrophenol	Dotzauer *et al.* (2006)
Polymer-impregnated ceramic TiO_2 filters	Polycyclic aromatic hydrocarbons (PAHs)	Arkas *et al.* (2006)
Polymer-impregnated ceramic alumina and silicon-carbon filters	Trihalogen methanes, PAHs, pesticide	Allabashi *et al.* (2007)

NT, not tested.

exceptional thermal and mechanical stability of nanotubes with the high surface area and cost-effective fabrication of the nanotube membranes, may in future allow them to compete with ceramic- and polymer-based separation membranes used commercially (Srivastava *et al.*, 2004). The use of nanofabrication methods to develop novel nanocapillary array (NCA) membranes with enhanced solute retention and reduced fouling tendency has also been reported (Chatterjee *et al.*, 2005). The novel NCA membranes consisted of poly(dimethoxysiloxane)-crossed microfluidic channels [14 mm × 100 μm (wide) × 60 μm (deep]) separated by a nanoporous polycarbonate track-etched membrane (30 nm pore diameter). It was proposed that the NCA membranes, which have tunable pore sizes and chemistry, could be used in future as functional materials and building blocks in a variety of applications, including water purification.

Nanoreactive membranes

In addition to the above nanostructured membranes, several reactive and functionalized membranes have also been developed for use in water filtration processes. DeFriend *et al.* (2003) reported the successful fabrication of alumina UF membranes, using alumina (A-alumoxane) nanoparticles that showed selectivity towards different synthetic dyes (i.e. Direct Red 81, Direct Blue 71 and Direct

Yellow 71). Subsequently, the fabrication of novel NF membranes by deposition of 4.5–5.0 layer pairs of poly(styrene sulphonate)/poly(allylamine hydrochloride) onto porous alumina has also been reported (Stanton *et al.*, 2003). These NF membranes exhibited high water flux, high retention of divalent cations [Ca(II) and Mg(II)] and Cl^-/SO_4^{2-} selectivity ratios up to 80. Polyelectrolytes or polyamines, which contain multifunctional chelating groups (e.g. amines, carboxylic acids or pyridines), form strong and stable interactions with metal ions (Rivas *et al.*, 2003), and procedures for their attachment on membrane internal pore surfaces have been studied previously for use in metal ions sorption (Bhattacharyya *et al.*, 1998; Ritchie *et al.*, 2001). Amino acids are a particularly interesting group of chelating species, since a primary amine and a carboxylic acid are available for metal ion binding. Consequently, homopolymers of amino acids may be used effectively for metal sorption (Autry *et al.*, 1995; Bhattacharyya *et al.*, 1998; Ritchie *et al.*, 1999, 2001). Novel membranes have been prepared by deposition of multilayers of charged polypeptides [poly(L-glutamic acid) or poly(L-lysine)] inside the pores of functionalized polycarbonate track-etched membranes, and these membranes were subsequently shown to display enhanced metal ion retention and permeate flux (Hollman and Bhattacharyya, 2004).

The application of electro-negative metals have been extensively studied for reductive dechlorination of halogenated organic compounds (HOCs) (Matheson and Tratnyek, 1994; Gotpagar *et al.*, 1997; Arnold and Roberts, 1998), but more recent publications have shown that much higher dehalogenation efficiencies can be obtained by using iron (Fe) nanoparticles (Zhang, 2003; Liu *et al.*, 2005a) or bimetallic nanoparticles (Zhang *et al.*, 1998a; Schrick *et al.*, 2002). Although the reactivity observed for especially Fe nanoparticles can be lost through rapid oxidation and hydrolysis (Giri *et al.*, 2001), the long-term stability of these nanoparticles can, however, be enhanced by immobilization in a support (Ponder *et al.*, 2000). Consequently, studies have been extended to the preparation of reactive membranes by incorporation of zero-valent iron (nZVI; Fe^0) nanoparticles into cellulose acetate films (Wu *et al.*, 2005). The nZVI nanoparticles were synthesized in a water–oil microemulsion, mixed with a cellulose acetate-acetone solution and then formed into a porous membrane by phase inversion. The reductive dehalogenation of trichloroethylene (TCE) in water by the nZVI laden cellulose acetate membranes was investigated and effective dechlorination of TCE could be demonstrated (Wu *et al.*, 2005). Moreover, Dotzauer *et al.* (2006) described a method for modifying alumina and polymeric membranes with gold nanoparticles through layer-by-layer adsorption of polyelectrolytes and citrate-stabilized gold nanoparticles. The modified membranes were subsequently shown to reduce more than 99% of 0.4 mM 4-nitrophenol to 4-aminophenol.

Redox-reactive membranes have also been prepared by incorporation of bimetallic Pt/Fe nanoparticles into cellulose acetate films (Meyer *et al.*, 2004), and through immobilization of Ni/Fe or Pd/Fe particles in a polyacrylic acid/polyether sulphone composite membrane matrix (Xu *et al.*, 2005a). In these studies, the metal laden membranes were shown to reduce TCE rapidly and completely with ethane as the only by-product. The enhanced performance of the bimetallic nanoparticles is attributed to Fe serving as the reductant for water to generate hydrogen, whilst the second metal acts as a catalyst (Schrick *et al.*, 2002). With the second catalytic metal, HOCs are preferentially dehalogenated by hydrogen reduction rather than electron transfer, which greatly increase the transformation rate and inhibits toxic intermediate product formation (Schrick *et al.*, 2002).

Polymer-supported filtration

Advances in macromolecular chemistry such as the synthesis of dendritic polymers have provided opportunities to refine, as well as to develop effective filtration processes for purification of water contaminated by different organic solutes and inorganic anions. Dendrimers are highly branched macromolecules with controlled composition and architecture, and monodispersity. They consist of a central core, repeating units and terminal functional groups (Bosman *et al.*, 1999; Tully and Frechet, 2001). The nature of the internal repeating units determines the microenvironment of the interior and consequently its solubilization properties, whilst the external groups determine the chemical behaviour of dendrimers in the external medium (Frechet and Tomalia, 2001; Klajnert and Bryszewska, 2001). Hyperbranched polymers maintain many of the architectural features and properties of dendrimers, but they are synthesized by one-step reactions forming polydisperse polymers of a non-perfectly symmetric shape (Jikei *et al.*, 2001; Gao and Yan, 2004; Yates and Hayes, 2004). In contrast, cyclodextrins are cyclic polysaccharide oligomers and have a truncated-cone shape that forms a well-defined cylindrical cavity (Schneiderman and Stalcup, 2000; Tick *et al.*, 2003; Del Valle, 2004). The interior of the cavity is lipophilic and can encapsulate various different organic compounds with suitable geometry and polarity, forming stable inclusion complexes. The outer surface of cyclodextrins has multiple hydroxyl groups, which are susceptible to further functionalization, thus leading to diversified molecules suitable for various applications, including the removal of pesticides (Sawicki and Mercier, 2006) and organic contaminants from water (Arkas *et al.*, 2006). Indeed, studies have established that modified poly(propylene imine) (Yates and Hayes, 2004) and poly(amidoamine) dendrimers (Diallo *et al.*, 2004), as well as polyethylene imine hyperbranched polymers (Arkas *et al.*, 2005) can effectively remove various organic molecules from water. Moreover, dendritic polymers have also been used successfully as delivery vehicles or scaffolds for antimicrobial agents such as ionic silver (Ag(I)) and quaternary ammonium chlorides (Balogh *et al.*, 2001; Chen and Cooper, 2002).

A dendrimer-enhanced ultrafiltration (DEUF) process for recovering metal ions from aqueous solutions has been reported (Diallo *et al.*, 2004; Diallo *et al.*, 2005) (Fig. 1.1). Using poly(amidoamine) dendrimers, separation of dendrimer–Cu(II) complexes from aqueous solutions was achieved simply by UF membranes with the appropriate molecular weight cut-off (MWCO). The metal ion laden dendrimers could be regenerated by decreasing the solution pH to 4.0, thus enabling the recovery of the bound Cu(II) ions and recycling of the dendritic polymer (Diallo *et al.*, 2005). Based on the very low tendency of the dendrimers to foul commercially available regenerated

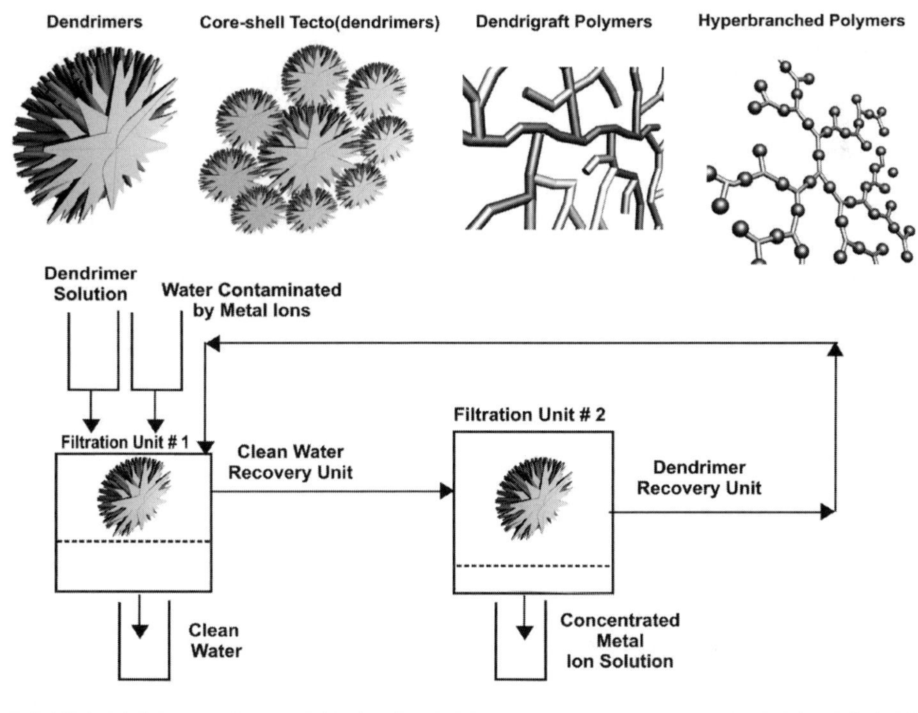

Figure 1.1 Recovery of metal ions from aqueous solutions by dendrimer-enhanced filtration.

cellulose membranes and on their small intrinsic viscosities, it was concluded that a comparatively smaller operating pressure and energy consumption could be achieved with dendritic polymers in tangential/cross-flow UF systems, which are typically used in water purification (Diallo *et al.*, 2005). The development of a method that permits removal of organic pollutants by employing a simple filtration step has been described by Arkas *et al.* (2006). This was pursued by utilizing titanium dioxide (TiO_2) porous ceramic filters of which the pores were impregnated with an alkylated poly(propylene imine) dendrimer, poly(ethylene imine) hyperbranched polymer or β-cyclodextrin, thus resulting in hybrid organic/inorganic filter modules of high mechanical strength and high surface area. These filters were subsequently tested for the purification of water by continuous filtration experiments, employing a variety of water pollutants. Polycyclic aromatic hydrocarbons (PAHs) could be removed very efficiently (more than 95%). Moreover, representatives of different pollutant groups [i.e. trihalogen methanes (THMs), monoaromatic hydrocarbons (benzene, toluene, xylene)] and pesticides (simazine) were also removed efficiently (in excess of 80%), but the filters became saturated considerably faster in such instances (Arkas *et al.*, 2006). These studies have been extended through the use of ceramic membranes made from alumina (A_2O_3) and silicon-carbon (SiC) impregnated with cross-linked silylated dendritic or cyclodextrin polymers (Allabashi *et al.*, 2007). A comparison of the silylated polymers performance with the aforementioned alkyl substituted polymers revealed that the latter were better suited for PAH sorption, but that silylated cyclodextrins were able to remove a wider range of pollutants.

Nanotechnologies for water remediation

The use of metals and chemicals in process industries has resulted in the generation of large quantities of effluent that contain high levels of toxic heavy metals, whereas mining and mineral processing operations also generate toxic liquid wastes (Coetser *et al.*, 2007). Moreover, groundwater is often contaminated by spills, agricultural practices, past waste disposal practices and leaking underground storage tanks (Kemper, 2004; Foley *et al.*, 2005). The presence of different organic and heavy metal contaminants in these, as well as other environmental water sources has a large environmental, public health and economic impact. In addition to highly toxic metallic cations such as arsenic (As), lead (Pb), chromium (Cr), cadmium (Cd) and mercury (Hg) (Järup, 2003; Lee *et al.*, 2005a), inorganic anions such as nitrate (NO_3^-) and perchlorate (ClO_4^-) are also of concern. Whereas high levels of nitrates in drinking water may be harmful to newborn babies and contribute to cancer (Kapoor and Viraraghavan, 1997; Yang and Lee, 2005), perchlorate, which has emerged as a high profile contaminant and has consequently received considerable regulatory attention (Urbansky and Schock, 1999; Cheng *et al.*, 2007), has been reported to lead to hypothyroidism in adults (Blount *et al.*, 2006). Moreover, almost all of the chlorinated aromatic compounds exhibit high toxicity and several of these compounds are carcinogenic (Xu *et al.*, 2005b). Notably, trichloroethene (TCE) is considered to be one of the most common hazardous organic contaminants found in groundwater and has been linked to liver damage and impaired pregnancies in humans (Nutt *et al.*, 2006). In addition, halogenated aromatic compounds have become a serious environment contamination problem because they have a long life, are chemically stable and are non-biodegradable (Keum and Li, 2004).

Most of the traditional technologies such as solvent extraction, activated carbon adsorption, and common chemical oxidation, whilst effective, very often are costly and time-consuming (Pattersom, 1985; Langwaldt and Puhakka, 2000; Schwarzenbach *et al.*, 2006). Biological degradation is environmentally friendly and cost-effective; but it is usually time-consuming (Pattersom, 1985; Ahluwalia and Goyal, 2007). Thus, the ability to remove toxic contaminants from these environments to a safe level and doing so rapidly, efficiently and within reasonable costs is important (Savage and Diallo, 2005). Nanotechnology could play an important role in this regard. An active emerging area of research is the development of novel nanomaterials with increased affinity, capacity and selectivity for heavy metals and other contaminants. The benefits from use of nanomaterials may derive from their enhanced reactivity, surface area and sequestration characteristics (Zhang, 2003; Li *et al.*, 2006b). A variety of nanomaterials are in various stages of research and development, each possessing unique functionalities that is potentially applicable to the remediation of industrial effluents, groundwater, surface water and drinking water (Table 1.2).

Zeolites

Zeolites are inorganic crystalline porous materials that have a highly ordered structure and are generally comprised of silicon (Si), aluminium (Al) and oxygen (Breck, 1974; Camblor *et al.*, 1998). Their physicochemical characteristics such as high mechanical and chemical resistance in addition to their high surface area, have formed the basis for their widespread use in catalysis, separation and ion-exchange (Vogel *et al.*, 2002; Sakaguchi *et al.*, 2005; Song *et al.*, 2005a; Tavolaro *et al.*, 2007). Although conventional synthesis methods produce zeolites on the scale of 1 to 10 μm, nanoscale zeolites, which comprise discrete, uniform crystals with dimensions ranging from 5 to 100 nm, have been synthesized successfully (Camblor *et al.*, 1998; Vogel *et al.*, 2002; Song *et al.*, 2004; 2005a; Bell *et al.*, 2006; Ding and Zeng, 2007). Compared to micron-scale zeolites, the nanocrystalline zeolites exhibit greater external surface areas, smaller diffusion path lengths and a greater aversion to coke formation (Vogel *et al.*, 2002; Song *et al.*, 2005a). The Si/Al ratio of zeolites can be varied either during synthesis or post-synthetically and, in general, zeolites have a low Si/Al ratio and therefore a high ion-exchange capacity. Conversely, if the Al content and thus the ion-exchange capacity of a zeolite is reduced, it becomes more hydrophobic or organophilic in its adsorptive characteristics (Tavolaro *et al.*, 2007).

In contrast to nanocrystalline zeolites, micron-scale zeolytes have been tested for a variety of environmental applications. NaP1 zeolites, which can be synthesized inexpensively by alkaline hydrothermal activation of coal fly ash in 1–2 M NaOH solutions (Moreno *et al.*, 2001), have been used successfully to remove heavy metals from acid mine wastewaters (Moreno *et al.*, 2001) and to remove Cr(III), Ni(II), Zn(II), Cu(II) and Cd(II) from metal electroplating wastewaters (Alvarez-Ayuso *et al.*, 2003). Moreover, surfactant-modified zeolites (SMZ) have also been used to not only remediate water containing the radioactive species ^{137}Cs and ^{90}Sr found in nuclear plant wastewater, but also to adsorb both tetrachloroethene and chromate (CrO_4^{2-}) from groundwater, and to remove petroleum

Table 1.2 Examples of nanoparticles and nanomaterials for use in water remediation

Nanoparticle/nanomaterial	Pollutant	Reference
Nanocrystalline zeolites	Toluene, nitrogen dioxide	Song *et al.* (2004, 2005b)
Carbonaceous nanomaterials		
Activated carbon fibres (ACFs)	Benzene, toluene, xylene, ethylbenzene	Mangun *et al.* (2001)
CeO_2-carbon nanotubes (CNTs)	Heavy metal ions	Peng *et al.* (2005), Di *et al.* (2006)
CNTs functionalized with polymers	*p*-Nitrophenol	Salipira *et al.* (2007
CNT functionalized with Fe	Benzene, toluene, dimethylbenzene	Jin *et al.* (2007)
Single-walled carbon nanotubes	Heavy metal ions	Lu and Chiu (2006)
	Trihalomethanes (THMs)	Lu *et al.* (2005)
Multiwalled carbon nanotubes	Heavy metal ions	Li *et al.* (2003, 2006b), Lu and Liu (2006)
	THMs	Lu *et al.* (2006a)

Table 1.2 Continued

Nanoparticle/nanomaterial	Pollutant	Reference
	Chlorophenols	Cai *et al.* (2005)
	Herbicides	Zhou *et al.* (2006a,b, 2007)
	Microcystin toxins	Yan *et al.* (2006)
Self-assembled monolayer on mesoporous supports (SAMMS)		
Anion-SAMMS	Inorganic ions	Kelly *et al.* (2001)
Thiol-SAMMS	Heavy metal ions	Mattigod (1999, 2006)
HOPO-SAMMS	Actinides and lanthanides	Fryxell *et al.* (2004, 2005), Yantasee *et al.* (2005), Lin *et al.* (2005)
Biopolymers	Heavy metal ions	Kostal *et al.* (2001, 2003)
Single-enzyme nanoparticles (SENs)	NT	Kim and Grate (2003, 2005)
Zero-valent iron nanoparticles (nZVI)	Polychlorinated biphenyls (PCBs)	He and Zhao (2005)
	Inorganic ions	Choe *et al.* (2000), Cao *et al.* (2005), Yang and Lee (2005)
	Chlorinated organic compounds	Wang and Zhang (1997), Zhang *et al.* (1998a). Cheng *et al.* (2007)
	Heavy metal ions	Ponder *et al.* (2000), Kanel *et al.* (2005, 2006), Li and Zhang (2006), Xu and Zhao (2007)
Bimetallic nanoparticles		
Pd/Fe nanoparticles	PCBs	Wang and Zhang (1997)
	Chlorinated ethene	Zhang *et al.* (1998a)
	Chlorinated methanes	Lien and Zhang (1999, 2001)
	TCE and PCBs	He and Zhao (2005)
	Dichlorophenol	Wei *et al.* (2006)
	Triclorobenzene	Zhu *et al.* (2006a)
Ni/Fe nanoparticles	Chlorinated ethene	Schrick *et al.* (2002)
	Brominated organic compounds (BOCs)	Lim *et al.* (2007)
Pd/Au nanoparticles	TCE	Nutt *et al.* (2005, 2006)
TiO_2 photocatalysts		
Nanocrystalline TiO_2	Heavy metal ions	Pena *et al.* (2005)
Nitrogen (N)-doped TiO_2	Azo dyes	Liu *et al.* (2005b)
Fe(III)-doped TiO_2	Phenol	Nahar *et al.* (2006)
Supported TiO_2 nanoparticles	Aromatic pollutants	López-Munoz *et al.* (2005)
TiO_2-based *p–n* junction nanotubes	Toluene	Chen *et al.* (2005)

NT, not tested.

hydrocarbons such as benzene, toluene, ethylbenzene and xylene from oilfield wastewater (Bowman, 2002). In contrast, nanocrystalline zeolites appear to be in the research phase and no specific water remediation processes have been proposed. The synthesis of nanocrystalline NaY zeolites capable of absorbing ca. 10% more toluene and ca. 30% more nitrogen dioxide relative to commercial NaY zeolites have, however, been described (Song *et al.*, 2005b). In addition, nanocrystalline ZSM-5 zeolites, with a particle size of 15 nm, have been reported to adsorb ca. 50% more toluene than ZSM-5 samples with larger particle sizes (Song *et al.*, 2004), furthermore suggesting that such nanoscale zeolites may be potentially superior to traditional micron-scale zeolite sorbents.

Carbon nanotubes

Carbon nanotubes (CNTs) are cylinder-shaped macromolecules of which the walls of the tubes are made up of a hexagonal lattice of carbon atoms and they are capped at their ends by one half of a fullerene-like molecule (Iijima, 1991). CNTs can be divided essentially into single-walled carbon nanotubes (SWCNTs) and multi-walled carbon nanotubes (MWCNTs) based on the principle of hybridized carbon atom layers in the walls of CNTs (Fig. 1.2). Whereas SWCNTs have diameters ranging from 0.3 to 3 nm, the MWCNTs are composed of a concentric arrangement of many cylinders and can reach diameters of up to 100 nm (Iijima and Ichihashi, 1993; Balasubramanian and Burghard, 2005). Since their discovery, carbon nanotubes have attracted considerable attention which stems from their outstanding structural, mechanical and chemical properties (Popov, 2004; Miyagawa *et al.*, 2005). In addition to their potential application in electronics, chemical and biological sensing, catalysis and reinforced composite materials (Balasubramanian and Burghard, 2005, 2006; Polizu *et al.*, 2006), their large surface area and tubular structure also make CNTs a promising adsorbent material.

Carbon nanotubes (CNTs) have been evaluated for the sorption of 1,2-dichlorobenzene (DCB) and it was reported that they can be used in a wide pH range of 3 to 10 and that sorption of DCB onto the CNTs took 40 min to reach equilibrium with a maximum sorption capacity of 30.8 mg/g (Peng *et al.*, 2003). More recently, novel sorbents with high surface areas (in excess of 189 m^2/g) have been developed that consists of cerium oxide supported on either carbon nanotubes (CeO_2-CNT) (Peng *et al.*, 2005) or on aligned carbon nanotubes (CeO_2-ACNTs) (Di *et al.*, 2006). Although the CeO_2-CNTs were effective sorbents for As(V), addition of the divalent cations Ca(II) and Mg(II)

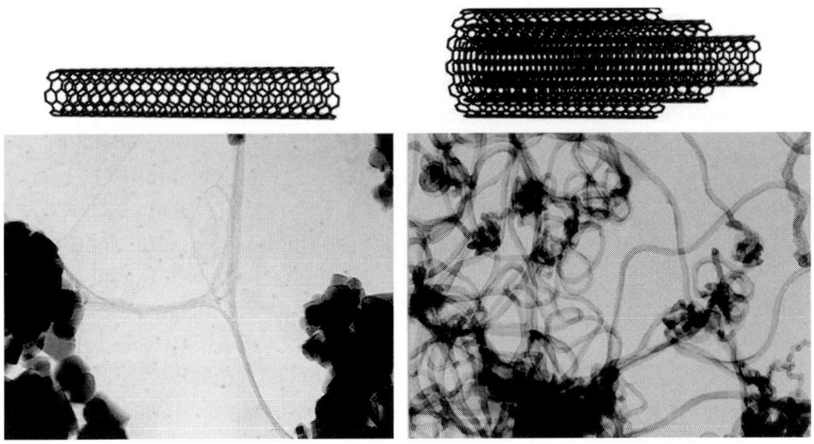

Figure 1.2 Molecular representations of single-walled carbon nanotubes (SWCNTs; top left) and multiwalled carbon nanotubes (MWCNTs; top right) with typical transmission electron micrographs below (Donaldson, K. *et al.*, Carbon nanotubes: A review of their properties in relation to pulmonary toxicology and workplace safety. Toxicol. Sci. 2006, *92*, 5–22, by permission of Oxford University Press).

at concentrations ranging between 1–10 mg/l was reported to increase the amount of sorbed As(V) (from 10 to 82 mg/g) (Peng *et al.*, 2005). The CeO_2-ACNTs exhibited high Cr(VI) adsorption efficiency from drinking water at a pH range of 3–7.4. The largest Cr(VI) adsorption capacity reached was 30.2 mg/g at pH 7.0, which was *ca.* two times higher than that of activated carbon and Al_2O_3 (Di *et al.*, 2006). Nanoporous activated carbon fibres (ACFs), prepared by electrospinning of CNTs, with surface areas ranging from 171 to 483 m^2/g, have been synthesized. Their organic sorption capability was compared to granular activated carbon and in all cases, the ACFs had much higher organic sorption equilibrium constants for benzene, toluene, xylene and ethylbenzene than granular activated carbon (Mangun *et al.*, 2001).

Multi-walled carbon nanotubes (MWCNTs), pre-treated with nitric acid, have been used successfully for the sorption of different heavy metal ions, including Pb(II) (97.08 mg/g), Cu(II) (24.49 mg/g) and Cd(II) (10.86 mg/g) from an aqueous solution. These sorption capacities were three and four times higher than those of powder and granular activated carbon, respectively (Li *et al.*, 2003). In addition, SWCNTs and MWCNTs were shown to have better Ni(II) sorption properties following their oxidation with NaClO. The treatment improved the surface properties of the CNTs, e.g. functional groups, total acidic sites and negatively charged carbons, thus resulting in them becoming more hydrophilic and therefore able to sorb more of the Ni(II) from the aqueous solution (Lu and Liu, 2006). Moreover, both SWCNTs (Lu *et al.*, 2005) and MWCNTs (Lu *et al.*, 2006a) have been investigated for their ability to remove trihalomethanes (THMs; $CHCl_3$, $CHBrCl_2$, $CHBr_2Cl$ and $CHBr_3$) from water. These compounds, which are recognized as potentially carcinogenic substances (Bull *et al.*, 1995), are formed during the chlorination of drinking water (Rook, 1974). Both CNTs displayed high adsorption capacities for the respective THMs at 5 and 15°C, which fluctuated very little in the pH range 3 to 7. Compared to activated carbon, the purified CNTs possessed two to three times higher adsorption capacities for $CHCl_3$, which accounts for a major portion of THMs in chlorinated drinking water (Lu *et al.*, 2006a). Recently, Salipira *et al.* (2007) reported that co-polymerization of cross-linked nanoporous polymers with functionalized CNTs resulted in novel polymers capable of efficient removal of organic pollutants from water. These novel polymers removed *p*-nitrophenol by 99% from a 10 mg/l spiked water sample, whereas granular activated carbon removed only 47%. Moreover, these polymers removed trichloroethylene (10 mg/l spiked sample) to non-detectable levels (detection limit <0.01 parts per billion (ppb)) compared to 55% for granular activated carbon (Salipira *et al.*, 2007).

In addition to their ability to sorb heavy metals and organic compounds, CNTs may also be useful as adsorbents for toxins and herbicides in environmental water. Based on a report by Long and Yang (2001), indicating that significantly higher dioxin removal efficiency is found with CNTs than with activated carbon, MWCNTs have subsequently been used successfully as the sorbent for the preconcentration and separation of chlorophenols (Cai *et al.*, 2005) and different herbicides, including triazine herbicides (Zhou *et al.*, 2006b), DDT and its metabolites (Zhou *et al.*, 2006a), as well as sulphonylurea herbicides (Zhou *et al.*, 2007), from river, tap and lake water samples. The ability of CNTs to adsorb MC-RR and LR, the most prevalent microcystin (MC) toxins produced by cyanobacterial bloom, has also been evaluated. Cyanobacteria blooms in natural waters have become a growing environmental problem worldwide due to the increased discharge of nitrogen- and phosphorus-containing wastewater into rivers and lakes, and MC-RR and LR have been found in fresh water bodies as well as in oceans (Haider *et al.*, 2003). Generally, MCs are very stable in the water body and resistant to removal from drinking water by traditional water treatment technology (Falconer, 1999; Haider *et al.*, 2003). Compared with wood-based activated carbon and clays (sepiolite, kaolinite and talc), the adsorption amounts of MCs in lake water by CNTs, of which the outer diameters ranged from 2 to 10 nm, were four times higher (Yan *et al.*, 2006), suggesting that CNTs may be a potentially promising nanomaterial for the removal of MCs from drinking water.

Whilst the above studies indicate that CNTs are potentially efficient adsorbents for a variety of pollutants in both drinking and environmental waters, their practical application may be hampered by their high cost, as well as their poor solubility and the difficulty in collecting them from their dispersing medium. It is therefore interesting to note that in a recent study, Lu *et al.* (2007) performed statistical analyses on the replacement cost of NaClO-oxidized SWCNTs and MWCNTs, both of which had been reported to be effective Zn(II) sorbents and could be re-used through 10 cycles of water treatment and regeneration (Lu and Chiu, 2006; Lu and Liu, 2006). The results of their analyses revealed the use of such re-usable carbonaceous sorbents can indeed be cost-effective in spite of their high unit cost at the present time (Lu *et al.*, 2007). Moreover, Jin *et al.* (2007) reported a simple method for fabricating magnetic Fe nanoparticle functionalized MWCNTs, which not only displayed improved water solubility, but could also be recovered easily from the water by magnetic separation. The functionalized water-soluble MWCNTs (Fe-MWCNT-CH_2COONa) were also tested as a sorbent using four model aromatic compounds. The results indicated that addition of 10 mg of Fe-MWCNT-CH_2COONa to aqueous solutions containing 8.6 ppb of the corresponding model compound, adsorbed benzene (79%), toluene (81%), dimethylbenzene (83%) and styrene (88%) very effectively. Notably, after being washed several times with methanol and dried in vacuum, the Fe-MWCNT-CH_2COONa could be re-used.

Self-assembled monolayer on mesoporous supports (SAMMS)

Self-assembled monolayers on mesoporous supports (SAMMS) have emerged as a powerful new class of environmental sorbent materials and can most simply be described as functionalized nanoporous ceramics (Mattigod, 1999). SAMMS are made via surfactant-templated synthesis and following calcination to remove the surfactant, nanoporous ceramic oxides are produced with very large surface areas (ca. 1 000 m^2/g) and a high density of sorption sites. These can be functionalized through molecular self-assembly to increase their selectivity towards target pollutants (Mattigod, 1999; Liu *et al.*, 2003; Fryxell *et al.*, 2005). The overall ceramic structure resembles a hexagonal honeycomb and the monolayers, formed within the porous surfaces, adsorb or bind molecules. This ability to alter the exposed functional group of the monolayer allows this class of sorbent materials to potentially remediate a broad range of contaminants (Fryxell *et al.*, 2005).

Different classes of SAMMS have been developed through fine-tuning of their surface chemistry to selectively sequester a specific target species. Anion-SAMMS or metal-capped ethylenediamine (EDA) SAMMS have been developed to absorb anions, such as chromate (CrO_4^{2-}) and arsenate (AsO_4^{3-}) (Kelly *et al.*, 2001). Cu-FC-EDA-SAMMS, a type of anion-SAMMS, has been further functionalized to bind radiocaesium, which is present in fission product waste solutions, by immobilization of copper ferrocyanide (Cu-FC) into the SAMMS matrix (Lin *et al.*, 2001). A variety of phosphonate and hydroxypyridinone (HOPO) functionalized SAMMS have also been synthesized and shown to sequester actinides and lanthanides (Fryxell *et al.*, 2004; 2005; Yantasee *et al.*, 2005; Lin *et al.*, 2005). Thiol-SAMMS have been designed mainly for the sequestration of mercury (Hg(II)), but can also bind metallic cations, such as Ag(I), Cd(II), Pb(II) and Cu(II) (Mattigod, 1999). Tests conducted using samples of different aqueous and non-aqueous waste streams indicated that thiol-SAMMS can achieve mercury loading as high as *ca.* 635 mg/g with a high affinity. The mercury was absorbed very rapidly, with about 99.9% adsorption occurring within the first five minutes. Experimental data furthermore indicated that mercury adsorption performance was not significantly affected by pH, ionic strength or the presence of other cations and complexing anions in the solution (Mattigod *et al.*, 2006).

Biopolymers

Advances in genetic and protein engineering techniques have enabled the construction of synthetic gene templates encoding biopolymers of which the sizes, compositions and functions at the molecular level can be controlled precisely. In this manner, it has become possible to specifically design protein-based nano-biomaterials with both metal-binding and tunable properties (Kostal *et al.*, 2005). Selected metal-binding proteins may also be fused readily to the tunable biopolymers in order to provide different metal-binding functionalities with enhanced specificity and affinity compared to chemical sorbents (Kostal *et al.*, 2001, 2003). Moreover, it provides for a cost-effective approach, since mass quantities of the biopolymer can be produced by overexpression in *Escherichia coli*. The use of tunable metal-binding biopolymers, based on elastin-like polypeptides (ELP), is emerging as a promising technology for heavy metal remediation from aqueous wastes (Kostal *et al.*, 2001, 2003). ELPs are biopolymers that consist of the repeating pentapeptide Val-Pro-Gly-Val-Gly and posses structurally similar characteristics to the mammalian protein elastin. The most interesting feature of ELPs is their ability to undergo reversible phase transition under a range of temperature, pH and ionic strength (Urry *et al.*, 1992), which are properties ideal for recovery of the metal–polymer complexes (Fig. 1.3). The phase transition temperature can be controlled precisely by varying the chain length and peptide sequence (Kostal *et al.*, 2001, 2003). Thus, tuning the process temperature allows for reversible network formation between the individual polymers that subsequently enables the recovery of sequestered metals by precipitation. After extraction of the metals, the aggregated polymers can be re-solubilized and re-used for subsequent remediation cycles (Kostal *et al.*, 2005).

Following synthesis of an ELP containing a terminal hexahistidine domain, Kostal *et al.* (2001) reported the biopolymers bound to cadmium (Cd(II)) at a 1:1 ratio. The addition of six extra histidines increased the cadmium-binding capacity to a ratio of 1.5:1, suggesting that the metal-binding capability of the biopolymers can be easily modified. Regeneration of the metal binding sites for repeated usage was achieved by incubating the biopolymer–cadmium complexes in acetate buffer (pH 4). The regenerated biopolymer aggregates remained fully functional even after four repeating cycles of cadmium binding (Kostal *et al.*, 2001). In addition, ELPs with either polyhistidine moieties (Prabhukumar *et al.*, 2004) or a synthetic phytochelatin metal-binding domain (Lao *et al.*, 2007) have been reported to remediate cadmium from contaminated soil *ex situ*. Compared with biosurfactants, which adsorb extensively to soil, less than 20% of the respective ELP biopolymers were adsorbed under the washing conditions, resulting in a significantly lower concentration of biopolymer required to achieve higher extraction efficiencies. In addition, Kostal *et al.* (2003) created a tunable biopolymer with a specific affinity for mercury (Hg(II)). A bacterial metallo-regulatory protein, MerR, which binds mercury with high affinity and selectivity (O'Halloran *et al.*, 1989), was fused to ELP to provide the highly selective nature of the biopolymers (Kostal *et al.*, 2003). The synthesized biopolymers demonstrated the ability to selectively remediate mercury in buffered solutions to less than 2 ppb levels with minimal binding of competing cations (cadmium, nickel and zinc) at 100-fold excess. Following separation, the mercury could be removed and the biopolymers recycled. The ability of the biopolymers to remediate mercury-contaminated lake water, which has a higher pH and turbidity, was also tested and it was found that the efficiency was maintained (Kostal *et al.*, 2003).

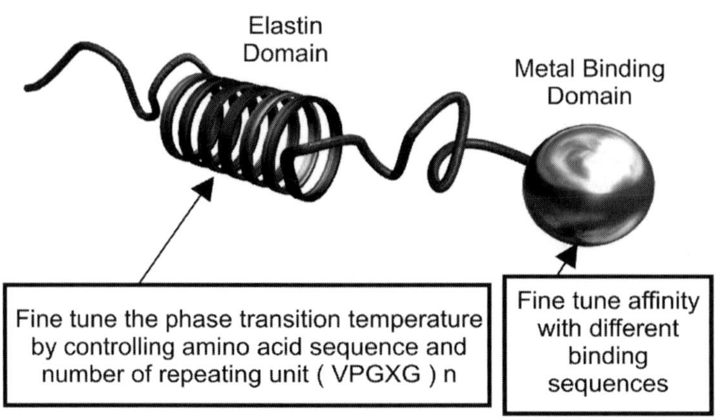

Figure 1.3 Customization of metal-binding and phase transition properties of biopolymers.

Single-enzyme nanoparticles (SENs)

Owing to their specificity and targeted effectiveness, enzymes may offer vast capabilities in the areas of chemical conversions, biosensing and bioremediation (Duran and Esposito, 2000; Schmid *et al.*, 2001). However, the lack of stability and relatively short catalytic lifetimes of enzymes may limit their usefulness as cost-effective alternatives to synthetic catalysts (Burton *et al.*, 2002). Although methods such as enzyme immobilization, enzyme modification and genetic modification have been used frequently to improve the stability and subsequent persistence of enzymes, nanotechnology has provided a new method whereby enzymes can be stabilized in the form of single-enzyme nanoparticles (SENs). The first SENs were assembled by Kim and Grate (2003), using chymotrypsin as a model enzyme. The synthesis of SENs involves enzyme surface modification, vinyl polymer growth from the enzyme surface, and finally an orthogonal polymerization by silanol condensation reactions that cross-link the attached polymer chains into a network around the enzyme (Kim and Grate, 2003). Not only does the hybrid organic/inorganic network of nanometre-scale thickness stabilize the enzyme, but it is sufficiently porous to allow substrates to access the active site of the enzyme. The yields of enzyme activity in the form of SENs was reported to be up to 73%, indicating that the catalytic stability and activity of chymotrypsin was not lost in the form of SENs. Moreover, results indicated that that the half-life of chymotrypsin activity in the form of SENs could reach up to 143 days (Kim and Grate, 2003). Although research has thus far only been conducted on the enzymes chymotrypsin and trypsin, the technology is transferable across all enzymes and it is thus envisaged that the use of SENs may offer potential decontamination methods for a broad range of organic contaminants (Kim and Grate, 2005). However, enzyme stability and lifetime may prove to be a cost-limiting factor for large-scale remedial purposes.

Zero-valent iron nanoparticles (nZVI)

Iron nanoparticle technology, which represents one of the first generation nanoscale environmental technologies (Wang and Zhang, 1997), has been receiving considerable attention for its potential applications in groundwater treatment and site remediation. The greatest interest, however, is in the use of zero-valent iron nanoparticles (Fe^0; nZVI) (Zhang, 2003; Tratnyek and Johnson, 2006; Zhang and Elliot, 2006) (Fig. 1.4). It is assumed that the reactivity of core-shell nanoparticles is driven by oxidation of the Fe^0 core (Liu *et al.*, 2005a; Nurmi *et al.*, 2005). Compared to their microscale counterparts, the higher reactivity of the nanoscale particles has been attributed to the greater density and

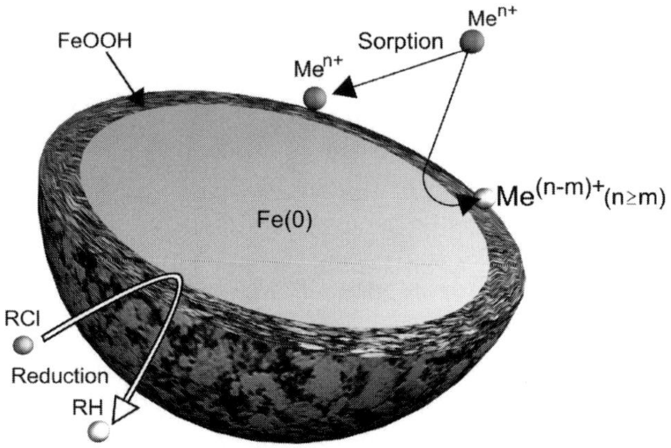

Figure 1.4 The core–shell model of zero-valent iron nanoparticles (nZVI). The core consists of mainly metallic iron, whilst the shell consists mostly of iron oxides and hydroxides. Thus, iron nanoparticles exhibit characteristics of both iron oxides (e.g. as a sorbent) and metallic iron (e.g. as a reductant).

higher intrinsic reactivity of their reactive surface sites (Klabunde *et al.*, 1996; Zhang, 2003; Nurmi *et al.*, 2005). Over the last few years, various methods have been developed for the production of nZVI particles (Li *et al.*, 2006a,b; Sun *et al.*, 2006) and the modification of their surface properties (Ponder *et al.*, 2000; He and Zhao, 2005). Extensive laboratory studies have demonstrated that nZVI particles are effective for the transformation of a wide array of common environmental contaminants such as chlorinated organic compounds (Wang and Zhang, 1997; Zhang *et al.*, 1998a; Cheng *et al.*, 2007), polychlorinated biphenyls (PCBs) (He and Zhao, 2005), inorganic ions such as nitrate (NO_3^-) and perchlorate (ClO_4^-) (Choe *et al.*, 2000, 2005) and heavy metal ions such as As(III), Pb(II), Cu(II), Ni(II) and Cr(VI) (Ponder *et al.*, 2000; Kanel *et al.*, 2005; Li and Zhang, 2006). Moreover, several field tests have demonstrated the promising prospective for *in situ* remediation (Elliot and Zhang, 2001; Li *et al.*, 2006b; Zhang and Elliot, 2006). Typically, nZVI particle-water slurries are injected under pressure and/or by gravity to the contaminated plume where treatment is needed. The nZVI particles can also remain in suspension for extended periods of time to establish an *in situ* treatment zone. As an alternative, nZVI particles may also be anchored onto a solid matrix such as activated carbon and/or zeolite for enhanced treatment of water and wastewater (Zhang, 2003).

Several studies have provided insights into nZVI particle properties associated with its potential to transform metal ions, as well as inorganic anions like perchlorate and nitrate. Similar to results obtained with As(III) (Kanel *et al.*, 2005), Kanel *et al.* (2006) reported that absorption of As(V) to nZVI particles was rapid, occurring within minutes. Approximately 25% of the As(V) was reduced to As(III) by the nZVI particles after 90 days. However, it was reported that in the presence of competing ions such as HCO_3^-, SiO_4^{2-} and PO_4^{3-}, the As(V) adsorption was reduced and thus a larger amount of nZVI particles may be required to remove As(III) and As(V) from aqueous solutions. Whilst reduction appears to be the predominant mechanism for metal ion removal by nZVI particles, it was recently reported that for Ni(II) removal both reduction and surface complex formation (sorption) occurred (Li and Zhang, 2006). At equilibrium, about 50% of Ni(II) is reduced as Ni^0 at the surface and 50% of Ni(II) remains adsorbed at the nZVI particle surface. Notably, the capacity [0.13 g Ni(II)/g nZVI] is more than 100% higher compared to sorbents such as zeolite. Cao *et al.* (2005) reported the near complete reduction of perchlorate to chloride in aqueous solutions by nZVI particles and no sequential degradation products were observed. In contrast, the reduction by microscale iron particles was negligibly slow. However, the high activation energy required for nZVI-mediated reduction of perchlorate (79 kJ/mol), suggests that the reaction is limited by slow kinetics. The use of nZVI particles for the chemical reduction of nitrate has also been investigated and, in contrast to microscale iron particles, which converted nitrate to ammonia, nZVI particles converted nitrate to nitrogen gas (Choe *et al.*, 2000). More recently, Yang and Lee (2005) reported that chemical reduction of nitrate by nZVI particles is primarily an acid-driven surface-mediated process. Thus, in addition to the requirement for a large surface area provided by the nanoparticles, both pH and dose of nZVI particles (i.e. the stoichiometric ratio of nZVI to nitrate (as N)) were identified as being important for efficient reduction of nitrate. In general, acidic pHs were found to be favourable and high doses of nZVI particles enhanced the reaction rate.

Due to their extremely high reactivity, nanoparticles prepared using current methods tend to either react with surrounding media or agglomerate, resulting in the formation of much larger flocs and a significant loss in reactivity (Liu *et al.*, 2004; He and Zhao, 2005). Since efficient dispersion of the nZVI particles is thus a critical factor to improving their efficiency, various particle-stabilizing strategies have been reported. Ponder *et al.* (2000) tested a class of resin-supported nZVI particles (Ferragels) to reduce Cr(VI) in aqueous solutions. The resin-supported nanoparticles had an average size of 10–30 nm and a specific surface area of 24 m^2/g of nanoparticles. Notably, reduction of Cr(VI) was found to be 20–30 times faster than the commercial iron filings or iron powder per unit mass of Fe applied. In addition, stabilization of nZVI particles in cetylpyridinium chloride (CPC) has been described, resulting in a specific surface area of 25.4 m^2/g of nanoparticles (Chen *et al.*, 2004), whilst immobilization of bimetallic Pd/Fe nanoparticles in water-soluble starch offered a surface area of 55 m^2/g of nanoparticles (He and Zhao, 2005). The feasibility of using carboxymethyl cellulose (CMC)-stabilized nZVI particles for *in situ* reductive immobilization of Cr(VI) in water has recently been investigated by Xu and Zhao (2007). Batch kinetic tests indicated that 0.08 g/l of nZVI particles were able to rapidly reduce 34 mg/l of Cr(VI). Compared to non-stabilized nanoparticles, the stabilized nZVI particles displayed much greater reactivity and reaction kinetics for chromate reduction. When the dose was increased from 0.04 g/l to 0.12 g/l (i.e. the Fe:Cr molar ratio was increased from 1.1 to 3.3), the reduction of Cr(VI) in water was increased from 24% to 90% (Xu and Zhao, 2007). The above results suggest that the use of stabilized nZVI particles may represent an effective means for *in situ* remediation of groundwater or industrial effluents.

Bimetallic nanoscale iron particles

Although nanoscale zero-valent iron particles (nZVI) have been proven to be effective in degrading chlorinated organics, its intrinsic structure may limit its reaction rate (Zhang *et al.*, 1998a; Cheng and Wu, 2000; Li *et al.*, 2003). The decrease of iron reactivity over time is probably due to the formation of oxide layers on the particle surface during the reaction or upon contact of the nZVI particles with air (Li *et al.*, 2006a). Subsequent efforts to enhance the reactivity and functionality of nZVI particles have led to the discovery that physical mixing of bimetallic preparations of iron (Fe^0) and either palladium (Pd^0) or nickel (Ni^0) yielded superior reductants for chlorinated organics (Cheng and Wu, 2000, 2001). Moreover, Zhang *et al.* (1998a) reported that dehalogenation rate constants of trichloroethylene (TCE) were 30 times greater when synthesized bimetallic Pd/Fe nanoparticles were used compared to nZVI particles alone. Due to the advantages associated with the use of nanoparticles (i.e. increased surface area resulting in increased surface reactivity), bimetallic nanoparticles have been synthesized by co-reduction of the precursors of the two metals (Schrick *et al.*, 2002) or post-deposition of the second metal on the surface of the iron particles (Wang and Zhang, 1997; Zhang *et al.*, 1998a). In the typical nZVI-based bimetallic particle system, Fe acts as the reducing agent, whilst the noble metals, such as Ni or Pd, act as a catalyst with hydrogen generated from water reduction (Matheson and Tratnyek, 1994; Zhang *et al.*, 1998a; Schrick *et al.*, 2002). Despite Ni having a weaker catalytic effect compared to Pd, its good corrosion stability and lower cost may make it more appealing for site remediation and wastewater treatment (Menini *et al.*, 2000; Simagina *et al.*, 2003; Keane *et al.*, 2004; Xu *et al.*, 2005b). Thus, the introduction of a second metal not only makes the particles more stable in air by inhibiting the oxidation in some cases but also increases the reactivity (Zhang *et al.*, 1998a; Lien and Zhang, 2001).

Most studies regarding the use of bimetallic nanoparticles are concerned with chlorinated organics, because they are widespread, mobile and represent important environmental contaminants. Moreover, almost all of the chlorinated aromatic compounds exhibit high toxicity (Shin and Keane, 1999). The application of bimetallic Pd/Fe or Ni/Fe nanoparticles to the hydrodechlorination of chlorinated aliphatics (Lien and Zhang, 1999, 2001; Schrick *et al.*, 2002; Zhu *et al.*, 2006a), chlorinated aromatics (Zhang *et al.*, 1998a; Wei *et al.*, 2006) and polychlorinated biphenyls (PCBs) (Wang and Zhang, 1997; He and Zhao, 2005) has been reported. Overall, faster and enhanced hydrodechlorination reactions have been reported with nanoscale bimetallic particles compared to macroscale iron. Although the exact mechanism of this enhanced reactivity is unknown yet, it has been suggested that galvanic cells are formed in the coupled bimetallic particles when nZVI is attached to a noble metal. Iron serves as the electron donor and preferably becomes reacted with the contaminants, whereas the noble metal is protected (Schrick *et al.*, 2002).

Brominated organic compounds (BOCs) belong to an important class of disinfection by-products formed during chlorination of water and wastewater (Hozalski *et al.*, 2001), and are biorefractory in the natural environment (Wever *et al.*, 2000). Using carbon tetrabromide (CTB), bromoform (BF) and dibromomethane (DBM) as target compounds, Lim *et al.*, (2007) reported that nanoscale bimetallic Ni/Fe nanoparticles (with a metal loading of 2.5 g/l) was more reactive for reductive dehalogenation of the brominated methanes compared to commercial microscale Fe (metal loading of 125 g/l). Analysis of the reduction kinetics showed that CTB and BF were mainly reduced through a hydrogenolysis reaction, while over 76% of DBM was reduced through concerted reductive debromination to methane. Moreover, it has been reported that TCE could also be removed from water by cellulose acetate (CA)-supported bimetallic Ni/Fe nanoparticles via sorption in the CA support and reduction by the bimetallic nanoparticles (Wu and Ritchie, 2006). The sorption of TCE by CA reached equilibrium in less than 15 min and the supported bimetallic Ni/Fe nanoparticles was shown to have a much higher reactivity than the CA-supported monometallic Ni^0 or Fe^0 nanoparticles. The observed reduction rate constant reached a maximum between 14.3 and 21.4 wt.% Ni, and decreased with further increase in Ni-content. Notably, post-coating was reported to be more effective than co-reduction as a means to prepare bimetallic Ni/Fe nanoparticles with high reactivity. As mentioned, He and Zhao (2005), in an effort to avoid aggregation of the nanoparticles and thus a reduction in reactivity, modified the conventional synthesis methods for bimetallic Pd/Fe nanoparticles by applying a water-soluble starch as a stabilizer (0.2% by weight). At an initial TCE concentration of 52 mg/l, the starch-stabilized Pd/Fe nanoparticles (0.1 g/l) reduced 98% TCE within 1 h, and no toxic by-products such as vinyl chloride, *cis-* or *trans*-dichloroethene were detected in the water. The starched nanoparticles at 1 g/l were able to transform over 80% of PCBs (initial concentration of 2.5 mg/l) in less than 100 h, as compared to only 24% with non-starched Pd/Fe nanoparticles.

Despite the success in using nZVI-based bimetallic nanoparticles, palladium-coated gold nanoparticles (Pd/Au) have emerged as a promising alternative. In this instance, the Au particles, which serve as the base metal, do not react with organic compounds nor do they act as catalysts, but they do, however, augment the catalytic ability of Pd. In comparative analyses regarding the TCE hydrodechlorination catalytic activity in water of unsupported Pd nanoparticles, Pd supported on alumina (Pd/Al_2O_3) and Pd black, Nutt *et al.* (2005) found that the Pd nanoparticles were the most active due to more efficient use of the metal, i.e. higher Pd dispersion. Subsequently, in an effort to increase the dispersion and thus catalytic activity, bimetallic Pd/Au nanoparticles were synthesized and it was reported that Pd catalytic activity was increased by as much as 70 times (943 l/g Pd/min) compared with Pd/Al_2O_3 (12 l/g Pd/min) (Nutt *et al.*, 2005). More recently, it was reported that bimetallic Pd/Au nanoparticles with a 4-nm Au core and controlled Pd loading were more active for the aqueous-phase TCE hydrodechlorination reaction than Pd-only catalysts. These bimetallic nanoparticles exhibited volcano-like catalytic activity, in which hydrodechlorination rate was maximum near 70% Pd surface coverage. Although it is not known yet how Au promotes Pd activity, it has been suggested that geometric effects could be responsible, i.e. that Pd ensembles and Pd–Au mixed sites can participate as adsorption and surface reaction sites (Nutt *et al.*, 2006). Moreover, trimetallic Pd–Sn–Au catalysts supported on silica were recently shown to be more active and selective in nitrate and nitrite reduction compared to the parent Pd–Sn catalyst, suggesting that the addition of gold improved the stability and selectivity towards nitrogen formation (Garron *et al.*, 2006).

Nanoscale semiconductor photocatalysts

Since the discovery of photocatalytic water splitting on titanium(IV) dioxide (TiO_2) electrodes by Fujishima and Honda (1972), interest in applying TiO_2 as homogeneous photocatalyst for the remediation of groundwater and wastewater has increased (Hsien *et al.*, 2001; Kositzi *et al.*, 2004; Prieto *et al.*, 2005; Cho *et al.*, 2005a, 2006; Liu *et al.*, 2006b), particularly due to its potential to degrade a wide range of organic and inorganic compounds (Linsebigler *et al.*, 1995; Herrmann, 1999; Kabra *et al.*, 2004). The effective photoexcitation of TiO_2 photocatalysts requires the application of light with energy higher than its band-gap energy, thus resulting in the formation of electrons (e^-) in the conduction band and positive holes (h^+) in the valence band. Hydroxyl ions and water are likely traps for holes, leading to the formation of hydroxyl radicals ($OH^\bullet$) that are strong oxidizing agents, whilst the traps for electrons are adsorbed oxygen species, leading to the formation of superoxide species ($O_2^\bullet$) that are unstable. The reactive oxygen species generated ($OH^\bullet$, $O_2^\bullet$) react with molecules adsorbed on the photocatalyst surface, resulting in its hydroxylation, oxidation and finally mineralization to carbon dioxide and water (Linsebigler *et al.*, 1995; Mills and LeHunte, 1997; Carp *et al.*, 2004).

Most of the semiconductor photocatalysts investigated is either metal oxides (e.g. TiO_2, ZnO, SnO_2, WO_3) or chalcogenides (e.g. CdS, ZnS, CdSe, ZnSe, CdTe) (Peral *et al.*, 1997; Carp *et al.*, 2004). Owing to its physical and chemical stability, lower cost and resistance to corrosion, TiO_2 is the most commonly used photocatalyst for environmental applications. However, the wide technological usage of TiO_2 is hampered by its wide band-gap (rutile: 3.02 eV, anastase: 3.2 eV), which thus requires ultraviolet (UV) light irradiation for photocatalytic activation (Dvoranova *et al.*, 2002). Since UV light accounts for only a small fraction (less than 5%) of solar irradiation compared to visible light (45%) (Romero *et al.*, 1999), any shift in the optical response of TiO_2 from the UV ($\lambda \leq 385$ nm) to the visible spectral range ($\lambda \geq 420$ nm) should have a beneficial effect on the photocatalytic efficiency of the material (Adesina, 2004). Several approaches have therefore been used to lower the band-gap energy of TiO_2 photocatalysts, including reductive hydrogen plasma treatment (Takeuchi *et al.*, 2000; Ihara *et al.*, 2001; Palmer *et al.*, 2002), sensitizing by adsorbed dyes (Hirano *et al.*, 2000; Bae and Choi, 2003; Moon *et al.*, 2003) or by semiconductors having lower band-gaps (Serpone *et al.*, 1995; Yu *et al.*, 2003a), as well as transition metal doping (Choi *et al.*, 1994; Di Paola *et al.*, 2001; Dvoranova *et al.*, 2002; Shah *et al.*, 2002; Yamashita *et al.*, 2003; Sakthivel *et al.*, 2004; Barakat *et al.*, 2005; Nahar *et al.*, 2006). Albeit a widely used approach, the photocatalytic activity of metal-doped TiO_2, however, depends on the nature and concentration of the dopant ion, preparation methods and calcination processes (Navio *et al.*, 1996; Bae and Choi, 2003; Oh *et al.*, 2003; Barakat *et al.*, 2005; Enríquez and Pichat, 2006). Consequently, anionic dopants such as nitrogen (Asahi *et al.*, 2001; Ihara *et al.*, 2003; Yang *et al.*, 2004; Liu *et al.*, 2006a), carbon (Irie *et al.*, 2003; Neumann *et al.*, 2005) and sulphur (Ohno *et al.*, 2003; Yu *et al.*, 2005; Jiang *et al.*, 2006), have been investigated as an alternative approach.

Commercially available TiO_2 Degussa P-25, consisting of 80% of anatase and 20% of rutile with an average particle size of 30 nm, is widely used in the treatment of contaminated wastewater and has become a research standard (Ohno *et al.*, 2001; Kirchnerova *et al.*, 2005). However, nanocrystalline TiO_2 (particle sizes of *ca.* 6–8 nm), synthesized mostly by using sol–gel techniques, has emerged as promising photocatalysts for water remediation and purification. The decrease in semiconductor particle size has two significant benefits. The high surface density of active sites available for substrate adsorption, consistent with the high surface to volume ratio typical of nanosized particles, makes it possible to increase the overall photoreaction rate. Also, the band-gap of the nanosized semiconductor particle becomes size-dependent due to quantization effects, allowing for fine tuning of the band-gap and, thus, also the electron–hole pair redox potentials (Zhang *et al.*, 1998b; Rao *et al.*, 2002). Consequently, a measure of selectivity in the associated photoreactions can be achieved. Moreover, the presence of crystal defects, which can be associated to trap states for the photogenerated carriers, is generally less when high quality nanocrystals are synthesized, thus allowing the electron-hole separation yield to be maximized (Amal *et al.*, 1999). In recent years, the synthesis of visible light-activated TiO_2 nanoparticles has attracted increasing interest. Bae and Choi (2003) have synthesized visible light-activated TiO_2 nanoparticles based on TiO_2 modified by ruthenium-complex sensitizers and platinum (Pt) deposits. The Pt/TiO_2/RuIIL3 nanoparticles drastically enhanced the rate of reductive dehalogenation of trichloroacetate and carbon tetrachloride in aqueous solutions under visible light. Pena *et al.* (2005) reported that, in the presence of sunlight and dissolved oxygen, As(III) (2 mg/l) was completely converted to As(V) in a suspension containing 0.2 g/l TiO_2 nanocrystals through photocatalytic oxidation within 25 min. The adsorption capacity of the TiO_2 for As(V) and As(III) was much higher than Degussa P-25 and granular ferric oxide. Using three azo dyes, i.e. acid orange 7 (AO7), procion red MX-5B (MX-5B) and reactive black 5 (RB5), nitrogen (N)-doped TiO_2 nanoparticles were reported to exhibit substantial photocatalytic activity under direct sunlight irradiation, with 70% of the dye colour removed in 1 h and complete decolourization within 3 h. In contrast, Degussa P-25 did not cause detectable dye decolourization under identical experimental conditions (Liu *et al.*, 2005b). Furthermore, Fe(III)-doped TiO_2 nanoparticles were reported to degrade phenol in water and although the doped TiO_2 was more active than Degussa P-25 under solar light irradiation, it was less active than Degussa P-25 under UV light irradiation (Nahar *et al.*, 2006).

For water treatment, the TiO_2 particles and nanocrystals can either be deposited as prepared or they can be incorporated in a suitable nanocomposite by using conventional grafting, precipitation and impregnation methods (Hsien *et al.*, 2001; Yu *et al.*, 2002; Cheng *et al.*, 2003; Tuel and Hubert-Pfalzgraf, 2003; Landau *et al.*, 2005; Zhang *et al.*, 2005a; Yang *et al.*, 2006). The latter is, however, preferred since it not only prevents agglomeration of the TiO_2 particles, which can decrease the active surface area and thus reduce the efficiency of the photocatalytic reactions, but it also facilitates greatly the separation and recovery of the catalysts following treatment (Yang *et al.*, 2006). López-Munoz *et al.* (2005) reported the use of tubular arrays of meso- and microporous molecular sieves composed of TiO_2 nanoparticles, supported by mesoporous silica, for water remediation of aromatic pollutants in the presence of UV light. Photocatalysts composed of nanostructured TiO_2, uniformly deposited onto Fe_2O_3, have also been incorporated into ultrafiltration membranes and were shown to successfully reduce the fouling burden and improve the permeate flux (Sun *et al.*, 2004). The ability to form films and nanotubes with semiconductor photocatalysts has also provided a means whereby photocatalytic structures with a high a surface area can be obtained. Recently, the synthesis of TiO_2-based *p–n* junction nanotubes have been described of which the outside of the tubes is comprised of TiO_2 and the inside is comprised of platinum (Chen *et al.*, 2005). The nature of the *p–n* junction allows the outside of the tube to act as an oxidizing surface, whilst the inside of the tube acts as a reductive surface. When these catalysts were exposed to UV light, they were able to degrade toluene and the photocatalytic destruction rates were higher compared to non-nanotube structured material and to nanotubes lacking a *p–n* junction (Chen *et al.*, 2005). Although still in early stages of research, this technology may in future find application in sensors, bifunctional catalysts and in nanofiltration membranes for water treatment. In addition, the use of nanostructured ZnO semiconductor films for simultaneous sensing and degradation of organic contaminants has been demonstrated (Kamat *et al.*, 2002). Using 4-chlorocatechol as model compound, its presence was detected from the quenching of visible emission of the ZnO semiconductor film. Not only was a detection sensitivity of 1 ppm of 4-chlorocatechol in water achieved, but it was also photocatalytically degraded by the ZnO film under UV irradiation. The coupling of these two features may greatly facilitate monitoring of the degradation process because as decontamination occurs, a direct change in the emission intensity follows (Kamat *et al.*, 2002).

Bioactive nanoparticles for water disinfection

It is estimated that infectious diseases cause about 40% of the ca. 50 million total annual deaths worldwide. Waterborne pathogens cause 10–20 million of these deaths and non-fatal infection of more than 200 million people each year (Leonard *et al.*, 2003). Disinfection of drinking water supplies is commonly accomplished by the addition of a strong oxidant such as chlorine. In addition to being able to inactivate viruses and bacteria, including coliforms, chlorine disinfectants also provide residual protection for some period of time against re-growth of bacteria and pathogens (Szewzyk *et al.*, 2000). However, the use of such disinfectants may lead to undesirable tastes and odours of the treated water (Suffet *et al.*, 1995). Moreover, these compounds tend to generate toxic disinfection by-products such as trihalomethanes, haloacetic acids and aldehydes (Rook, 1974; Becher, 1999; Gopal *et al.*, 2007). The trend towards replacing free chlorine with chloramine to meet more stringent disinfection by-product regulations in drinking water may impact on bacterial re-growth in distribution systems. Although chloramine produces a more stable residual than free chlorine and provides lasting protection against bacterial re-growth (Neden *et al.*, 1992; Norton and Lechevallier, 1997; Zhang and Digiano, 2002), it does, however, potentially add a nutrient, ammonia, to the distribution system (Skadsen, 1993). Nanotechnology may present a reasonable alternative for development of new chlorine-free biocides. Among the most promising nanomaterials with antimicrobial properties are metallic and metal-oxide nanoparticles, as well as TiO_2 photocatalysts.

Metal and metal oxide nanoparticles

Silver (Ag) ions and silver compounds have long been known to exhibit strong toxicity to a wide range of microorganisms (Spadaro *et al.*, 1974; Zhao and Stevens, 1998) and have thus been used as antimicrobial compounds in various biomedical products and applications (Liau *et al.*, 1997). Consequently, several investigators have begun evaluating the use of Ag nanoparticles as biocides (Sondi and Salopek-Sondi, 2004; Baker *et al.*, 2005; Panáček *et al.*, 2006; Kim *et al.*, 2007; Shrivastava *et al.*, 2007). In general, the effect appears to be dose-dependent and is more pronounced against Gram-negative (*Escherichia coli*) than against Gram-positive (*Staphylococcus aureus*) bacteria, possibly as a consequence of the differences between the structure of their respective cell walls (Madigan and Martinko, 2005). The bactericidal properties of the Ag nanoparticles appear to be dependent on both their size and shape (Morones *et al.*, 2005; Pal *et al.*, 2007). Analysis regarding the effect of silver nanoparticles in the range of 1–100 nm on the growth of four Gram-negative bacteria (*E. coli, Vibrio cholerae, Pseudomonas aeruginosa* and *Salmonella typhus*) indicated that growth of the bacterial cultures [5×10^7 colony-forming units (cfu)/ml] was inhibited at concentrations above 75 μg/ml of the nanoparticles. However, subsequent characterization of the interactions of the Ag nanoparticles with the bacteria by high angle annular dark field (HAADF) scanning transmission electron microscopy (STEM) showed that nanoparticles attached to the surface of the cell membrane or located inside the bacteria were mainly in the range of 1–10 nm (Morones *et al.*, 2005). Moreover, despite magnesium oxide (MgO; periclase) having been reported to be an effective biocide against *E. coli, Bacillus megaterium* and *B. subtilis* spores (Koper *et al.*, 2002; Stoimenov *et al.*, 2002), it was subsequently shown that the antibacterial activities of MgO nanoparticles are also dependent on their size. Results indicated that smaller nanoparticles (8 nm) were the most active in killing *E. coli* and *S. aureus*, whilst a gradual decrease in the bactericidal activity was observed with an increase in particle size (11–23 nm) (Makhluf *et al.*, 2005). The shape dependence of the bactericidal activity of Ag nanoparticles against *E. coli* was recently investigated using Ag nanoparticles synthesized with different shapes (Pal *et al.*, 2007). Although the respective nanoparticles were each capable of interacting with *E. coli*, the inhibition results indicated that truncated triangular silver nanoplates displayed the strongest biocidal action, compared with spherical and rod-shaped nanoparticles. It was proposed that differences in the effective surface areas in terms of active facets may influence the killing activity or relate to the bacterial killing capacity of the differently shaped Ag nanoparticles.

The stabilization and immobilization of metallic nanoparticles in different matrixes has also gained increased importance since such nanoparticles purportedly present high antibacterial activity, low toxicity, chemical stability, a long-lasting action period and thermal resistance (Esteban-Cubillo *et al.*, 2006). In this regard, Sondi and Salopek-Sondi (2004) have reported the preparation of stable Ag nanoparticles of narrow size distribution (12 nm) by reducing silver nitrate solutions with ascorbic acid in the presence of Daxad 19 (sodium salt of a high-molecular-weight naphthalene sulphonate formaldehyde condensate) as stabilizing agent. The Ag nanoparticles were shown to be effective biocides against *E. coli*. In an alternative approach, Son *et al.* (2004) prepared ultrafine cellulose acetate (CA) fibres by direct electrospinning of a CA solution with 0.05 wt.% $AgNO_3$, followed by UV irradiation photoreduction. The CA fibres with embedded Ag nanoparticles (average diameters 15.4 nm) were reported to be effective biocides against Gram-positive (*S. aureus*) and Gram-negative (*E. coli, Klebsiella pneumoniae* and *P. aeruginosa*) bacteria. Recently, copper monodispersed nanoparticles (2–5 nm) incorporated into sepiolite (a magnesium phylosilicate), which has a low ion exchange capacity but a large surface area (300 m^2/g), has been shown be an effective biocide against *E. coli* and *S. aureus*, reducing their concentrations ($2.5–10 \times 10^5$ cfu/ml) by 99.9% after 24 h of incubation (Esteban-Cubillo *et al.*, 2006). Despite the above results indicating that stabilized nanoparticles display enhanced biocide activity, it may, however, not always be the case since a report by Top and Ülkü (2004) indicated that incorporation of Cu(II) into zeolite had negligible antibacterial activity.

Although the antibacterial activity of silver ions has been studied extensively, the effects of Ag nanoparticles on the bacteria and bactericidal mechanism are only partially understood. Based on studies having shown that Ag nanoparticles anchor to and penetrate the bacterial cell wall (Sondi and Salopek-Sondi, 2004; Morones *et al.*, 2005), it is reasonable to suggest that the resultant structural changes in the cell membrane could cause an increase in cell permeability, leading to uncontrolled transport through the plasma membrane and ultimately cell death. It has also been proposed, based on electronspin resonance spectroscopy studies of Ag nanoparticles, that damage of the bacterial membranes may furthermore be induced by the formation of free radicals (Danilczuk *et al.*, 2006; Kim *et al.*, 2007). Moreover, based on the greater tendency of silver (soft acid) to react with sulphur- and phosphorus-containing soft bases (Hatchett and Henry, 1996) and on the presence of Ag nanoparticles inside the cells (Morones *et al.*, 2005), it is likely that further damage could thus be caused by interactions with sulphur-containing proteins and phosphorus-containing compounds such as DNA. These interactions may prevent cell division and DNA replication from occurring, and also ultimately lead to the death of the cell. In addition, evidence has recently been obtained suggesting that the Ag nanoparticles may modulate the phosphotyrosine profile of putative bacterial peptides that could affect cellular signalling and therefore inhibit the growth of the organisms (Shrivatava *et al.*, 2007).

Photocatalytic disinfection by TiO_2 catalysts

Since Matsunaga *et al.* (1985) first reported that bacteria (*E. coli* and *Lactobacillus acidophilus*) and yeasts (*Saccharomyces cerevisae*) in water could be killed by contact with a TiO_2–Pt catalyst upon illumination with near-UV light, interest in using TiO_2-mediated photocatalysis for water disinfection has grown steadily. The efficiency of photocatalytic TiO_2 substrates (mostly Degussa P-25), induced primarily by UV light irradiation, in killing bacteria, fungi, algae and viruses has been studied extensively (Watts *et al.*, 1995; Bekbolet and Araz, 1996; Bekbolet, 1997; Butterfield *et al.*, 1997; Mills and LeHunte, 1997; Lee *et al.*, 1998; Shephard *et al.*, 2002; Ibanez *et al.*, 2003; Liu and Yang, 2003; Cho *et al.*, 2004, 2005b). Studies, using *E. coli* as model organism, have also indicated that the bactericidal activity of TiO_2/UV photocatalysts can be enhanced by using TiO_2 nanorods (Joo *et al.*, 2005), iron (Fe(III))-doped TiO_2 (Egerton *et al.*, 2006) and upon addition of silver (Ag) to the TiO_2 photocatalysts (Sokmen *et al.*, 2001; Vamathevan *et al.*, 2004; Zhang *et al.*, 2005b). Moreover, novel biocidal photocatalytic composite materials, composed of TiO_2 and multiwalled carbon nanotubes (MWCNTs) (Lee *et al.*, 2005b), or of 5 wt.% Ag-TiO_2 supported on hydroxyapatite (HAP) (Reddy *et al.*, 2007), have been reported to successfully inactivate *B. cereus* endospores and *E. coli*, respectively, under UV light irradiation.

Since visible light comprises *ca.* 45% of the solar spectrum, visible light-induced photocatalytic disinfection appears to be a promising technology from the viewpoint of using solar energy. To date, comparatively few studies have been conducted to investigate visible light-induced antimicrobial activity (Yu *et al.*, 2003b; Goswami *et al.*, 2004; McLouglin *et al.*, 2004; Hu *et al.*, 2006). In the majority of cases, the Gram-negative bacterium *E. coli* has been used as model organism, although the application of TiO_2 photodisinfection to the destruction of more resistant microbes has also been described. Lonnen *et al.* (2005) reported that although TiO_2, after 8 h of simulated solar exposure, reduced the viability of several waterborne pathogens such as protozoa (the trophozoite stage of *Acanthamoeba polyphaga*), fungi (*Candida albicans* and *Fusarium solani*) and Gram-negative bacteria (*P. aeruginosa* and *E. coli*), it was inefficient in inactivating spores of *B. subtilis* and ineffective against the cyst stage of *A. polyphaga*. In a recent study, Mitoraj *et al.* (2007) reported that carbon (C)-doped TiO_2, when illuminated by visible light, inactivated *E. coli*, but Gram-positive bacteria (*S. aureus* and *Enterococcus faecalis*) appeared to be less sensitive, whilst the viability of *Candida albicans* was slightly reduced and no growth inhibition of *Aspergillus niger* was observed. The differences reported in TiO_2-mediated photocatalytic killing may be related to differences in the complexity and density of the respective microbial cell walls. In contrast to the Gram-negative bacterium *E. coli*, which has a thin peptidoglycan layer in its cell wall, Gram-positive bacteria have thicker and more compact cell walls, whilst eukaryotes and fungi are surrounded by a rigid cell wall composed mainly of polysaccharide polymers (Madigan and Martinko, 2005). Notably, several studies have indicated that nitrogen (N)-doped TiO_2 has better visible light photocatalytic bactericidal activity than C-doped TiO_2, as was evidenced by drastically reduced *E. coli* survival rates following irradiation (Liu *et al.*, 2006a; Shieh *et al.*, 2006; Wong *et al.*, 2006). However, cultures (2.5×10^6 cfu/ml) of different waterborne Gram-positive bacterial pathogens (*Shigella flexneri*, *Listeria monocytogenes* and *Vibrio parahaemolyticus*) have been reported to be more resistant to N-doped TiO_2-mediated killing than *E. coli*, with a killing efficiency of approximately 50% compared with 80–95% (Wong *et al.*, 2006). In this regard, it is interesting to note that Yu *et al.* (2005) reported that sulphur (S)-doped TiO_2 nanoparticles were highly effective in killing the Gram-positive bacterium *Micrococcus lylae* (3×10^7 cfu/ml) under visible light irradiation (97% reduction in 1 h), and results from electronspin resonance spectroscopy confirmed the formation of hydroxyl radicals. TiO_2 modified with platinum (Pt(IV)) chloride complexes (4% $H_2[PtCl_6]$/TH-0) and immobilized at the surface of plastic plates has been reported to exhibit a high inactivation effect on *E. coli* (10^6 cfu/ml) in both the presence (99% reduction after 30 min) and absence of visible light (complete reduction after 60 min). This enhanced toxicity of the catalyst was ascribed to the presence of Pt(II) and Pt(IV), both of which are known inhibitors of bacterial DNA, RNA and protein synthesis. Their presence resulted from the desorption of Pt(IV) from the photocatalyst surface and, in the presence of reducing agents in the medium, Pt(IV) can be partially reduced to Pt(II) (Mitoraj *et al.*, 2007).

Although a wealth of information has demonstrated the efficacy of the biocidal actions of the TiO_2 photocatalyst, the fundamental mechanism underlying the photocatalytic killing process has not been well established yet. Although the biocidal action of the TiO_2 photocatalysts has been frequently ascribed to reactive oxygen species, such as $OH^\bullet$, $O_2^\bullet$ and hydrogen peroxide (H_2O_2) that are generated on irradiated TiO_2 surfaces (Linsebigler *et al.*, 1995; Mills and LeHunte, 1997; Herrmann, 1999; Konaka *et al.*, 1999), there are, however, differing opinions regarding the bactericidal mechanism of these photocatalysts (Fig. 1.5). In early studies, Matsunaga *et al.* (1985, 1988) proposed that the direct oxidation of coenzyme A (CoA) in TiO_2-treated cells, which led to decreased respiratory activities, was the reason for cell death. However, several studies have proposed that destruction of the cell membrane structure and membrane functions is the root cause of cell death. Whereas Saito *et al.* (1992) reported that decomposition of the cell membrane and loss of its permeability, as was evidenced by leakage of intracellular potassium ions, are the main factors responsible for bacterial cell death, Maness *et al.* (1999) presented results showing that irradiated TiO_2 induces peroxidation of polyunsaturated phospholipid components of the lipid membrane. Sunada *et al.* (2003) demonstrated that decomposition of the outer membrane caused by peroxidation processes results in cytoplasmic membrane disorder and afterwards in cellular death. More recently, Kiwi and Nadtochenko (2005) reported that photocatalytic degradation of lipopolysaccharide, phosphatidyl-ethanolcholine and peptidoglycan, which are the main wall structure elements, occurred at the TiO_2 interface. However, Gogniat *et al.* (2006) reported that adsorption of the bacterial cells onto TiO_2 aggregates alters bacterial membrane integrity and that the effect is subsequently greatly amplified when illumination occurred, thus suggesting that adsorption onto TiO_2 is a prerequisite for bacterial inactivation. In addition, Hirakawa *et al.* (2004) reported that photoirradiated TiO_2 catalysed site-specific DNA damage via generation of H_2O_2. All of these studies reported on the mechanism of TiO_2 lethality under UV light and it is not known yet whether photocatalytic degradation of the bacterial cell wall leads to death under visible light.

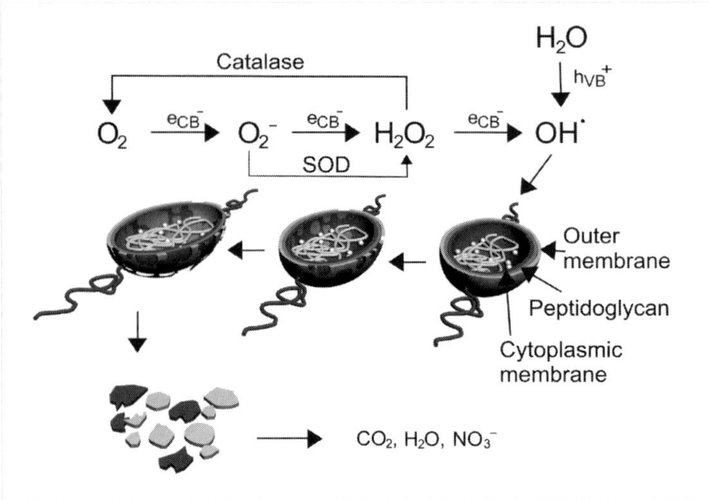

Figure 1.5 Selected processes occurring during the photocatalytic inactivation of *E. coli* in the presence TiO_2, as proposed by Mitoraj *et al.* (2007). Hydrogen peroxide (H_2O_2) is generated at the surface of irradiated TiO_2 as a result of a two-step reduction of an adsorbed O_2 molecule with O_2^- as an intermediate. The electrons participating in this process come from the conduction band of photoexcited TiO_2 (e_{CB}^-). In addition, O_2^- may be reduced to H_2O_2 also by superoxide dismutase (SOD). The formation of H_2O_2 competes with its decomposition in the presence of catalase. H_2O_2 in a photocatalytic system acts as an irreversible electron acceptor, which reduces significantly the electron-hole recombination and leads to the formation of $OH^•$ radicals. Highly reactive $OH^•$ is generated either as a result of the described multi-step reduction of O_2 or in the process of direct water oxidation by photogenerated holes in the valence band of titania, h_{VB}^+. The latter reaction can only take place under direct TiO_2 excitation with UV light. $OH^•$ radicals together with O_2^- are responsible for cellular wall damage, cell death and final mineralization of organic matter.

Toxic potential of engineered nanoparticles and nanomaterials

The use of engineered nanoparticles and nanomaterials for water treatment and groundwater remediation, as highlighted in the previous sections, has raised concerns for human exposure which stems from the absence of specific technologies aimed at the removal of engineered nanomaterials from the water, and the safety of the new nanoparticles and nanomaterials that may be used by the water industry. These concerns are often based on the assumption that nanoparticles will be highly mobile in porous media because of their small size; thus implying a greater potential for exposure as they are dispersed over greater distances and their effective persistence in the environment increases. However, Wiesner *et al.* (2006) suggested that smaller nanoparticles should not be very mobile, since their relatively large diffusivity would enable them to produce more frequent contacts with the surfaces of porous media such as groundwater aquifers or the sand filters used in potable water treatment. Moreover, retention of nanoparticles by porous media has been reported to be increased by conditions such as high ionic strength and the presence of small quantities of divalent ions (Brandt *et al.*, 2005). Because groundwater aquifers and surface waters typically have ionic strengths of greater than 10^{-4} M, in addition to significant concentrations of calcium or magnesium, conditions should thus favour deposition of the nanomaterial. Consequently, Wiesner *et al.* (2006) concluded that there is no evidence to support the scenario of a new nanoparticulate contaminant that current water treatment infrastructure cannot handle.

Although the use of engineered nanoparticles and nanomaterials may have a positive health benefit on humans from the viewpoint of drinking water quality, various studies have nevertheless commented on their potential adverse effects on human health (Hoet *et al.*, 2004; Borm *et al.*, 2006; Nel *et al.*, 2006; Maynard, 2007; Oberdörster *et al.*, 2007) and the environment (Colvin, 2003; Biswas and Wu, 2005; Moore, 2006). Since it is likely that waste generated during the production and use of these nanoparticles will appear eventually in various environments, there is a possibility that humans could be exposed to these nanoparticles through inhalation, dermal contact or ingestion and absorption through the digestive tract. Several hypotheses concerning physiochemical characteristics of nanoparticles responsible for adverse health effects have been proposed (Lanone and Boczkowski, 2006; Nel *et al.*, 2006; Maynard, 2007). These include properties such as surface area and reactivity (Oberdörster *et al.*, 2000), particle size, shape and structure (Watanabe *et al.*, 2002; Chen *et al.*, 2004; Maynard and Kuempel, 2005; Oberdörster *et al.*, 2005; Donaldson *et al.*, 2006; Lam *et al.*, 2006), chemical composition (Shvedova *et al.*, 2003; Donaldson *et al.*, 2006), as well as surface treatments (coatings) and the degree to which nanoparticles aggregate or agglomerate (Warheit *et al.*, 2003; Bermudez *et al.*, 2004; Heckel *et al.*, 2004; Ryman-Rasmussen *et al.*, 2006; Wang *et al.*, 2006). Despite the wide variety of different nanoparticles and nanomaterials, the toxicological effects, however, of only a limited range of nanomaterials have been investigated. Studies have focused mostly on the effects of carbon-based nanomaterials and metal-oxide nanoparticles, including TiO_2 nanoparticles, of which the results of selected *in vivo* and *in vitro* studies are summarized in Tables 1.3 and 1.4, respectively. Whereas results from *in vivo* studies, which have focussed on respiratory exposure in

Table 1.3 Examples of the toxic effects of nanomaterials in *in vivo* studies

Nanomaterial	Dose and exposure route	Species	Effects	Reference
SWCNT	0.1 or 0.5 mg; intratracheal instillation	Mice	56% mortality in 0.5 mg dose. Weight loss, lung lesions in 0.5-mg group, necrosis, macrophages and granulomas, interstitial and peribronchial inflammation	Lam *et al.* (2004)
SWCNT	2-mg implant; subcutaneous implantation	Mice	Activation of the histocompatibility complex in $CD4^+/CD8^+$ T-cells	Koyama *et al.* (2006)
Intact or ground MWCNT	0.5, 2 or 5 mg; intratracheal instillation	Rats	Intact MWCNT induced collagen-rich granulomas and surrounding alveolitis. Ground MWCNT induced inflammatory and fibrotic responses	Muller *et al.* (2005)
MWCNT	12.5 mg; intratracheal instillation	Guinea pigs	Pneumonitis and pulmonary lesions	Grubek-Jaworska *et al.* (2006)
Ultrafine TiO_2	0.5, 2 or 10 mg/ml; aerosol inhalation	Mice, rats, hamsters	Pulmonary particle overload and inflammation in rats and mice exposed to 10 mg/ml. Inflammation included increased number of macrophages and neutrophils, progressive fibrosis in rats, elevated protein and lactate dehydrogenase levels	Bermudez *et al.* (2004)
TiO_2 particles, nanoscale rods or dots	1 or 5 mg; intratracheal instillation	Rats	TiO_2 rods or dots produced transient inflammatory and cell injury effects, and were not different from the effects of larger sized TiO_2 particles	Warheit *et al.* (2006)

SWCNT, single-walled carbon nanotubes; MWCNT, multiwalled carbon nanotubes.

Table 1.4 Examples of the toxic effects of nanomaterials in *in vitro* studies

Nanomaterial	Dose	Cells	Effects	Reference
SWCNT	0.06–0.24 mg/ml	HEK cells in culture (human epidermal keratinocytes)	Cells showed a dose-dependent decrease in viability. Oxidative stress indicated	Shvedova *et al.* (2003)
SWCNT	0.78–200 mg/ml	Human HEK293 (human embryo kidney cells)	Inhibition of cell proliferation and cell apoptosis was induced	Cui *et al.* (2005)
MWCNT	0.1–0.2 mg/ml	HEK	MWCNT present within cytoplasmic vacuoles of HEK cells, initiated irritation response and release of the proinflammatory cytokine interleukin 8	Monteiro-Riviere *et al.* (2005)
MWCNT	40–400 µg/ml	Human T lymphocytes	Dose- and time-dependent decrease in cell viability, and induced cell apoptosis. Only oxidized MWCNT induced an increase in tyrosine phosphorylation. Oxidized MWCNT more toxic than pristine MWCNT	Bottini *et al.* (2006)
TiO_2 nanoparticles (anastase form: 10 or 20 nm; rutile form: 200 nm)	10 mg/ml	BEAS 2B (human bronchial epithelial cells)	Anastase forms caused DNA damage, lipid peroxidation and micronuclei formation. Rutile form was generally much less toxic	Gurr *et al.* (2005)
TiO_2 nanoparticles (anastase form: 30–40 nm)	5 mg/ml	Human red blood cells	Aggregates of nanoparticles present in the cells. Red cell morphology was not generally affected, suggesting that the red cells were not damaged	Rothen-Rutishauser *et al.* (2006)
TiO_2 nanoparticles (different sizes in both anastase or rutile form)	3–30 mg/ml	A549 cells (human lung epithelial cells)	Cells showed a dose-dependent decrease in viability. Interleukin 8 was induced at highest doses	Sayes *et al.* (2006)

SWCNT, single-walled carbon nanotubes; MWCNT, multiwalled carbon nanotubes.

mostly laboratory rodents, raises concerns about lung injury, inflammation and possible tumorigenesis, the results from *in vitro* studies, using different tissue and cell cultures, suggest that oxidative stress (resulting from the production of reactive oxygen species) may be a possible mechanism of toxicity. However, it should be noted that the doses or exposure concentrations of nanomaterials used in *in vitro* and *in vivo* toxicological studies are most often extraordinarily high, and also delivered as a bolus (extremely high dose rate), so as to have no or limited relevancy to real or anticipated human exposures (Borm *et al.*, 2006; Oberdörster *et al.*, 2007). Moreover, many of the *in vivo* studies have been conducted in rats, which are known to be especially sensitive when it comes to adverse lung responses (Donaldson and Tran, 2004), thus furthermore complicating the extrapolation of rodent sensitivity to human responses. In addition, there is only limited information available regarding the uptake, translocation and distribution of nanoparticles in the body (Kreyling *et al.*,

2002; Oberdörster *et al.*, 2004; Elder *et al.*, 2006; Singh *et al.*, 2006). Without this basic knowledge, it becomes difficult to predict what all the major toxic effects might be.

Despite the current lack of knowledge relating to which environments, and thus which organisms are most likely to be exposed to nanoparticles, some literature regarding the toxicological impact of nanoparticles on different aquatic organisms has started to emerge. Studies have been published regarding the effects of nanoparticles (Buckminster fullerenes (C_{60})) on fish such as the largemouth bass (Oberdörster, 2004) and the fathead minnow (Zhu *et al.*, 2006b), as well as the effects of C_{60} and TiO_2 (Degussa P-25) on aquatic invertebrates such as water fleas (*Daphnia magna*) (Lovern and Klaper, 2006; Oberdörster *et al.*, 2006; Zhu *et al.*, 2006b). To aid solubilization of the nanoparticles used in these latter studies, they were prepared in the organic solvent tetrahydrofuran (THF), which is a neurotoxin. Consequently, it is unclear as to whether the high toxicity observed in these instances is due to increased particle dispersion or because of THF-induced toxicity. However, preparation of the nanoparticles by stirring in water only indicated a reduced toxicity in different aquatic species compared with THF-solubilized C_{60} (Oberdörster *et al.*, 2006). The toxicity of nanoparticles to microorganisms has been studied more extensively (e.g. silver nanoparticles, MgO and TiO_2), albeit in relation to the development of biocidal agents, as highlighted previously. Since microorganisms are the foundation of all known ecosystems, serving as the basis of food webs and the primary agents for global biogeochemical cycles, an understanding of nanomaterial–microbe interactions is especially important. Currently, there is insufficient evidence to suggest that all nanoparticles have antimicrobial effects or that all nanoparticles are toxic to any organism encountered in an exposed environment (Oberdörster *et al.*, 2007). Consequently, more research is needed before generalized statements can be made regarding nanoparticle ecotoxicology.

Conclusion

Although water plays a critical role in every facet of human activity, it is becoming an increasingly scarce resource in many parts of the world. Besides utilization of non-traditional sources for production of high-quality freshwater and the conservation and protection of water bodies from pollution, equally important is the development of innovative new technologies and materials whereby challenges associated with the provision of safe potable water can be addressed. It is widely recognized that nanotechnology and applications thereof may play an important role in resolving issues relating to water shortage and water quality. Owing to their large surface areas and their size- and shape-dependent catalytic properties, considerable efforts are under way to explore uses of nanomaterials in applications such as membrane separations, catalysis and adsorption. Moreover, nanomaterials can be functionalized with various different chemical groups to increase their affinity towards a given compound, thus resulting in ligands that are not only recyclable but also have a high capacity and selectivity for organic and inorganic solutes, as well as toxic metal ions and inorganic anions in aqueous solutions.

While much attention has been focused on the development and potential benefits of nanomaterials in water treatment processes, concerns have also been raised regarding their potential human and environmental toxicity. Indeed, studies have indicated that the selfsame properties of nanomaterials that make them attractive (e.g. size, shape, structure, reactivity) may also cause them to be toxic. It is difficult, however, to assess the effect of nanomaterials on health and the environment because the methods and tools for such a task have not been well developed yet. In addition, common frameworks for risk research, risk assessment, and risk management are lacking at present. It is vital that these processes be developed and investigated to ensure that nanomaterials are as safe as possible, whilst reaching their full potential. Notwithstanding these information gaps, it is certain that new nanomaterials, particularly in water and wastewater treatment, will play key roles in ensuring sufficient and good quality water to meet the ever-increasing demand for potable water.

Acknowledgements

Reprinted with permission from *Critical Reviews in Microbiology*. J. Theron, J.A. Walker and T.E. Cloete (2008). Nanotechnology and water treatment: applications and emerging opportunities. Crit. Rev. in Microbiol., *34*, 43–69.

References

Adesina, A.A. (2004). Industrial exploitation of photocatalysis: Progress, perspectives and prospects. Catal. Surv. Asia *8*, 265–273.

Ahluwalia, S.S., and Goyal, D. (2007). Microbial and plant derived biomass for removal of heavy metals from wastewater. Bioresour. Technol. *98*, 2243–2257.

Ahmad, A.L., Ooi, B.S., Mohammad, A.W., and Choudhury, J.P. (2004). Development of a highly hydrophobic nanofiltration membrane for desalination and water treatment. Desalination. *168*, 215–221.

Allabashi, R., Arkas, M., Hörmann, G., and Tsiourvas, D. (2007). Removal of some organic pollutants in water employing ceramic membranes impregnated with cross-linked silylated dendritic and cyclodextrin polymers. Water Res. *41*, 476–486.

Alvarez-Ayuso, E., Garcia-Sanchez, A., and Querol, X. (2003). Purification of metal electroplating waste waters using zeolites. Water Res. *37*, 4855–4862.

Amal, R., McEvoy, S., Beydoun, D., and Low, G. (1999). Role of nanoparticles in photocatalysis. J. Nanopart. Res. *1*, 439–458.

Arkas, M., Allabashi, R., Tsiourvas, D., Mattausch, E.-M., and Perfler, R. (2006). Organic/inorganic hybrid filters based on dendritic and cyclodextrin 'nanosponges' for the removal of organic pollutants from water. Environ. Sci. Technol. *40*, 2771–2777.

Arkas, M., Paleos, C.M., Eleades, L., and Tsiourvas, D. (2005). Alkylated hyperbranched polymers as molecular nanosponges for the purification of water from polycyclic aromatic hydrocarbons. J. Appl. Polym. Sci. *97*, 2299–2305.

Arnold, W.A., and Roberts, A.L. (1998). Pathways of chlorinated ethylene and chlorinated acetylene reaction with Zn(0). Environ. Sci. Technol. *32*, 3017–3025.

Asahi, R., Morikawa, T., Ohwaki, T., Aoki, K., and Taga, Y. (2001). Visible light photocatalysis in nitrogen-doped titanium oxides. Science *293*, 269–271.

Ashbolt, N.J. (2004). Microbial contamination of drinking water and disease outcomes in developing regions. Toxicology *198*, 229–238.

Autry, H.A., and Holcombe, J.A. (1995). Cadmium, copper and zinc-complexes of poly-L-cysteine. Analyst *120*, 2643–2647.

Bae, E., and Choi, W. (2003). Highly enhanced photoreductive degradation of perchlorinated compounds on dye-sensitized metal/TiO_2 under visible light. Environ. Sci. Technol. *37*, 147–152.

Baker, C., Pradhan, A., Pakstis, L., Pochan, D.J., and Shah, S.I. (2005). Synthesis and antibacterial properties of silver nanoparticles. J. Nanosci. Technol. *5*, 244–249.

Balasubramanian, K., and Burghard, M. (2005). Chemically functionalized carbon nanotubes. Small *1*, 180–192.

Balasubramanian, K., and Burghard, M. (2006). Biosensors based on carbon nanotubes. Anal. Bioanal. Chem. *385*, 452–468.

Balogh, L., Swanson, D.R., Tomalia, D.A., Hagnauer, G.L., and McManus, A.T. (2001). Dendrimer–silver complexes and nanocomposites as antimicrobial agents. Nano. Lett. *1*, 18–21.

Barakat, M.A., Schaeffer, H., Hayes, G., and Ismat-Shah, S. (2005). Photocatalytic degradation of 2-chlorophenol by Co-doped TiO_2 nanoparticles. Appl. Catal. B: Environ. *57*, 23–30.

Becher, G. (1999). Drinking water chlorination and health. Acta Hydrochem. Hydrobiol. *27*, 100–102.

Bekbolet, M. (1997). Photocatalytic bactericidal activity of TiO_2 in aqueous suspensions of *E. coli*. Water Sci. Technol. *35*, 95–100.

Bekbolet, M., and Araz, C. (1996). Inactivation of *Escherichia coli* by photocatalytic oxidation. Chemosphere *32*, 959–965.

Bell, J.L., Sarin, P., and Kriven, W.M. (2006). Formation of nanocrystalline zeolites in geopolymer gels. Microsc. Microanal. *12*, 738–739.

Bermudez, E., Mangum, J.B., Wong, B.A., Asgharian, B., Hext, P.M., Warheit, D.B., and Everitt, J.I. (2004). Pulmonary responses of mice, rats, and hamsters to subchronic inhalation of ultrafine titanium dioxide particles. Toxicol. Sci. *77*, 347–357.

Bhattacharyya, D., Hestekin, J.A., Brushaber, P., Cullen, L., Bachas, L.G., and Sikdar, S.K. (1998). Novel poly-glutamic acid functionalized microfiltration membranes for sorption of heavy metals at high capacity. J. Membr. Sci. *141*, 121–135.

Biswas, P., and Wu, C.-Y. (2005). Nanoparticles and the environment. J. Air Waste Manage. Assoc. *55*, 708–746.

Blount, B.C., Pirkle, J.L., Osterloh, J.D., Valentin-Blasini, L., and Caldwell, K.L. (2006). Urinary perchlorate and thyroid hormone levels in adolescent and adult men and women living in the United States. Environ. Health Pers. *114*, 1865–1871.

Borm, P.J.A., Robbins, D., Haubold, S., Kuhlbusch, T., Fissan, H., Donaldson, K., Schins, R., Stone, V., Kreyling, W., Lademann, J., Krutmann, J., Warheit, D., and Oberdörster, E. (2006). The potential risks of nanomaterials: A review carried out for ECETOC. Particle and Fiber Toxicology *3*, (doi: 10.1186/1743–8977–3–11).

Bosman, A.W., Janssen, H.M., and Meijer, E.W. (1999). About dendrimers: Structure, physical properties, and applications. Chem. Rev. *99*, 1665–1688.

Bottero, J.Y., Rose, J., and Wiesner, M.R. (2006). Nanotechnologies: Tools for sustainability in a new wave of water treatment processes. Integr. Environ. Assess. Manag. *2*, 391–395.

Bottini, M., Bruckner, S., Nika, K., Bottini, N., Bellucci, S., Magrini, A., Bergamaschi, A., and Mustelin, T. (2006). Multi-walled carbon nanotubes induce T lymphocyte apoptosis. Toxicol. Lett. *160*, 121–126.

Bowman, R.S. (2002). Applications of surfactant-modified zeolites to environmental remediation. Micropor. Mesopor. Mater. *61*, 43–56.

Brant, J.A., Lecoanet, H., and Wiesner, M.R. (2005). Aggregation and deposition characteristics of fullerene nanoparticles in aqueous systems. J. Nanopart. Res. *7*, 545–553.

Breck, D.W. (1974). Zeolite Molecular Sieves. (Wiley and Sons Publishers, New York).

Bull, R.J., Brinbaum, L.S., Cantor, K.P., Rose, J.B., Butterworth, B.E., Pegram, R., and Tuomisto, J. (1995). Water chlorination: Essential process and cancer hazard. Fund. Appl. Toxicol. *28*, 155–166.

Burton, S.G., Cowan, D.A., and Woodley, J.M. (2002). The search for the ideal biocatalyst. Nature Biotechnol. *20*, 37–45.

Butterfield, I., Christensen, P., Curtis, T., and Gunlazuardi, J. (1997). Water disinfection using an immobilized titanium dioxide film in a photochemical reactor with electric field enhancement. Water Res. *31*, 675–677.

Cai, Y.Q., Cai, Y., Mou, S.F., and Lu, Y.Q. (2005). Multi-walled carbon nanotubes as a solid-phase extraction adsorbent for the determination of chlorophenols in environmental water samples. J. Chromatogr. A *1081*, 245–247.

Camblor, M.A., Corma, A., and Valencia, S. (1998). Synthesis in fluoride media and characterization of aluminosilicate zeolite beta. J. Mater. Chem. *8*, 2137–2145.

Cao, J., Elliott, D., and Zhang, W.-X. (2005). Perchlorate reduction by nanoscale iron particles. J. Nanopart. Res. *7*, 499–506.

Carp, O., Huisman, C.L., and Reller, A. (2004). Photoinduced reactivity of titanium dioxide. Prog. Solid State Chem. *32*, 33–177.

Chatterjee, A.N., Cannon, D.M., Gatimu, E.N., Sweedler, J.V., Aluru, N.R., and Bohn, P.W. (2005). Modelling and simulation of ionic currents in three-dimensional microfluidic devices with nanofluidic interconnects. J. Nanopart. Res. *7*, 507–516.

Chen, Y., Chen, J., Dong, J., and Jin, Y. (2004). Comparing study of the effect of nanosized silicon dioxide and microsized silicon dioxide on fibrogenesis in rats. Toxicol. Ind. Health *20*, 21–27.

Chen, C.Z.S., and Cooper, S. (2002). Interactions between dendrimer biocides and bacterial membranes. Biomaterials *23*, 3359–3368.

Chen, Y., Crittenden, J., Hackney, S., Sutter, L., and Hand, D. (2005). Preparation of a novel TiO_2-based *p–n* junction nanotube photocatalyst. Environ. Sci. Technol. *39*, 1201–1208.

Chen, S.-S., Hsu, H.-D., and Li, C.-W. (2004). A new method to produce nanoscale iron for nitrate removal. J. Nanopart. Res. *6*, 639–647.

Cheng, R., Wang, J.-L., and Zhang, W.-X. (2007). Comparison of reductive dechlorination of *p*-chlorophenol using Fe^0 and nano-sized Fe^0. J. Hazard. Mater. *144*, 334–339.

Cheng, S.-F., and Wu, S.-C. (2000). The enhancement methods for the degradation of TCE by zero-valent metals. Chemosphere *41*, 1263–1270.

Cheng, S.-F., and Wu, S.-C. (2001). Feasibility of using metals to remediate water containing TCE. Chemosphere *43*, 1023–1028.

Cheng, P., Zheng, M., Jin, Y., Huang, Q., and Gu, M. (2003). Preparation and characterization of silica-doped titania photocatalyst through sol–gel method. Mater. Lett. *57*, 2989–2994.

Cho, I.-H., Kim, Y.-G., Yang, J.-K., Lee, N.-H., and Lee, S.-M. (2006). Solar-chemical treatment of groundwater contaminated with petroleum at gas station sites: *Ex situ* remediation using solar/TiO_2 photocatalysis and solar photofenton. J. Environ. Sci. Health A *41*, 457–473.

Cho, I.-H., Park, J.-H., and Kim, Y.-G. (2005a). Oxidative degradation and toxicity reduction of trichloroethylene (TCE) in water using TiO_2/solar light: Comparative study of TiO_2 slurry and immobilized systems. J. Environ. Sci. Health A *40*, 1033–1044.

Cho, M., Chung, H., Choi, W., and Yoon, J. (2005b). Different inactivation behaviours of MS-2 phage and *Escherichia coli* in TiO_2 photocatalytic disinfection. Appl. Environ. Microbiol. *71*, 270–275.

Cho, M., Chung, H., Choi, W., and Yoon, J. (2004). Linear correlation between inactivation of *E. coli* and OH radical concentration in TiO_2 photocatalytic disinfection. Water Res. *38*, 1069–1077.

Choe, S., Chang, Y.Y., Hwang, K.Y., and Khim, J. (2000). Kinetics of reductive denitrification by nanoscale zero-valent iron. Chemosphere *41*, 1307–1311.

Choi, W., Termin, A., and Hoffmann, M.R. (1994). The role of metal ion dopants in quantum-sized TiO_2: Correlation between photoreactivity and charge carrier recombination dynamics. J. Phys. Chem. *98*, 13669–13679.

Coetser, S.E., Heath, R.G., and Ndombe, N. (2007). Diffuse pollution associated with the mining sectors in South Africa: A first-order assessment. Water Sci. Technol. *55*, 9–16.
Cui, D.X., Tian, F.R., Ozkan, C.S., Wang, M., and Gao, H.J. (2005). Effect of single wall carbon nanotubes on human HEK293 cells. Toxicol. Lett. *155*, 73–85.
Colvin, V.L. (2003). The potential environmental impact of engineered nanomaterials. Nature Biotechnol. *21*, 1166–1170.
Danilczuk, M., Lund, A., Saldo, J., Yamada, H., and Michalik, J. (2006). Conduction electron spin resonance of small silver particles. Spectrochim. Acta: A *63*, 189–191.
DeFriend, K.A., Wiesner, M.R., and Barron, A.R. (2003). Alumina and aluminate ultrafiltration membranes derived from alumina nanoparticles. J. Membr. Sci. *224*, 11–28.
Del Valle, E.M.M. (2004). Cyclodextrins and their uses: A review. Process Biochem. *39*, 1033–1046.
Di, Z.-C., Ding, J., Peng, X.-J., Li, Y.-H., Luan, Z.-K., and Liang, J. (2006). Chromium adsorption by aligned carbon nanotubes supported ceria nanoparticles. Chemosphere *62*, 861–865.
Di Paola, A., Ikeda, S., Marci, G., and Ohtani, B. (2001). Transition metal doped TiO_2: Physical properties and photocatalytic behavior. Intern. J. Photoenergy *3*, 171–176.
Diallo, M.S., Christie, S., Swaminathan, P., Balogh, L., Shi, X., Um, W., Papelis, C., Goddard, W.A., and Johnson, J.H. (2004). Dendritic chelating agents 1. Cu(II) binding to ethylene diamine core poly(amidoamine) dendrimers in aqueous solutions. Langmuir *20*, 2640–2651.
Diallo, M.S., Christie, S., Swaminathan, P., Johnson, J.H., and Goddard, W.A. (2005). Dendrimer-enhanced ultrafiltration. 1. Recovery of Cu(II) from aqueous solutions using Gx-NH_2PAMAM dendrimers with ethylene diamine core. Environ. Sci. Technol. *39*, 1366–1377.
Ding, L., and Zheng, Y. (2007). Nanocrystalline zeolite beta: The effect of template agent on crystal size. Mater. Res. Bull. *42*, 584–590.
Donaldson, K., Aitken, R., Tran, L., Stone, V., Duffin, R., Forrest, G., and Alexander, A. (2006). Carbon nanotubes: A review of their properties in relation to pulmonary toxicology and workplace safety. Toxicol. Sci. *92*, 5–22.
Donaldson, K., and Tran, C.L. (2004). An introduction to the short-term toxicology of respirable industrial fibres. Mutat. Res. *553*, 5–9.
Dotzauer, D.M., Dai, J., Sun, L., and Bruening, M.L. (2006). Catalytic membranes prepared using layer-by-layer adsorption of polyelectrolyte/metal nanoparticle films in porous supports. Nano Lett. *6*, 2268–2272.
Duran, N., and Esposito, E. (2000). Potential applications of oxidative enzymes and phenoloxidase-like compounds in wastewater and soil treatment: A review. Appl. Catal. B: Environ. *28*, 83–99.
Dvoranova, D., Brezova, V., and Malati, M.A. (2002). Investigations of metal-doped titanium dioxide photocatalysts. Appl. Catal. B: Environ. *37*, 91–105.
Egerton, T.A., Kosa, S.A., and Christensen, P.A. (2006). Photoelectrocatalytic disinfection of *E. coli* suspensions by iron doped TiO_2. Phys. Chem. Chem. Phys. *8*, 398–406.
Elder, A., Gelein, R., Silva, V., Feikert, T., Opanashuk, L., Carter, J., Potter, R., Maynard, A., Ito, Y., Finkelstein, J., and Oberdörster, G. (2006). Translocation of inhaled ultrafine manganese oxide particles to the central nervous system. Environ. Health Perspect. *114*, 1172–1178.
Elliott, D.W., and Zhang, W.-X. (2001). Field assessment of nanoscale bimetallic particles for groundwater treatment. Environ. Sci. Technol. *35*, 4922–4926.
Enríquez, R., and Pichat, P. (2006). Different net effect of TiO_2 sintering temperature on the photocatalytic removal rates of 4-chlorophenol, 4-chlorobenzoic acid and dichloroacetic acid in water. J. Environ. Sci. Heal A *41*, 955–966.
Eshelby, K. (2007). Dying for a drink. BMJ *334*, 610–612.
Esteban-Cubillo, A., Pecharromán, C., Aguilar, E., Santarén, J., and Moya, J.S. (2006). Antibacterial activity of copper monodispersed nanoparticles into sepiolite. Mater. Sci. *41*, 5208–5212.
Falconer, I.R. (1999). An overview of problems caused by toxic blue-green algae (cyanobacteria) in drinking and recreational water. Environ. Toxicol. *14*, 5–12.
Falconer, I.R., and Humpage, A.R. (2005). Health risk assessment of cyanobacterial (blue-green algal) toxins in drinking water. Int. J. Environ. Res. Public Health *2*, 43–50.
Fawell, J., and Nieuwenhuijsen, M.J. (2003). Contaminants in drinking water. Br. Med. Bull. *68*, 199–208.
Foley, J.A., DeFries, R., Asner, G.P., Barford, C., Bonan, G., Carpenter, S.R., Chapin, F.S., Coe, M.T., Daily, G.C., Gibbs, H.K., Helkowski, J.H., Holloway, T., Howard, E.A., Kucharik, C.J., Monfreda, C., Patz, J.A., Prentice, I.C., Ramankutty, N., and Snyder, P.K. (2005). Global consequences of land use. Science *309*, 570–574.
Frechet, J.M.J., and Tomalia, D.A. (2001). Dendrimers and other dendritic polymers. (Wiley and Sons Publishers, New York).
Fryxell, G.E., Lin, Y., Fiskum, S., Birnbaum, J.C., and Wu, H. (2005). Actinide sequestration using self-assembled monolayers on mesoporous supports. Environ. Sci. Technol. *39*, 1324–1331.
Fryxell, G.E., Wu, H., Lin, Y., Shaw, W.J., Birnbaum, J.C., Linehan, J.C., Nie, Z., Kemner, K., and Kelly, S. (2004). Lanthanide selective sorbents: Self-assembled monolayers on mesoporous supports (SAMMS). J. Mater. Chem. *14*, 3356–3363.
Fujishima, A., and Honda, K. (1972). Electrochemical photolysis of water at a semiconductor electrode. Nature *238*, 37–38.

Gao, C., and Yan, D. (2004). Hyperbranched polymers: From synthesis to applications. Prog. Polym. Sci. *29*, 183–275.

Garron, A., Lázár, K., and Epron, F. (2006). Characterization by Mössbauer spectroscopy of trimetallic Pd-Sn-Au/Al_2O_3 and Pd-Sn-Au/SiO_2 catalysts for denitration of drinking water. Appl. Catal. B: Environ. *65*, 240–248.

Giri, S., Ganguli, S., and Bhattacharya, M. (2001). Surface oxidation of iron nanoparticles. Appl. Surf. Sci. *182*, 345–349.

Gogniat, G., Thyssen, M., Denis, M., Pulgarin, C., and Dukan, S. (2006). The bactericidal effect of TiO_2 photocatalysis involves adsorption onto catalyst and the loss of membrane integrity. FEMS Microbiol. Lett. *258*, 18–24.

Gopal, K., Tripathy, S.S., Bersillon, J.L., and Dubey, S.P. (2007). Chlorination by-products, their toxicodynamics and removal from drinking water. Hazard, J. . Mater. *140*, 1–6Goswami, D.Y., Vijayaraghavan, S., Lu, S., and Tamm, G. (2004). New and emerging developments in solar energy. Solar Energy *76*, 33–43.

Gotpagar, J.K., Grulke, E.A., Tsang, T., and Bhattacharyya, D. (1997). Reductive dehalogenation of trichloroethylene using zero-valent iron. Environ. Progr. *6*, 137–143.

Grubek-Jaworska, H., Nejman, P., Czuminska, K., Przybylowski, T., Huczko, A., Lange, H., Bystrzejewski, M., Baranowski, P., and Chazan, R. (2006). Preliminary results on the pathogenic effects of intratracheal exposure to one-dimensional nanocarbons. Carbon *44*, 1057–1063.

Gurr, J.R., Wang, A.S.S., Chen, C.H., and Jan, K.Y. (2005). Ultrafine titanium dioxide particles in the absence of photoactivation can induce oxidative damage to human bronchial epithelial cells. Toxicology *213*, 66–73.

Haider, S., Naithani, V., Viswanathan, P.N., and Kakkar, P. (2003). Cyanobacterial toxins: A growing environmental concern. Chemosphere *52*, 1–21.

Hatchett, D.W., and Henry, S. (1996). Electrochemistry of sulfur adlayers on the low-index faces of silver. J. Phys. Chem. *100*, 9854–9859.

He, P., and Zhao, D.Y. 2005. Preparation and characterization of a new class of starch-stabilized bimetallic nanoparticles for degradation of chlorinated hydrocarbons in water. Environ. Sci. Technol. *39*, 3314–3320.

Heckel, K., Kiefmann, R., Dorger, M., Stoeckelhuber, M., and Goetz, A.E. (2004). Colloidal gold particles as a new *in vivo* marker of early acute lung injury. Am. J. Physiol. Lung Cell. Mol. Physiol. *287*, 867–878.

Herrmann, J.M. (1999). Heterogeneous photocatalysis: Fundamentals and applications to the removal of various types of aqueous pollutants. Catal. Today *53*, 115–129.

Hirakawa, K., Mori, M., Yoshida, M., Oikawa, S., and Kawanishi, S. (2004). Photo-irradiated titanium dioxide catalyzes site-specific DNA damage via generation of hydrogen peroxide. Free Radic. Res. *38*, 439–447.

Hirano, K., Suzuki, E., Ishikawa, A., Moroi, T., Shiroishi, H., and Kaneko, M. (2000). Sensitization of TiO_2 particles by dyes to achieve H_2 evolution by visible light. J. Photochem. Photobiol. A: Chem. *136*, 157–161.

Hoet, P.H.M., Bruske-Hohlfeld, I., and Salata, O.V. (2004). Nanoparticles – known and unknown health risks. J. Nanobiotechnol. *2*, 12–27.

Hollman, A.M., and Bhattacharyya, D. (2004). Pore assembled multilayers of charged polypeptides in microporous membranes for ion separation. Langmuir *20*, 5418–5424.

Hozalski, R.M., Zhang, L., and Arnold, W.A. (2001). Reduction of haloacetic acids by Fe^0: Implications for treatment and fate. Environ. Sci. Technol. *35*, 2258–2263.

Hsien, Y.-H., Chang, C.-F., Chen, Y.-H., and Cheng, S. (2001). Photodegradation of aromatic pollutants in water over TiO_2 supported on molecular sieves. Appl. Catal. B: Environ. *31*, 241–249.

Hu, C., Hu, X., Guo, J., and Qu, J. (2006). Efficient destruction of pathogenic bacteria with NiO/$SrBi_2O_4$ under visible light irradiation. Environ. Sci. Technol. *40*, 5508–5513.

Ibanez, J., Litter, M., and Pizarro, R. (2003). Photocatalytic bactericidal effect of TiO_2 on *Enterobacter cloacae* – Comparative study with other Gram (-) bacteria. J. Photochem. Photobiol. A: Chem. *157*, 81–85.

Ihara, T., Miyoshi, M., Ando, M., Sugihara, S., and Iriyama, Y. (2001). Preparation of a visible light-active TiO_2 photocatalyst by RF plasma treatment. J. Mater. Sci. *36*, 4201–4207.

Ihara, T., Miyoshi, M., Iriyama, Y., Matsumoto, O., and Sugihara, S. (2003). Visible light-active titanium oxide photocatalyst realized by an oxygen-deficient structure and by nitrogen doping. Appl. Catal. B: Environ. *42*, 403–409.

Iijima, S. (1991). Helical microtubules of graphitic carbon. Nature *354*, 56–58.

Iijima, S., and Ichihashi, T. (1993). Single-shell carbon nanotubes of 1-nm diameter. Nature *363*, 603–605.

Irie, H., Watanabe, Y., and Hashimoto, K. (2003). Carbon-doped anatase as a visible light-sensitive photocatalyst. Chem. Lett. *32*, 772–773.

Järup, L. (2003). Hazards of heavy metal contamination. Br. Med. Bull. *68*, 167–182.

Jiang, F., Zheng, Z., Xu, Z., Zheng, S., Guo, Z., and Chen, L. (2006). Aqueous Cr(VI) photo-reduction catalyzed by TiO_2 and sulfated TiO_2. J. Hazard. Mater. *134*, 94–103.

Jikei, M., and Kakimoto, M.-A. (2001). Hyperbranched polymers: A promising new class of materials. Prog. Polym. Sci. *26*, 1233–1285.

Jin, J., Li, R., Wang, H., Chen, H., Liang, K., and Ma, J. (2007). Magnetic Fe nanoparticle functionalized water-soluble multi-walled carbon nanotubules: Towards the preparation of sorbent for aromatic compounds removal. Chem. Commun. *4*, 386–388.

Johnson, D.M., Hokanson, D.R., Zhang, Q., Czupinski, K.D., and Tang, J. (2007). Feasibility of water purification technology in rural areas of developing countries. J. Env. Man. *88*, 416–427.

Joo, J., Kwon, S.G., Yu, T., Cho, M., Lee, J., Yoon, J., and Hyeon, T. (2005). Large-scale synthesis of TiO_2 nanorods via nonhydrolytic sol–gel ester elimination reaction and their application to photocatalytic inactivation of *E. coli*. J. Phys. Chem. B *109*, 15297–15302.

Kabra, K., Chaudhary, R., and Sawhney, R.L. (2004). Treatment of hazardous organic and inorganic compounds through aqueous-phase photocatalysis: A review. Ind. Eng. Chem. Res. *43*, 7683–7696.

Kamat, P., Huehn, R., and Nicolaescu, R. (2002). A 'sense and shoot' approach for photocatalytic degradation of organic contaminants in water. J. Phys. Chem. B *106*, 788–794.

Kanel, S.R., Greneche, J.M., and Choi, H. (2006). Arsenic(V) removal from groundwater using nano-scale zero-valent iron as a colloidal reactive barrier material. Environ. Sci. Technol. *40*, 2045–2050.

Kanel, S.R., Manning, B., Charlet, L., and Choi, H. (2005). Removal of arsenic(III) from groundwater by nanoscale zero-valent iron. Environ. Sci. Technol. *39*, 1291–1298.

Kapoor, A., and Viraraghavan, T. (1997). Nitrate removal from drinking water. J. Environ. Eng. *123*, 371–380.

Keane, M.A., Pina, G., and Tavoularis, G. (2004). The catalytic hydrodechlorination of mono-, di- and trichlorobenzenes over supported nickel. Appl. Catal. B: Environ. *48*, 275–286.

Kelly, S.D., Kemmer, K.M., Fryxell, G.E., Liu, J., Mattigod, S.V., and Ferris, K.F. (2001). X-ray-absorption fine-structure spectroscopy study of the interactions between contaminant tetrahedral anions and self-assembled monolayers on mesoporous supports. J. Phys. Chem. B *105*, 6337–6346.

Kemper, K.E. (2004). Groundwater – from development to management. Hydrogeol. J. *12*, 3–5.

Keum, Y.S., and Li, Q.X. (2004). Reduction of nitroaromatic pesticides with zero-valent iron. Chemosphere *54*, 255–263.

Kim, J., and Grate, J.W. (2003). Single-enzyme nanoparticles armored by a nanometer-scale organic/inorganic network. Nano. Lett. *3*, 1219–1222.

Kim, J., and Grate, J.W. (2005). Nano-biotechnology in using enzymes for environmental remediation: Single-enzyme Nanoparticles. In ACS Symposium Series, Nanotechnology and the Environment: Applications and Implications, Karn, B., Masciangioli, T., Zhang, W., Colvin, V., Alivisatos, P., eds. (American Chemical Society, Washington, D.C), pp. 220–225.

Kim, J.S., Kuk, E., Yu, K.M., Kim, J.-H., Park, S.J., Lee, H.J., Kim, S.H., Park, Y.K., Park, Y.H., Hwang, C.-Y., Kim, Y.-K., Lee, Y.-S., Jeong, D.-H., and Cho, M.-H. (2007). Antimicrobial effects of silver nanoparticles. Nanomedicine *3*, 95–101.

Kirchnerova, J., Herrera Cohen, M.-L., Guy, C., and Klvana, D. (2005). Photocatalytic oxidation of *n*-butanol under fluorescent visible light lamp over commercial TiO_2 (Hombicat UV100 and Degussa P25). Appl. Catal. A: Gen. *282*, 321–332.

Kiwi, J., and Nadtochenko, V. (2005). Evidence for the mechanism of photocatalytic degradation of the bacterial wall membrane at the TiO_2 interface by ATR-FTIR and laser kinetic spectroscopy. Langmuir *21*, 4631–4641.

Klabunde, K.J., Stark, J., Koper, O., Mohs, C., Park, D.G., Decker, S., Jiang, Y., Lagadic, I., and Zhang, D. (1996). Nanocrystals as stoichiometric reagents with unique surface chemistry. J. Phys. Chem. *100*, 12142–12153.

Klajnert, B., and Bryszewska, M. (2001). Dendrimers: Properties and applications. Acta Biochim. Pol. *48*, 199–208.

Konaka, R., Kasahara, E., Dunlap, W.C., Yamamoto, Y., Chien, K.C., and Inoue, M. (1999). Irradiation of titanium dioxide generates both singlet oxygen and superoxide anion. Free Radical Biol. Med. *27*, 294–300.

Koper, O.B., Klabunde, J.S., Marchin, G.L., Klabunde, K.J., Stoimenov, P., and Bohra, L. (2002). Nanoscale powders and formulations with biocidal activity towards spores and vegetative cells of *Bacillus* species, viruses, and toxins. Curr. Microbiol. *44*, 49–55.

Kositzi, M., Poulios, I., Malato, S., Caceres, J., and Campos, A. (2004). Solar photocatalytic treatment of synthetic municipal wastewater. Water Res. *38*, 1147–1154.

Kostal, J., Mulchandani, A., and Chen, W. 2001. Tunable biopolymers for heavy metal removal. Macromolecules *34*, 2257–2261.

Kostal, J., Mulchandani, A., Gropp, K., and Chen, W. (2003). A temperature-responsive biopolymer for mercury remediation. Environ. Sci. Technol. *37*, 4457–4462.

Kostal, J., Prabhukumar, G., Lao, U.L., Chen, A., Matsumoto, M., Mulchandani, A., and Chen, W. (2005). Customizable biopolymers for heavy metal remediation. J. Nanopart. Res. *7*, 517–523.

Koyama, S., Endo, M., Kim, Y.A., Hayashi, T., Yanagisawa, T., Osaka, K., Koyama, H., Haniu, H., and Kuroiwa, N. (2006). Role of systemic T-cells and histopathological aspects after subcutaneous implantation of various carbon nanotubes in mice. Carbon *44*, 1079–1092.

Kreyling, W., Semmler, M., Erbe, M.P., Takenaka, S., Schulz, H., Oberdörster, G., and Ziesenis, A. (2002). Translocation of ultrafine insoluble iridium particles from lung epithelium to extrapulmonary organs is size dependent but very low. J. Toxicol. Environ. Health A *65*, 1513–1530.

Lam, C.W., James, J.T., McCluskey, R., and Hunter, R.L. (2004). Pulmonary toxicity of single-wall carbon nanotubes in mice 7 and 90 days after intratracheal instillation. Toxicol. Sci. *77*, 126–134.

Lam, C.W., James, J., McCluskey, R., Sivaram, A., and Hunter, R. (2006). A review of carbon nanotube toxicity and assessment of potential occupational and environmental health risks. Crit. Rev. Toxicol. *36*, 189–217.

Landau, M.V., Vradman, L., Wang, X., and Titelman, L. (2005). High loading TiO_2 and ZrO_2 nanocrystals ensembles inside the mesopores of SBA-15: Preparation, texture and stability. Micropor. Mesopor. Mater. *78*, 117–129.

Langwaldt, J.H., and Puhakka, J.A. (2000). On-site biological remediation of contaminated groundwater: A review. Environ. Pollut. *107*, 187–197.
Lanone, S., and Boczkowski, J. (2006). Biomedical applications and potential health risks of nanomaterials: Molecular mechanisms. Curr. Mol. Med. *6*, 651–663.
Lao, U.L., Chen, A., Matsumoto, M.R., Mulchandani, A., and Chen, W. (2007). Cadmium removal from contaminated soil by thermally-responsive elastin (ELPEC20) biopolymers. Biotechnol. Bioeng. *98*, 349–355.
Lee, J.-S., Chon, H.-T., and Kim, K.-W. (2005a). Human risk assessment of As, Cd, Cu and Zn in the abandoned metal mine site. Environ. Geochem. Health *27*, 185–191.
Lee, S., Nakamura, M., and Ohgaki, S. (1998). Inactivation of phage Qβ by 254 nm UV light and titanium dioxide photocatalyst. J. Environ. Sci. Health A *33*, 1643–1655.
Lee, S.-H., Pumprueg, S., Moudgil, B., and Sigmund, W. (2005b). Inactivation of bacterial endospores by photocatalytic nanocomposites. Colloids and Surfaces B: Biointerfaces *40*, 93–98.
Lee, E.J., and Schwab, K.J. (2005). Deficiencies in drinking water distribution systems in developing countries. J. Water Health *3*, 109–127.
Leonard, P., Hearty, S., Brennan, J., Dunne, L., Quinn, J., Chakraborty, T., and O'Kennedy, R. (2003). Advances in biosensors for detection of pathogens in food and water. Enzyme Microb. Technol. *32*, 3–13.
Li, L., Fan, M., Brown, R.C., and van Leeuwen, J. (2006a). Synthesis, properties, and environmental applications of nanoscale iron-based materials: A review. Crit. Rev. Environ. Sci. Tech. *36*, 405–431.
Li, X.-Q., Elliott, D.W., and Zhang, W.-X. (2006b). Zero-valent iron nanoparticles for abatement of environmental pollutants: Materials and engineering aspects. Crit. Rev. Solid State Mater. Sci. *31*, 111–122.
Li, Y.-H., Ding, J., Luan, Z.K., Di, Z.C., Zhu, Y.F., Xu, C.L., Wu, D.H., and. Wei, B.Q. (2003). Competitive adsorption of Pb^{2+}, Cu^{2+} and Cd^{2+} ions from aqueous solutions by multi-walled carbon nanotubes. Carbon *41*, 2787–2792.
Li, X.-Q., and Zhang, W.-X. (2006). Iron nanoparticles: The core-shell structure and unique properties for Ni(II) sequestration. Langmuir *22*, 4638–4642.
Liau, S.Y., Read, D.C., Pugh, W.J., Furr, J.R., and Russell, A.D. (1997). Interaction of silver nitrate with readily identifiable groups: Relationship to the antibacterial action of silver ions. Lett. Appl. Microbiol. *25*, 279–283.
Lien, H.L., and Zhang, W.-X. (1999). Transformation of chlorinated methanes by nanoscale iron particles. J. Environ. Eng. *125*, 1042–1047.
Lien, H.L., and Zhang, W.-X. (2001). Nanoscale iron particles for complete reduction of chlorinated ethenes. Colloids Surf. A: Physicochem. Eng. Aspects *191*, 97–105.
Lim, T.-T., Feng, J., and Zhu, B.-W. (2007). Kinetic and mechanistic examinations of reductive transformation pathways of brominated methanes with nano-scale Fe and Ni/Fe particles. Water Res. *41*, 875–883.
Lin, Y., Fiskum, S.K., Yantasse, W., Wu, H., Mattigod, S.V., Vorpagel, E., and Fryxell, G.E. (2005). Incorporation of hydroxypyridinone ligands into self-assembled monolayers on mesoporous supports for selective actinide sequestration. Environ. Sci. Technol. *39*, 1332–1337.
Lin, Y., Fryxell, G.E., Wu, H., and Engelhard, M. (2001). Selective sorption of cesium using self-assembled monolyaers on mesoporous supports. Environ. Sci. Technol. *35*, 3962–3966.
Linsebigler, A.L., Lu, G., and Yates, J.T. (1995). Photocatalysis on TiO_2 surfaces: Principles, mechanisms, and selected results. Chem. Rev. *95*, 735–758.
Liu, Y., Chen, X., Li, J., and Burda, C. (2005b). Photocatalytic degradation of azo dyes by nitrogen-doped TiO_2 nanocatalysts. Chemosphere *61*, 11–18.
Liu, Z., He, Y., Li, F., and Liu, Y. (2006b). Photocatalytic treatment of RDX wastewater with nano-sized titanium dioxide. Environ. Sci. Pollut. Res. Int. *13*, 328–332.
Liu, Y., Li, J., Qiu, X., and Burda, C. (2006a). Novel TiO_2 nanocatalysts for wastewater purification – Tapping energy from the sun. Water Practice and Technology *1*, 1 (doi: 10.2166/WPT.2006072).
Liu, Y., Majetich, S.A., Tilton, R.D., Sholl, D.S., and Lowry, G.V. (2005a). TCE dechlorination rates, pathways, and efficiency of nanoscale iron particles with different properties. Environ. Sci. Technol. *39*, 1338–1345.
Liu, J., Shin, Y., Wang, L.-Q., Exarhos, G.J., Chang, J.H., Fryxell, G.E., Nie, Z., Zemanian, T., and Samuels, W.D. (2003). Strategies for the design and synthesis of hybrid multi-functional nanoporous materials. J. Nanostructured Catalysts *19*, 297–310.
Liu, Z.L., Wang, H.B., Lu, Q.H., Du, G.H., Peng, L., Du, Y.Q., Zhang, S.M., and Yao, K.L. (2004). Synthesis and characterization of ultrafine well-dispersed magnetic nanoparticles. J. Magn. Mater. *283*, 258–262.
Liu, H., and Yang, T. (2003). Photocatalytic inactivation of *Escherichia coli* and *Lactobacillus helveticus* by ZnO and TiO_2 activated with ultraviolet light. Process Biochem. *39*, 475–481.
Long, Q.R., and Yang, R.T. (2001). Carbon nanotubes as superior sorbent for dioxin removal. J. Am. Chem. Soc. *123*, 2058–2059.
Lonnen, J., Kilvington, S., Kehoe, S.C., Al-Touati, F., and McGuigan, K.G. (2005). Solar and photocatalytic disinfection of protozoan, fungal and bacterial microbes in drinking water. Water Res. *39*, 877–883.
López-Munoz, M.J., Van Grieken, R., Aguado, J., and Marugán, J. (2005). Role of the support on the activity of silica-supported TiO_2 photocatalysts: Structure of the TiO_2/SBA-15 photocatalysts. Catal. Today *101*, 307–314.
Lovern, S.B., and Klaper, R.D. (2006). *Daphnia magna* mortality when exposed to titanium dioxide and fullerene (C_{60}) nanoparticles. Environ. Toxicol. Chem. *25*, 1132–1137.

Lu, C., and Chiu, H. (2006). Adsorption of zinc(II) from water with purified carbon nanotubes. Chem. Eng. Sci. *61*, 1138–1145.

Lu, C., Chiu, H., and Bai, H. (2007). Comparisons of adsorbent cost for the removal of zinc(II) from aqueous solution by carbon nanotubes and activated carbon. J. Nanosci. Nanotechnol. *7*, 1647–1652.

Lu, C., Chiu, H., and Liu, C. (2006b). Removal of zinc(II) from aqueous solution by purified carbon nanotubes: Kinetic and equilibrium studies. Ind. Eng. Chem. Res. *45*, 2850–2855.

Lu, C., Chung, Y.-L., and Chang, K.-F. (2005). Adsorption of trihalomethanes from water with carbon nanotubes. Water Res. *39*, 183–189.

Lu, C., Chung, Y.-L., and Chang, K.-F. (2006a). Adsorption thermodynamic and kinetic studies of trihalomethanes on multiwalled carbon nanotubes. J. Hazard. Mater. *138*, 304–310.

Lu, C., and Liu, L. (2006). Removal of nickel(II) from aqueous solution by carbon nanotubes. J. Chem. Technol. Biotechnol. *81*, 1932–1940.

Madigan, M., and Martinko, J. (2005). Brock Biology of Microorganisms, 11th ed. (Prentice Hall Publishers, Englewood Cliffs).

Makhluf, S., Dror, R., Nitzan, Y., Abramovich, A., Jelinek, R., and Gedanken, A. (2005). Microwave-assisted synthesis of nanocrystalline MgO and its use as a bacteriocide. S. Adv. Funct. Mater. *15*, 1708–1715.

Maness, P.-C., Smolinski, S., Blake, D.M., Huang, Z., Wolfrum, E.J., and Jacoby, W.A. (1999). Bactericidal activity of photocatalytic TiO_2 reaction: Toward an understanding of its killing mechanism. Appl. Environ. Microbiol. *65*, 4094–4098.

Mangun, C.L., Yue, Z.R., Economy, J., Maloney, S., Kemme, P., and Cropek, D. (2001). Adsorption of organic contaminants from water using tailored ACFs carbon. Chem. Mater. *13*, 2356–2360.

Mara, D.D. (2003). Water, sanitation and hygiene for the health of developing nations. Public Health *117*, 452–456.

Masciangioli, T., and Zhang, W.-X. (2003). Environmental technologies at the nanoscale. Environ. Sci. Technol. *37*, 102–108.

Matheson, L.J., and Tratnyek, P.G. (1994). Reductive dehalogenation of chlorinated methanes by iron metal. Environ. Sci. Technol. *28*, 2045–2053.

Matsunaga, T., Tomada, R., Nakajima, T., and Wake, H. (1985). Photochemical sterilization of microbial cells by semiconductor powders. FEMS Microbiol. Lett. *29*, 211–214.

Matsunaga, T., Tomoda, R., Nakajima, Y., Nakamura, N., and Komine, T. (1988). Continuous-sterilization system that uses photosemiconductor powders. Appl. Environ. Microbiol. *54*, 1330–1333.

Mattigod, S.V., Feng, X., Fryxell, G.E., Liu, J., and Gong, M. (1999). Separation of complexed mercury from aqueous wastes using self-assembled mercaptan on mesoporous silica. Sep. Sci. Technol. *34*, 2329–2345.

Mattigod, S.V., Fryxell, G.E., Feng, X., Parker, K.E., and Piers, E.M. (2006). Removal of mercury from aqueous streams of fossil fuel power plants using novel functionalized nanoporous sorbents. In Coal combustion by-products and environmental issues, Sajwan, K.S., Twardowska, I., Punshon, T. and Alva, A.K., ed. (Springer, New York), pp 99–104.

Maynard, A.D. (2007). Nanotechnology: The next big thing, or much ado about nothing? Ann. Occup. Hyg. *51*, 1–12.

Maynard, A.D., and Kuempel, E.D. (2005). Airborne nanostructured particles and occupational health. J. Nanopart. Res. *7*, 587–614.

McLoughlin, O.A., Ibanẽz, F.P., Gernjak, W., Rodríguez, S.M., and Gill, L.W. (2004). Photocatalytic disinfection of water using low cost compound parabolic collectors. Solar Energy *77*, 625–633.

Menini, C., Park, C., Shin, E.-J., Tavoularis, G., and Keane, M.A. (2000). Catalytic hydrodehalogenation as a detoxification methodology. Catal. Today *62*, 355–366.

Meyer, D.E., Wood, K., Bachas, L.G., and Bhattacharyya, D. (2004). Degradation of chlorinated organics by membrane-immobilized nanosized metals. Environ. Prog. *23*, 232–242.

Miller, S.A., Young, V.Y., and Martin, C.R. (2001). Electro-osmotic flow in template-prepared carbon nanotube membranes. J. Am. Chem. Soc. *123*, 12335–12342.

Mills, A., and LeHunte, S. (1997). An overview of semiconductor photocatalysis. J. Photochem. Photobiol. A: Chem. *108*, 1–35.

Mitoraj, D., Jánczyk, A., Strus, M., Kisch, H., Stochel, S., Heczko, P.B., and Macyk, W. (2007). Visible light-inactivation of bacteria and fungi by modified titanium dioxide. Photochem. Photobiol. Sci. *6*, 642–648.

Miyagawa, H., Misra, M., and Mohanty, A.K. (2005). Mechanical properties of carbon nanotubes and their polymer nanocomposites. J. Nanosci. Nanotechnol. *5*, 1593–1615.

Moe, C.L., and Rheitigans, R.D. (2006). Global challenges in water, sanitation and health. J. Water Health. *4*, 41–57.

Mohsen, M.S., Jaber, J.O., and Afonso, M.D. (2003). Desalination of brackish water by nanofiltration and reverse osmosis. Desalination *157*, 167.

Monteiro-Riviere, N.A., Nemanich, R.J., Inman, A.O., Wang, Y.Y.Y., and Riviere, J.E. (2005). Multi-walled carbon nanotube interactions with human epidermal keratinocytes. Toxicol. Lett. *155*, 377–384.

Montgomery, M.A., and Elimelech, M. (2007). Water and sanitation in developing countries: Including health in the equation. Environ. Sci. Technol. *41*, 17–24.

Moon, J., Yun, C.Y., Chung, K.-W., Kang, M.S., and Yi, J. (2003). Photocatalytic activation of TiO_2 under visible light using Acid Red 44. Catal. Today *87*, 77–86.

Moore, M.N. (2006). Do nanoparticles present ecotoxicological risks for the health of the aquatic environment? Environ. Int. *32*, 967–976.

Moore, M., Gould, P., and Keary, B.S. (2003). Global urbanization and impact on health. Int. J. Hyg. Environ. Health *206*, 269–278.

Moreno, N., Querol, X., and Ayora, C. (2001). Utilization of zeolites synthesized from coal fly ash for the purification of acid mines water. Environ. Sci. Technol. *35*, 3526–3534.

Morones, J.R., Elechiguerra, J.L., Camacho, A., Holt, K., Kouri, J.B., Ramírez, J.P., and Yacaman, M.Y. (2005). The bactericidal effect of silver nanoparticles. Nanotechnology *16*, 2346–2353.

Muller, J., Huaux, F., Moreau, N., Misson, P., Heilier, J.F., Delos, M., Arras, M., Fonseca, A., Nagy, J.B., and Lison, D. (2005). Respiratory toxicity of multi-walled carbon nanotubes. Toxicol. Appl. Pharmacol. *207*, 221–231.

Nahar, S., Hasegawa, K., and Kagaya, S. (2006). Photocatalytic degradation of phenol by visible light-responsive iron-doped TiO_2 and spontaneous sedimentation of the TiO_2 particles. Chemosphere *65*, 1976–1982.

Navio, J.A., Colon, G., Litter, M.I., and Bianco, G.N. (1996). Synthesis, characterization and photocatalytic properties of iron-doped titanium semiconductors prepared from TiO_2 and iron(III) acetylacetonate. J. Mol. Catal. A: Chem. *106*, 267–276.

Neden, D.G., Jones, R.J., Smith, J.R., Kirmeyer, G.J., and Foust, G.W. (1992). Comparing chlorination and chloramination for controlling bacterial re-growth. J. Am. Water Works Assoc. *84*, 80–88.

Nel, A., Xia, T., Mädler, L., and Li, N. (2006). Toxic potential of materials at the nanolevel. Science *311*, 622–627.

Neumann, B., Bogdanoff, P., Tributsch, H., Sakthivel, S., and Kisch, H. (2005). Electrochemical mass spectroscopic and surface photovoltage studies of catalytic water photooxidation by undoped and carbon-doped titania. J. Phys. Chem. B *109*, 16579–16586.

Norton, C.D., and Lechevallier, M.W. (1997). Chloramination: Its effect on distribution system water quality. J. Am. Water Works Assoc. *89*, 66–77.

Nurmi, J.T., Tratnyek, P.G., Sarathy, V., Baer, D.R., and Amonette, J.E. (2005). Characterization and properties of metallic iron nanoparticles: Spectroscopy, electrochemistry, and kinetics. Environ. Sci. Technol. *39*, 1221–1230.

Nutt, M.O., Heck, K.N., Alvarez, P., and Wong, M.S. (2006). Improved Pd-on-Au bimetallic nanoparticle catalysts for aqueous-phase trichloroethene hydrodechlorination. Appl. Catal. B: Environ. *69*, 115–125.

Nutt, M.O., Hughes, J.B., and Wong, M.S. (2005). Designing Pd-on-Au bimetallic nanoparticle catalysts for trichloroethene hydrodechlorination. Environ. Sci. Technol. *39*, 1346–1353.

O'Halloran, T.V., Frantz, B., Shin, M.K., Ralston, D.M., and Wright, J.G. (1989). The MerR heavy metal receptor mediates positive activation in a topologically novel transcription complex. Cell *56*, 119–129.

Oberdörster, E. (2004). Manufactured nanomaterials (Fullerenes, C-60) induce oxidative stress in the brain of juvenile largemouth bass. Environ. Health Perspect. *112*, 1058–1062.

Oberdörster, G., Finkelstein, J.N., Johnston, C., Gelein, R., Cox, C., Baggs, R., and Elder, A.C. (2000). Acute pulmonary effects of ultrafine particles in rats and mice. Res. Rep. Health Eff. Inst. *96*, 5–86.

Oberdörster, G., Oberdörster, E., and Oberdörster, J. (2005). Nanotoxicology: An emerging discipline evolving from studies of ultrafine particles. Environ. Health Perspect. *113*, 823–839.

Oberdörster, G., Sharp, Z., Atudorei, V., Elder, A., Gelein, R., Kreyling, W., and Cox, C. (2004). Translocation of inhaled ultrafine particles to the brain. Inhalation Toxicol. *16*, 437–445.

Oberdörster, G., Stone, V., and Donaldson, K. (2007). Toxicology of nanoparticles: A historical perspective. Nanotoxicology *1*, 2–25.

Oberdörster, E., Zhu, S., Blickley, T.M., McClellan-Green, P., and Haasch, M.L. (2006). Ecotoxicology of carbon-based engineered nanoparticles: Effects of fullerene (C_{60}) on aquatic organisms. Carbon *44*, 1112–1120.

Oh, S.-M., Kim, S.-S., Lee, J.E., Ishigaki, T., and Park, D.-W. (2003). Effect of additives on photocatalytic activity of titanium dioxide powders synthesized by thermal plasma. Thin Solid Films *435*, 252–258.

Ohno, T., Mitsui, T., and Matsumura, M. (2003). Photocatalytic activity of S-doped photocatalyst under visible light. Chem. Lett. *32*, 364–365.

Ohno, T., Sarukawa, K., Tokieda, K., and Matsumura, M. (2001). Morphology of a TiO_2 photocatalyst (Degussa P-25) consisting of anatase and rutile crystalline phases. J. Catal. *203*, 82–86.

Pal, S., Tak, Y.K., and Song, J.M. (2007). Does the antibacterial activity of silver nanoparticles depend on the shape of the nanoparticle? A study of the Gram-negative bacterium *Escherichia coli*. Appl. Environ. Microbiol. *73*, 1712–1720.

Palmer, R.A., Doan, T.M., Lloyd, P.G., Jarvis, B.L., and Ahmed, N.U. (2002). Reduction of TiO_2 with hydrogen plasma. Plasma Chem. Plasma Process 3, 335–350.

Panáček, A., Kvítek, L., Prucek, R., Kolář, M., Večeřová, R., Pizrová, N., Sharma, V.K., Nevěčná, T., and Zbořil, R. (2006). Silver colloid nanoparticles: Synthesis, characterization, and their antibacterial activity. J. Phys. Chem. B *110*, 16248–16253.

Pattersom, J.W. (1985). Wastewater Treatment Technology. (Ann Arbor: Ann Arbor Science Publishers Inc).

Peltier, S., Cotte, E., Gatel, D., Herremans, L., and Cavard, J. (2003). Nanofiltration improvements of water quality in a large distribution system. Water Supply 3, 193–200.

Pena, M.E., Korfiatis, G.P., Patel, M., Lippincott, L., and Meng, X. (2005). Adsorption of As(V) and As(III) by nanocrystalline titanium dioxide. Water Res. *39*, 2327–2337.

Peng, X.J., Li, Y.H., Luan, Z.K., Di, Z.C., Wang, H.Y., Tian, B.H., and Jia, Z.P. (2003). Adsorption of *1*,2-dichlorobenzene from water to carbon nanotubes. Chem. Phys. Lett. *376*, 154–158.
Peng, X.J., Luan, Z.K., Ding, J., Di, Z.C., Li, Y.H., and Tian, B.H. (2005). Ceria nanoparticles supported nanotubes for the removal of arsenate from water. Mater. Lett. *59*, 399–403.
Peral, J., Domenech, X., and Ollis, D.F. (1997). Heterogeneous photocatalysis for purification, decontamination and deodorization of air. J. Chem. Technol. Biotechnol. *70*, 117–140.
Polizu, S., Savadogo, O., Poulin, P., and Yahia, L. (2006). Applications of carbon nanotubes-based biomaterials in biomedical nanotechnology. J. Nanosci. Nanotechnol. *6*, 1883–1904.
Ponder, S.M., Darab, J.G., and Mallouk, T.E. (2000). Remediation of Cr(VI) and Pb(II) aqueous solutions using supported, nanoscale zero-valent iron. Environ. Sci. Technol. *34*, 2564–2569.
Popov, V.N. (2004). Carbon nanotubes: Properties and application. Mater. Sci. Eng. *43*, 61–102.
Prabhukumar, G., Matsumoto, M., Mulchandani, A., and Chen, W. (2004). Cadmium removal from contaminated soil by tunable biopolymers. Environ. Sci. Technol. *38*, 3148–3152.
Prieto, O., Fermoso, J., Nuẽez, Y., Del Valle, J.L., and Irusta, R. (2005). Decolouration of textile dyes in wastewaters by photocatalysis with TiO_2. Solar Energy *79*, 376–383.
Qin, J.J., Oo, M.H., and Kekre, K.A. (2007). Nanofiltration for recovering wastewater from a specific dyeing facility. Sep. Pur. Tech. *56*, 199–203.
Rao, C.N.R., Kulkarni, G.U., Thomas, P.J., and Edwards, P.P. (2002). Size-dependent chemistry: Properties of nanocrystals. Chem. Eur. J. *8*, 28–35.
Reddy, M.P., Venugopal, A., and Subrahmanyam, M. (2007). Hydroxyapatite-supported Ag-TiO_2 as *Escherichia coli* disinfection photocatalysts. Water Res. *41*, 379–386.
Rickerby, D.G., and Morrison, M. (2007). Nanotechnology and the environment: A European perspective. STAM *8*, 19–24.
Ritchie, S.M.C., Bachas, L.G., Olin, T., Sikdar, S.K., and Bhattacharyya, D. (1999). Surface modification of silica- and cellulose-based microfiltration membranes with functional polyamino acids for heavy metal sorption. Langmuir *15*, 6346–6357.
Ritchie, S.M.C., Kissick, K.E., Bachas, L.G., Sikdar, S.K., Parikh, C., and Bhattacharyya, D. (2001). Polycysteine and other polyamino acid functionalized microfiltration membranes for heavy metal capture. Environ. Sci. Technol. *35*, 3252–3258.
Ritter, L., Solomon, K., Sibley, P., Hall, K., Keen, P., Mattu, G., and Linton, B. (2002). Sources, pathways, and relative risks of contaminants in surface water and groundwater: A perspective prepared for the Walkerton inquiry. J. Toxicol. Environ. Health *65*, 1–142.
Riu, J., Maroto, A., and Rius, F.X. (2006). Nanosensors in environmental analysis. Talanta *69*, 288–301.
Rivas, B.L., Pereira, E.D., and Moreno-Villoslada, I. (2003). Water soluble polymer-metal ion interactions. Prog. Polym. Sci. *28*, 173–208.
Rodriguez-Mozaz, S., López de Alda, M.J., and Barceló, D. (2004). Monitoring of estrogens, pesticides and bisphenol A in natural waters and drinking water treatment plants by solid-phase extraction-liquid chromatography-mass spectrometry. J. Chromatogr. A *1045*, 85–92.
Romero, M., Blanco, J., Sanchez, B., Vidal, A., Malato, S., Cardona, A.I., and Garcia, E. (1999). Solar photocatalytic degradation of water and air pollutants: Challenges and perspectives. Solar Energy *66*, 169–182.
Rook, J.J. (1974). Formation of haloforms during chlorination of natural waters. Water Treat. Exam. *23*, 234–243.
Rothen-Rutishauser, B.M., Schurch, S., Haenni, B., Kapp, N., and Gehr, P. (2006). Interaction of fine particles and nanoparticles with red blood cells visualized with advanced microscopic techniques. Environ. Sci. Technol. *40*, 4353–4359.
Ryman-Rasmussen, J.P., Riviere, J.E., and Monteiro-Riviere, N.A. (2006). Penetration of intact skin by quantum dots with diverse physiochemical properties. Toxicol. Sci. *91*, 159–165.
Saito, T., Iwase, T., and Morioka, T. (1992). Mode of photocatalytic bactericidal action of powdered semiconductor TiO_2 on mutans streptococci. J. Photochem. Photobiol. B: Biol. *14*, 369–379.
Sakaguchi, K., Matsui, M., and Mizukami, F. (2005). Applications of zeolite inorganic composites in biotechnology: Current state and perspectives. Appl. Microbiol. Biotechnol. *67*, 306–311.
Sakthivel, S., Shankar, V.M., Palanichamy, M., Arabindoo, B., Bahnemann, D.W., and Murugesan, V. (2004). Enhancement of photocatalytic activity by metal deposition: Characterization and photonic efficiency of Pt, Au and Pd deposited on TiO_2 catalyst. Water Res. *38*, 3001–3008.
Salipira, K.L., Mamba, B.B., Krause, R.W., Malefetse, T.J., and Durbach, S.H. (2007). Carbon nanotubes and cyclodextrin polymers for removing organic pollutants from water. Environ. Chem. Lett. *5*, 13–17.
Savage, N., and Diallo, M.S. (2005). Nanomaterials and water purification: Opportunities and challenges. J. Nanopart. Res. *7*, 331–342.
Sawicki, R., and Mercier, L. (2006). Evaluation of mesoporous cyclodextrin-silica nanocomposites for the removal of pesticides from aqueous media. Environ. Sci. Technol. *40*, 1978–1983.
Sayes, C.M., Wahi, R., Kurian, P.A., Liu, Y.P., West, J.L., Ausman, K.D., Warheit, D.B., and Colvin, V.L. (2006). Correlating nanoscale titania structure with toxicity: A cytotoxicity and inflammatory response study with human dermal fibroblasts and human lung epithelial cells. Toxicol. Sci. *92*, 174–185.

Schmid, A., Dordick, J.S., Hauer, B., Kiener, A., Wubbolts, M., and Witholts, B. (2001). Industrial biocatalysis today and tomorrow. Nature *409*, 258–268.

Schneiderman, E., and Stalcup, A.M. (2000). Cyclodextrins: A versatile tool in separation science. J. Chromatogr. B *745*, 83–102.

Schrick, B., Blough, J., Jones, D., and Mallouk, T. (2002). Hydrodechlorination of trichloroethylene to hydrocarbons using bimetallic nickel-iron nanoparticles. Chem. Mater. *14*, 5140–5147.

Schwarzenbach, R.P., Escher, B.I., Fenner, K., Hofstetter, T.B., Johnson, C.A., von Gunten, U., and Wehrli, B. (2006). The challenge of micropollutants in aquatic systems. Science *313*, 1072–1077.

Serpone, N., Maruthamuthu, P., Pichat, P., Pelizzetti, E., and Hidaka, H. (1995). Exploiting the interparticle electron transfer process in the photocatalysed oxidation of phenol, 2-chlorophenol and pentachlorophenol: Chemical evidence for electron and hole transfer between coupled semiconductors. Photochem. Photobiol. A: Chem. *85*, 247–255.

Shah, S.I., Li, W., Huang, C.P., Jung, O., and Ni, C. (2002). Study of Nd^{3+}, Pd^{2+}, Pt^{4+} and Fe^{3+} dopant effect on photoreactivity of TiO_2 nanoparticles. Proc. Natl. Acad. Sci. USA *99*, 6482–6486.

Shephard, G.S., Stockenstrom, S., de Villiers, D., Engelbrecht, W.J., and Wessels, G.F.S. (2002). Degradation of microcystin toxins in a falling film photocatalytic reactor with immobilized titanium dioxide catalyst. Water Res. *36*, 140–146.

Shieh, K.-J., Li, M., Lee, Y.-H., Sheu, S.-D., Liu, Y.-T., and Wang, Y.-C. (2006). Antibacterial performance of photocatalyst thin film fabricated by defection effect in visible light. Nanomed. Nanotechnol. Biol. Med. *2*, 121–126.

Shin, E.J., and Keane, M.A. (1999). Detoxification of dichlorophenols by catalytic hydrodechlorination using a nickel/silica catalyst. Chem. Eng. Sci. *54*, 1109–1120.

Shrivastava, S., Bera, T., Arnab Roy, A., Singh, G., Ramachandrarao, P., and Dash, D. (2007). Characterization of enhanced antibacterial effects of novel silver nanoparticles. Nanotechnology *18*, 1–9.

Shvedova, A.A., Castranova, V., Kisin, E.R., Schwegler-Berry, D., Murray, A.R., Gandelsman, V.Z., Maynard, A., and Baron, P. (2003). Exposure to carbon nanotube material: Assessment of nanotube cytotoxicity using human keratinocyte cells. J. Toxicol. Environ. Health A *66*, 1909–1926.

Simagina, V., Likholobov, V., Bergeret, G., Gimenez, M.T., and Renouprez, A. (2003). Catalytic hydrodechlorination of hexachlorobenzene on carbon supported Pd-Ni bimetallic catalysts. Appl. Catal. B: Environ. *40*, 293–304.

Singh, R., Pantarotto, D., Lacerda, L., Pastorin, G., Klumpp, C., Prato, M., Bianco, A., and Kostarelos, K. (2006). Tissue biodistribution and blood clearance rates of intravenously administered carbon nanotube radiotracers. Proc. Natl. Acad. Sci. USA *103*, 3357–3362.

Skadsen, J. (1993). Nitrification in a distribution system. J. Am. Water Works Assoc. *85*, 95–103.

Sokmen, M., Candan, F., and Sumer, Z. (2001). Disinfection of *E. coli* by the Ag-TiO_2/UV system: Lipidperoxidation. J. Photochem. Photobiol. A: Chem. *143*, 241–244.

Son, W.K., Youk, J.H., Lee, T.S., and Park, W.H. (2004). Preparation of antimicrobial ultrafine cellulose acetate fibers with silver nanoparticles. Macromol. Rapid Commun. *25*, 1632–1637.

Sondi, I., and Salopek-Sondi, B. (2004). Silver nanoparticles as antimicrobial agent: A case study on *E. coli* as a model for Gram-negative bacteria. J. Colloid Interface Sci. *275*, 1770–1782.

Song, W., Grassian, V.H., and Larsen, S.C. (2005a). High yield method for nanocrystalline zeolite synthesis. Chem. Commun. *20*, 2951–2953.

Song, W., Justice, R.E., Jones, C.A., Grassian, V.H., and Larsen, S.C. (2004). Synthesis, characterization, and adsorption properties of nanocrystalline ZSM-5. Langmuir *20*, 8301–8306.

Song, W., Li, G., Grassian, V.H., and Larsen, S.C. (2005b). Development of improved materials for environmental applications: Nanocrystalline NaY zeolites. Environ. Sci. Technol. *39*, 1214–1220.

Spadaro, J.A., Berger, T.J., Barranco, S.D., Chapin, S.E., and Becker, R.O. (1974). Antibacterial effects of silver electrodes with weak direct current. Microb. Agents Chemother. *6*, 637–642.

Srivastava, A., Srivastava, O.N., Talapatra, S., Vajtai, R., and Ajayan, P.M. (2004). Carbon nanotube filters. Nature Mater. *3*, 610–614.

Stanton, B.W., Harris, J.J., Miller, M.D., and Bruening, M.L. (2003). Ultrathin, multilayered polyelectrolyte films as nanofiltration membranes. Langmuir *19*, 7038–7042.

Stoimenov, P.K., Klinger, R.L., Marchin, G.L., and Klabunde, K.J. (2002). Metal oxide nanoparticles as bactericidal agents. Langmuir *18*, 6679–6686.

Strathmann, H. (2001). Membrane separation processes: Current relevance and future opportunities. AIChE Journal *47*, 1077–1087.

Suffet, I.H., Ho, J., Chou, D., Khiari, D., and Mallevialle, J. (1995). Taste-and-odor problems observed during drinking water treatment. In Advances in taste and odor treatment and control, Suffet, I.H., Mallevialle, J., and Kawczynski, E., eds. (AwwaRF, Denver), pp. 1–21.

Sun, Y.-P., Li, X.-Q., Cao, J., Zhang, W.-X., and Wang, H.P. (2006). Characterization of zero-valent iron nanoparticles. Adv. Colloid Interface Sci. *120*, 47–56.

Sun, D., Meng, T.T., Loong, T.H., and Hwa, T.J. (2004). Removal of natural organic matter from water using a nanostructured photocatalyst coupled with filtration membrane. Wat. Sci. Technol. *49*, 103–110.

Sunada, K., Watanabe, T., and Hashimoto, K. (2003). Studies on photokilling of bacteria on TiO_2 thin film. J. Photochem. Photobiol. A: Chem. *156*, 227–233.

Szewzyk, U., Szewzyk, R., Manz, W., and Schleifer, K.H. (2000). Microbiological safety of drinking water. Annu. Rev. Microbiol. *54*, 81–127.
Takeuchi, K., Nakamura, I., Matsumoto, O., Sugihara, S., Ando, M., and Ihara, T. (2000). Preparation of visible light-responsive titanium oxide photocatalysts by plasma treatment. Chem. Lett. *29*, 1354–1355.
Tavolaro, A., Tavolaro, P., and Drioli, E. (2007). Zeolite inorganic supports for BSA immobilization: Comparative study of several zeolite crystals and composite membranes. Colloids Surf. B: Biointerfaces *55*, 67–76.
Theron, J., and Cloete, T.E. (2002). Emerging waterborne infections: Contributing factors, agents, and detection tools. Crit. Rev. Microbiol. *28*, 1–26.
Tick, G.R., Lourenso, F., Wood, A.L., and Brusseau, M.L. (2003). Pilot-scale demonstration of cyclodextrin as a solubility-enhancement agent for remediation of a tetrachloroethene-contaminated aquifer. Environ. Sci. Technol. *37*, 5829–5834.
Top, A., and Ülkü, S. (2004). Silver, zinc and copper exchange in a Na-clinoptilolite and resulting effect on antibacterial activity. Appl. Clay Sci. *27*, 13–19.
Tratnyek, P.G., and Johnson, R.L. (2006). Nanotechnologies for environmental cleanup. Nanotoday *1*, 44–48.
Tuel, A., and Hubert-Pfalzgraf, L.G. (2003). Nanometric monodispersed titanium oxide particles on mesoporous silica: Synthesis, characterization, and catalytic activity in oxidation reactions in the liquid phase. J. Catal. *217*, 343–353.
Tully, D.C., and Frechet, J.M.J. (2001). Dendrimers at surfaces and interfaces: Chemistry and applications. J. Chem. Commun. *14*, 1229–1239.
Urbansky, E.T., and Schock, M.R. (1999). Issues in managing the risks associated with perchlorate in drinking water. J. Environ. Manage. *56*, 79–95.
Urry, D.W., Gowda, D.C., Parker, T.M., and Luan, C.H. (1992). Hydrophobicity scale for protein based on inverse temperature transitions. Biopolymers *32*, 1243–1250.
Vamathevan, V., Amal, R., Beydoun, D., Low, G., and McEvoy, S. (2004). Silver metallisation of titania particles: Effects on photoactivity for the oxidation of organics. Chem. Eng. J. *98*, 127–139.
Van der Bruggen, B., and Vandecasteele, C. (2002). Distillation vs. membrane filtration: Overview of process evolutions in seawater desalination. Desalination *143*, 207–218.
Van der Bruggen, B., and Vandecasteele, C. (2003). Removal of pollutants from surface water and groundwater by nanofiltration: Overview of possible applications in the drinking water industry. Environ. Pollut. *122*, 435–445.
Vaseashta, A., Vaclavikova, M., Vaseashta, S., Gallios, G., Roy, P., and Pummakarnchana, O. (2007). Nanostructures in environmental pollution detection, monitoring, and remediation. STAM *8*, 47–59.
Viessman, W., and Hammer, M.J. (1998). Water Supply and Pollution Control. (Addison-Wesley Longman Publishers, Menlo Park).
Vogel, B., Schneider, C., and Klemm, E. (2002). The synthesis of cresol from toluene and N_2O on H[Al]ZSM-5: Minimizing the product diffusion limitation by the use of small crystals. Catal. Lett. *79*, 107–112.
Vörösmarty, C.J., Green, P., Salisbury, J., and Lammers, R.B. (2000). Global water resources: Vulnerability from climate change and population growth. Science *289*, 284–288.
Walha, K., Amar, B.R., Firdaous, L., Quéméneur, F., and Jaouen, P. (2007). Brackish groundwater treatment by nanofiltration, reverse osmosis and electrodialysis in Tunisia: Performance and cost comparison. Desalination *207*, 95–106.
Wang, B., Feng, W.Y., Wang, T.C., Jia, G., Wang, M., Shi, J.W., Zhang, F., Zhao, Y.L., and Chai, Z.F. (2006). Acute toxicity of nano- and microscale zinc powder in healthy adult mice. Toxicol. Lett. *161*, 115–123.
Wang, C.-B., and Zhang, W.-X. (1997). Synthesizing nanoscale iron particles for rapid and complete dechlorination of TCE and PCBs. Environ. Sci. Technol. *31*, 2154–2156.
Warheit, D.B., Reed, K.L., and Webb, T.R. (2003). Development of pulmonary bridging studies: Pulmonary toxicity studies in rats with triethoxyoctylsilane (OTES)-coated pigment-grade titanium dioxide particles. Exp. Lung Res. *29*, 593–606.
Warheit, D.B., Webb, T.R., Sayes, C.M., Colvin, V.L., and Reed, K.L. (2006). Pulmonary instillation studies with nanoscale TiO_2 rods and dots in rats: Toxicity is not dependent upon particle size and surface area. Toxicol. Sci. *91*, 227–236.
Watanabe, M., Okada, M., Kudo, Y., Tonori, Y., Niitsuya, M., Sato, T., Aizawa, Y., and Kotani, M. (2002). Differences in the effects of fibrous and particulate titanium dioxide on alveolar macrophages of Fischer 344 rats. J. Toxicol. Environ. Health A *5*, 1047–1060.
Watts, R.J., Kong, S., Orr, M.P., Miller, G.C., and Henry, B.E. (1995). Photocatalytic inactivation of coliform bacteria and viruses in secondary wastewater effluent. Water Res. *29*, 95–100.
Weber, W.J. (2002). Distributed optimal technology networks: A concept and strategy for potable water sustainability. Water Sci. Technol. *46*, 241–246.
Wei, J., Xu, X., Liu, Y., and Wang, D. (2006). Catalytic hydrodechlorination of 2,4-dichlorophenol over nanoscale Pd/Fe: Reaction pathway and some experimental parameters. Water Res. *40*, 348–354.
Wever, H.D., Cole, J.R., Fettig, M.R., Hogan, D.A., and Tiedje, J.M. (2000). Reductive dehalogenation of trichloroacetic acid by *Trichlorobacter thiogenes* gen. nov., sp. nov. Appl. Environ. Microbiol. *66*, 2297–2301.

Wiesner, M.R., Lowry, G.V., Alvarez, P., Dionysiou, D., and Biswas, P. (2006). Assessing the risks of manufactured nanomaterials. Environ. Sci. Technol. *40*, 4336–4345.
Wong, M.-S., Chu, W.-C., Sun, D.-S., Huang, H.-S., Chen, J.-H., Tsai, P.-J., Lin, N.-T., Yu, M.-S., Hsu, S.-F., Wang, S.-L., and Chang, H.-H. (2006). Visible light-induced bactericidal activity of a nitrogen-doped titanium photocatalyst against human pathogens. Appl. Environ. Microbiol. *72*, 6111–6116.
World Health Organization (W.H.O). (2004). Meeting the MDG drinking water and sanitation target: A mid-term assessment of progress. Geneva: WHO.
Wu, L., and Ritchie, S.M.C. (2006). Removal of trichloroethylene from water by cellulose acetate supported bimetallic Ni/Fe nanoparticles. Chemosphere *63*, 285–292.
Wu, L., Shamsuzzoha, M., and Ritchie, S.M.C. (2005). Preparation of cellulose acetate supported zero-valent iron nanoparticles for the dechlorination of trichloroethylene in water. J. Nanopart. Res. *7*, 469–476.
Xu, J., Dozier, A., and Bhattacharyya, D. (2005a). Synthesis of nanoscale bimetallic particles in polyelectrolyte membrane matrix for reductive transformation of halogenated organic compounds. J. Nanopart. Res. *7*, 449–461.
Xu, X., Zhou, M., He, P., and Hao, Z. (2005b). Catalytic reduction of chlorinated and recalcitrant compounds in contaminated water. J. Hazard. Mater. *123*, 89–93.
Xu, Y., and Zhao, D. (2007). Reductive immobilization of chromate in water and soil using stabilized iron nanoparticles. Water Res. *41*, 2101–2108.
Yamashita, H., Harada, M., Misaka, J., Takeuchi, M., Neppolian, B., and Anpo, M. (2003). Photocatalytic degradation of organic compounds diluted in water using visible light-responsive metal ion-implanted TiO_2 catalysts: Fe ion-implanted TiO_2. Catal. Today *84*, 191–196.
Yan, H., Gong, A., He, H., Zhou, J., Wei, Y., and Lu, L. (2006). Adsorption of microcystins by carbon nanotubes. Chemosphere *62*, 142–148.
Yang, G.C.C., and Lee, H.-L. (2005). Chemical reduction of nitrate by nano-sized iron: Kinetics and pathways. Water Res. *39*, 884–894.
Yang, M.C., Yang, T.S., and Wong, M.S. (2004). Nitrogen-doped titanium oxide films as visible light photocatalyst by vapor deposition. Thin Solid Films *469*, 1–5.
Yang, J., Zhang, J., Zhu, L., Chen, S., Zhang, Y., Tang, Y., Zhua, Y., and Li, Y. (2006). Synthesis of nano-titania particles embedded in mesoporous SBA-15: Characterization and photocatalytic activity. J. Hazard. Mater. *137*, 952–958.
Yantasee, W., Fryxell, G.E., Lin, Y., Wu, H., Raymond, K.N., and Xu, J. (2005). Hydroxypyridinone functionalized self-assembled monolayers on nanoporous silica for sequestering lanthanide cations. J. Nanosci. Nanotechnol. *5*, 527–529.
Yates, C.R., and Hayes, W. (2004). Synthesis and applications of hyperbranched polymers. Eur. Polym. J. *40*, 1257–1281.
Yu, J.C., Ho, W., Yu, J., Yip, H., Wong, P.K., and Zhao, J. (2005). Efficient visible light-induced photocatalytic disinfection on sulfur-doped nanocrystalline titania. Environ. Sci. Technol. *39*, 1175–1179.
Yu, J.C., Wu, L., Lin, J., Li, P., and Li, Q. (2003a). Microemulsion-mediated solvo-thermal synthesis of nanosized CdS-sensitized TiO_2 crystalline photocatalyst. Chem. Commun. *8*, 1552–1553.
Yu, J.C., Xie, Y., Tang, H.Y., Zhang, L., Chan, H.C., and Zhao, J. (2003b). Visible light-assisted bactericidal effect of metalphtalocyanine-sensitized titanium dioxide films. J. Photochem. Photobiol. A *156*, 235–241.
Yu, J.C., Yu, J.G., and Zhao, J.C. (2002). Enhanced photocatalytic activity of mesoporous and ordinary TiO_2 thin films by sulfuric acid treatment. Appl. Catal. B: Environ. *36*, 31–43.
Zhang, W.-X. (2003). Nanoscale iron particles for environmental remediation: An overview. J. Nanopart. Res. *5*, 323–332.
Zhang, W., and Digiano, F.A. (2002). Comparison of bacterial re-growth in distribution systems using free chlorine and chloramine: A statistical study of causative factors. Water Res. *36*, 1469–1482.
Zhang, W.-X., and Elliott, D.W. (2006). Applications of iron nanoparticles for groundwater remediation. Remediation *16*, 7–21.
Zhang, W.-X., Wang, C.B., and Lien, H.L. (1998a). Treatment of chlorinated organic contaminants with nanoscale bimetallic particles. Catal. Today *40*, 387–395.
Zhang, Z., Wang, C.-C., Zakaria, R., and Ying, J.Y. (1998b). Role of particle size in nanocrystalline TiO_2-based photocatalysts. J. Phys. Chem. B *102*, 10871–10878.
Zhang, X., Zhang, F., and Chan, K.Y. (2005a). Synthesis of titania-silica mixed oxide mesoporous materials, characterization and photocatalytic properties. Appl. Catal. A *284*, 193–198.
Zhang, X., Zhou, M., and Lei, L. (2005b). Preparation of an Ag-TiO_2 photocatalyst coated on activated carbon by MOCVD meter. Chem. Phys. *91*, 73–79.
Zhao, G.J., and Stevens, S.E. (1998). Multiple parameters for the comprehensive evaluation of the susceptibility of *Escherichia coli* to the silver ion. Biometals *11*, 27–32.
Zhou, Q., Xiao, J., and Wang, W. (2006a). Using multi-walled carbon nanotubes as solid phase extraction adsorbents to determine dichlorodiphenyltrichloroethane and its metabolites at trace level in water samples by high performance liquid chromatography with UV detection. J. Chromatogr. A *1125*, 152–158.
Zhou, Q., Xiao, J., and Wang, W. (2007). Comparison of multi-walled carbon nanotubes and a conventional absorbent on the enrichment of sulfonylurea herbicides in water samples. Anal. Sci. *23*, 189–192.

Zhou, Q., Xiao, J., Wang, W., Li, G., Shi, Q., and Wang, J. (2006b). Determination of atrazine and simazine in environmental water samples using multi-walled carbon nanotubes as the adsorbents for preconcentration prior to high performance liquid chromatography with diode array detector. Talanta *68*, 1309–1315.
Zhu, B.-W., Lim, T.-T., and Feng, J. (2006a). Reductive dechlorination of 1,2,4-trichlorobenzene with palladized nanoscale Fe^0 particles supported on chitosan and silica. Chemosphere *65*, 1137–1145.
Zhu, S., Oberdörster, E., and Haasch, M.L. (2006b). Toxicity of an engineered nanoparticle (fullerene, C_{60}) in two aquatic species, *Daphnia* and fathead minnow. Mar. Environ. Res. *62*, S5-S9.

Current Molecular and Emerging Nanobiotechnology Approaches for the Detection of Microbial Pathogens

2

Jacques Theron, T. Eugene Cloete and Michele de Kwaadsteniet

Abstract

An adequate supply of safe drinking water is one of the major prerequisites for a healthy life, but waterborne diseases is still a major cause of death in many parts of the world, particularly in young children, the elderly, or those with compromised immune systems. As the epidemiology of waterborne diseases is changing, there is a growing global public health concern about new and re-emerging infectious diseases that are occurring through a complex interaction of social, economic, evolutionary, and ecological factors. An important challenge is therefore the rapid, specific and sensitive detection of waterborne pathogens. Presently, microbial tests are based essentially on time-consuming culture methods. However, newer enzymatic, immunological and genetic methods are being developed to replace and/or support classical approaches to microbial detection. Moreover, innovations in nanotechnology and nanosciences are having a significant impact in biodiagnostics, where a number of nanoparticle-based assays and nanodevices have been introduced for biomolecular detection. Accordingly, current and emerging molecular approaches for the detection of microbial pathogens as well as nanobiotechnologies that that will extend the limits of current molecular diagnostics are discussed.

Introduction

Waterborne microbial diseases, once expected to be eliminated as public health problems, not only remains the leading cause of death worldwide, but the spectrum of disease is expanding and the incidence of many waterborne microbial diseases once thought conquered is increasing (Ford, 1999; Leclerc *et al.*, 2002; Nwachcuku and Gerba, 2004). These infections, which may be transmitted by contaminated recreational waters, surface water and ground water, intended for drinking, place entire communities at risk. Even though major outbreaks of waterborne disease are comparatively rare, there is substantial evidence that human enteric pathogens that are frequently present in domestic water supplies are responsible for low-level incidence of waterborne microbial disease. Although these diseases are rarely debilitating to healthy adults for more than a few hours to a few days, enteric pathogens can cause severe illness, even death, for young children, the elderly or those with compromised immune systems (Ford and MacKenzie, 2000; Theron and Cloete, 2002; Ashbolt, 2004). Owing to the numerous infectious agents (including enteric bacteria, viruses and protozoa), reservoirs and asymptomatic infected individuals, complete eradication of waterborne diseases may not be possible. However, it should be possible to control these agents as long as they can be detected and monitored.

Traditionally, prediction of the presence of human enteric pathogens in water has been achieved by monitoring for established microbial 'indicators' of faecal pollution. Not necessarily pathogenic themselves, faecal coliforms, total coliforms, *E. coli*, enterococci and bacteriophages are all examples of organisms that when present are viewed as predictive of the potential presence of enteric pathogens, since they have the same faecal source as the pathogenic organisms (Toranzos and McFeters, 1996; Savichtcheva and Okabe, 2006). Tests for coliform bacteria are standardized and relatively easy and inexpensive to use. Consequently, they are more rapidly administered than tests determining the presence of individual pathogenic microorganisms in water. Despite being successful in predicting possible health risks in many circumstances, there are many flaws in using microbial indicators (Anderson *et al.*, 2005; Harwood *et al.*, 2005). Research has established an inability of many of these indicators to predict the presence of disease-causing viruses (such as hepatitis A and E, coxsackieviruses, echoviruses, adenoviruses and Norwalk viruses), indigenous bacteria (such as *Legionella* and *Helicobacter*), as well as parasites (such as *Cryptosporidium* and *Giardia*) (Moe, 1996; Lemarchand *et al.*, 2004).

Since it would appear that conventional detection methods do not adequately assess the risk of waterborne disease, the need for powerful new tools for the detection of pathogenic microorganisms in water is becoming increasingly more important. Although several different molecular methodologies are available, these have only recently been applied in the field of water science and technology. Moreover, the rapid progress of nanotechnology and advanced nanomaterials production offers significant opportunities not only for the detection and remediation of a broad range of environmental contaminants (Theron *et al.*, 2008; Andreescu *et al.*, 2009), but also for the development of new diagnostic assays that may serve as an appealing alternative to current molecular diagnostic techniques.

Molecular methodologies for detection of pathogenic microorganisms

Owing to advances in areas such as genomics and biotechnology, powerful methodologies have been developed to detect both specific pathogens and indicator organisms. Because many microorganisms are not easily cultured or can enter a viable but non-culturable (VBNC) state, the current methods focus on immunological or genetic characteristics to detect the presence of specific waterborne pathogenic microorganisms. Not only do these methods increase the rapidity of analysis, but they are also able to achieve a high degree of sensitivity and specificity without the need for complex cultivation and additional confirmation steps. Consequently, some of these methods permit the detection of specific culturable and/or non-culturable microorganisms within hours, instead of the days required with the traditional methods. A summary of available technologies for molecular detection and differentiation of microorganisms, as well as their limitations is shown in Table 2.1.

Immunological methods

Immunological methods are based on the specific recognition between antibodies and antigens, and the high affinity that is characteristic of this recognition reaction. Consequently, many different immunoassay methods have become available for both quantitative and qualitative analysis of pathogenic bacteria in water (Bottger *et al.*, 1987; Faude and Hofle, 1997; Buswell *et al.*, 1998). These include immunocapture of cells or antigens by enzyme-linked immunosorbent assay (ELISA or EIA), or detection of targeted cells by immunofluorescence (IFA). These assays can be performed by a direct or indirect manner. In a direct immunoassay, the monoclonal or polyclonal antibodies, directed against antigens located on the surface of the target pathogen (such as capsid proteins, cell wall or flagellar antigens), are conjugated with a fluorochrome or fluorescent dye. Alternatively, secondary enzymatically or fluorescently labelled antibodies directed against the primary antibodies (now serving as antigens) can be used in an indirect immunoassay. The advantage of this procedure is that the secondary antibodies can easily be obtained from a commercial supplier with a range of conjugated fluorochromes and it leads to signal amplification as several labelled secondary antibodies can bind to a single unlabelled primary antibody (Hoff, 1993). The antigen–antibody complex is detected and quantified by the ability of the enzyme to react with a substrate that produces either a coloured product for colorimetry or emits light for luminometry. The immunoassays are often performed on a solid phase to which the pathogen antigens have been applied, such as a membrane filter

Table 2.1 Molecular techniques for use in detection of pathogenic microorganisms (adapted from Rose and Grimes, 2001)

Molecular tool	Characteristics and advantages	Limitations
Immunomagnetic capture	Relatively specific for target organism Can be used for recovery from complex environmental samples	Currently unable to discern viable from non-viable organisms
Nucleic acid hybridization	Highly specific Can be used to differentiate virulent from non-virulent strains	Only applicable to culturable microbes
Polymerase chain reaction (PCR)	Rapid and specific Can be used for quantitative assays Can be applied to expressed genes (i.e. RT-PCR on mRNA) Can be used to detect non-culturable viruses and other microbes	Unless conditions are optimized, inconsistencies in performance can produce erroneous results Currently unable to directly discern viable from non-viable organisms
DNA microarrays	Rapid detection Very sensitive High throughput design Various applications	Low sensitivity for environmental samples Sample processing complexities

or the bottom of a microtitre plate well (De Silva *et al.*, 1992; Hasan *et al.*, 1994; Quadri *et al.*, 1994; Noah *et al.*, 1996).

Studies have shown that solid-phase enzyme immunoassays generally are too insensitive for direct detection of microbial pathogens in water, as they require a minimum of 10^3 to 10^4 target microbes (or their antigens) for detection (Faude and Hofle, 1997; Hanai *et al.*, 1997). In most situations drinking water and its sources rarely contain high enough levels of most target pathogens for direct immunoenzymatic detection. Nevertheless, enumeration of diluted specific cells can be obtained by means of immunomagnetic separation (IMS). Immunomagnetic separation, also termed immunocapture or antibody capture, is a method that uses paramagnetic synthetic beads or other magnetic particles that have been coated with monoclonal or polyclonal antibodies directed against the target microbes to recover the microbes from the sample by antigen–antibody reactions (Olsvik *et al.*, 1994). The retained microbes can be analysed directly or after they or their nucleic acids have been released or extracted from the antibody and solid phase by various physical or chemical methods. IMS methods have the advantage of selecting, separating and purifying specific target microbes from other microbes and from solutes, based on the specificity of the antigen-antibody reaction. This is a powerful approach for recovering, enriching, purifying and concentrating the target viruses, bacteria and parasites from the sample matrix (Jothikumar *et al.*, 1998; Sasaki *et al.*, 1999; Lowery *et al.*, 2000). However, it is not applicable to some pathogens because of the lack of antisera or the antigenic diversity of a large pathogen group lacking a common antigen and thus requiring many antisera.

As an alternative to the above assays, agglutination methods can be used to detect pathogens by combining dispersed microorganisms with antibodies (on a slide, for example) and allowing for antigen-antibody reactions to produce agglutination (clumping) that can be scored as negative or various degrees of positive (Colwell *et al.*, 1992; Quadri *et al.*, 1994). One modification is latex bead agglutination in which antibodies against a specific microbial antigen are attached to latex beads. The beads are reacted with the sample and should the sample contain the specific antigen, agglutination occurs by the reaction of antigens with antibodies on the beads resulting in the beads clumping together. As with enzyme immunoassays, agglutination tests are too insensitive to directly detect and quantify most waterborne pathogens in drinking water and other aquatic samples (Ito *et al.*,

1983). The target microbes must first be cultured in order to obtain a sufficient number of them or a sufficient amount of antigen to detect and antigenically characterize them by agglutination methods.

The use of immunological methods for the detection of specific microorganisms is a rapid and simple technique, the accuracy of which mainly depends on the specificity of the antibody (Kfir and Genthe, 1993). Nevertheless, its application to the detection of specific microorganisms from environmental water samples is limited (Hanai *et al.*, 1997). While IFA allows specific identification and detection at a single-cell level, it does not provide information on the physiological status or viability of the detected cells. The ELISA is a rapid, simple and quite sensitive test. However, assay limitations are often associated with the specificity of the antibody used, the concentration of both antibody and antigen, and the solid matrix often leads to non-specific binding of the antigen or of the secondary antibody (Kfir and Genthe, 1993).

Monoclonal antibodies are better suited for biosensors because of their higher specificity. Polyclonal antibodies recognize different epitopes on the same pathogen. False positives can be generated when these antigens are present in other closely related non-pathogenic microorganisms (Dover *et al.*, 2009). The thermal instability of antibodies, in particular monoclonal antibodies, is another drawback when applying them in environmental biosensors. Single domain antibodies (also referred to as nanobodies) have, however, been developed that are thermostable, even at temperatures as high as 90°C. Their small size, high solubility and refolding capacity are other features that make them ideally situated for biosensing applications (Goldman *et al.*, 2006a).

Nucleic acid-based methods

Nucleic acid hybridization techniques

The easiest way of detecting specific nucleic acid sequences is through direct hybridization of a probe to microbial nucleic acid extracts. These hybridization techniques rely on the specific binding of nucleic acid probes to complementary DNA or RNA (target nucleic acid). The probes are single strands of nucleic acid with the potential of carrying detectable marker molecules highly specific to complementary target sequences, even if these sequences account for only a small fraction of the target nucleic acid. Either DNA or RNA can serve as a nucleic acid probe, but for a number of reasons (e.g. ease of synthesis and stability), most studies have employed DNA probes (Holben and Tiedje, 1988). The probes may be used to detect genes in the bacterial genome (Southern blots) or to detect mRNA or rRNA (Northern blots) (Schneegurt and Kulpa, 1998).

According to their length, DNA probes can be grouped as either polynucleotide probes (more than 50 nucleotides) or as oligonucleotide probes (fewer than 20 nucleotides) (Amann *et al.*, 1997). The latter are more frequently used since they are short enough to allow for single mismatch discrimination of target nucleic acids and large quantities of oligonucleotides can be rapidly and inexpensively produced. During oligonucleotide synthesis a variety of marker or linker sequences can be introduced to the 5′ end of the oligonucleotide. The principal conjugate fluorochromes are derivatives of fluorescein or rhodamine, although labels such as digoxigenin and biotin have also been used (Zarda *et al.*, 1991; Amann *et al.*, 1993, 1995). For probes to be effective in nucleic acid hybridization, they must be specific and selective. This requires that sequence data be available. Although such data are available for some waterborne pathogens, the data are not comprehensive for all pathogens of concern. However, the use of bacterial 16S rRNA sequences to develop determinative hybridization probes is now well established (Olsen *et al.*, 1986; Pace *et al.*, 1986). The occurrence of 16S rRNA molecules in high copy numbers of usually more than 1000 in any living cell, as well as the fact that 16S rRNA sequences have been determined for a large fraction of bacterial species (Olsen *et al.*, 1986; Amann *et al.*, 1995; 1997), currently confers on these genes a high informative value at phylogenetic level. Comparative analyses of rRNA sequences have indicated that the rRNA molecules comprise highly conserved sequence domains interspersed with more variable regions (Gutell *et al.*, 1994; Van de Peer *et al.*, 1996). Consequently, so-called signature sequence motifs on various taxonomic levels, from domains to subspecies, can be identified (Woese, 1987; Giovannoni *et al.*, 1988; Manz *et al.*, 1992; Amann *et al.*, 1995). Whereas species-specific probes complement the most variable regions, more general probes target more conserved regions of the molecule (Ward *et al.*, 1990; Stahl and Amann, 1991). Although several oligonucleotide probes are commercially available, new specific probes may also be designed. The design of such 16S rRNA-targeted probes is greatly facilitated by computer-aided sequence comparisons of large rRNA sequence databases (Maidak *et al.*, 1994, 1997). The principal steps involved in the design of probes are: the alignment of rRNA gene sequences; the identification of sequence idiosyncrasies; the synthesis and labelling of complementary nucleic acid probes; and the experimental evaluation and optimization of the probe specificities and assay sensitivities using cultured reference strains (Manz *et al.*, 1992; Amann, 1995).

Nucleic acid hybridization can be performed in a variety of formats. The procedure typically entails the extraction of whole-cell DNA or RNA from the sample that is subsequently fixed to a nylon or nitrocellulose membrane. Alternatively, bacterial colonies can be replica-plated from agar plates to membranes and their nucleic acids exposed following lysis for subsequent hybridization. Whereas the former direct approach is usually inadequate due to high detection limits and large sample volumes that are impractical for most hybridization protocols without further pathogen concentration, the latter approach is time-consuming and cannot be applied to non-culturable organisms. Consequently, the use of whole-cell fluorescence *in situ* hybridization (FISH) as a method for the direct detection of bacteria in water samples has increased in popularity (Moter and Gobel, 2000; Amann *et al.*, 2001). The method typically combines membrane filtration, resuscitation of the bacteria on membranes and hybridization with rRNA-targeted probes. During FISH, the morphology of the cells in the sample has to be stabilized in order to maintain the morphological integrity of the cells under harsh hybridization conditions. The cell walls and membranes have to be permeabilized to allow free penetration of fluorescent oligonucleotides to the intracellular rRNA. This can be achieved with fixatives such as aldehydes and/or alcohols (Amann *et al.*, 1995; Bottari *et al.*, 2006). The membranes are hybridized in hybridization solution containing a fluorescently labelled oligonucleotide. Following incubation at the hybridization temperature for one to several hours, to allow the probe to bind to complementary rRNA sequences, washing steps are used to remove unbound or the non-specifically bound fluorescent probe, and the hybridized cells are then detected by epifluorescence microscopy. Probes will only bind correctly under defined hybridization conditions and the optimization of hybridization and washing conditions is as important as the probe design (Stahl and Amann, 1991; Manz *et al.*, 1992; Bottari *et al.*, 2006). Several probes labelled with spectrally different fluorochromes can be simultaneously used on one sample (Lee *et al.*, 1993; Amann *et al.*, 1997), while counterstaining of the fixed cells with DAPI (4′,6-diamidino-2-phenylindole) allows total counts to be made (Hicks *et al.*, 1992; Meier *et al.*, 1997). Alternatively, depending on the concentration of targeted cells in the sample and to increase resolution, FISH hybridization can be performed in suspension and detection can be performed by means of flow cytometry, which enables quantification of the fluorescence intensities for each target–probe hybrid (Fuchs *et al.*, 1998; Wagner *et al.*, 2003).

Although FISH is currently considered a highly specific cellular detection method and is relatively easy to perform, the procedure may have some limitations when applied to the detection of nutrient-starved bacterial cells disseminated in drinking waters. Generally, probes give a strong signal only if cells are metabolically active and contain large numbers of ribosomes and target rRNA. Conversely, low cellular ribosome contents may result in weak fluorescent hybridization signals (Amann *et al.*, 1995; Lebaron *et al.*, 1997). Another area of concern relates to the viability of the detected cells. Although it has been reported that the rRNA content of bacterial cells can be directly correlated with the growth rate (Amann *et al.*, 1995), some reports have indicated that a small number of rRNA molecules can remain for a relatively long period after the loss of culturability, thus resulting in false positives (McKillip *et al.*, 1998; Sheridan *et al.*, 1998). One possible way to overcome this issue is to couple direct viable count techniques with FISH detection (Nishimura *et al.*, 1993; Kalmbach *et al.*, 1997; Meier *et al.*, 1997).

Polymerase chain reaction (PCR)

As a consequence of the speed, specificity and low cost of the PCR, the procedure has become one of the most widely used assays for direct detection of low levels of pathogenic microbes in environmental samples. The PCR assay can be used to selectively amplify, to detectable levels, nucleic acid sequences associated with pathogens that might be present in low numbers in water samples. PCR is a process in which target DNA, synthetic oligonucleotide primers, a thermostable DNA polymerase and the DNA subunits are combined in a microcentrifuge tube and subjected to the temperature changes needed for the DNA duplication to occur. During the PCR process, different temperatures are used to facilitate DNA denaturation, annealing of the oligonucleotide primers to the target DNA and extension of the primers across the target sequence. These cycles are repeated many times, thus resulting in increasingly greater quantities of target sequence. Under ideal conditions, the PCR can generate millions of copies of a single DNA molecule in just 20 to 30 repetitions of the temperature cycle, with each cycle requiring only a few minutes (Saiki *et al.*, 1988). The PCR-amplified products can be detected by means of techniques such as electrophoresis on agarose gels and after staining of the amplification products by a fluorochrome dye or by hybridization with a labelled probe.

There are several essential steps in the development and application of PCR for successful detection of pathogens in water samples. The key steps include: identification and selection of oligonucleotide primers for target genomic sequences; testing of selected primers for sensitivity, specificity and selectivity; purification and concentration of the pathogens in environmental sample concentrates to enable efficient and reliable enzymatic amplification of low numbers of target genomic sequences; and testing of the methods for their applicability to natural pathogenic strains and actual field samples. By using PCR, a selected gene sequence specific to a group of organisms or a single species can be selectively amplified. Various primers have thus been described to amplify fragments of rRNA operons in order to detect specific organisms or groups of organisms in environmental samples (Dorsch *et al.*, 1994; Covert *et al.*, 1999). As in the case of designing oligonucleotide probes for hybridization purposes, selection of oligonucleotide primers for target pathogens requires that sequence data be available. Of particular importance is the type and function of the nucleic acid target; its length and location; and the extent to which its sequence is related to that of other, non-target but genetically related microbes. It is essential to select oligonucleotides having an appropriate length (20–30 nucleotides); desired sequence composition for specificity and selectivity; and appropriate melting and annealing temperatures to prevent the formation of undesirable secondary structures, primer dimers and other artefacts that would interfere with successful PCR (Saiki *et al.*, 1988; Steffan and Atlas, 1991).

There exist many variations of the basic PCR technique. The sensitivity and specificity of the PCR may be improved by adopting a nested approach. Nested PCR involves two consecutive rounds of PCR amplification. The first round of amplification is performed with group- or organism-specific primers, whilst the second round of amplification uses the initially amplified product as the template for another round of annealing and extension with different primers (Arnheim and Erlich, 1992). The use of nested primers provides an additional level of specificity since the second round of PCR amplification can only be performed if the correct sequence (complementary to the nested primers) was amplified during the first round. Alternatively, nucleic acid hybridization can be performed using highly specific oligoprobes that would hybridize only with amplicons from a single pathogen type or strain. However, nested PCR permits a more rapid detection (6–8 h) than confirmation of the correct sequence amplification by probe hybridization (few days). Another variation of the PCR technique, namely multiplex PCR, allows the simultaneous detection of more than one target organism in a single PCR using multiple pairs of primers designed to be specific for different target organisms (Bej *et al.*, 1991; Rochelle *et al.*, 1997). However, multiplex PCR may not perform well with all primer blends as the composition and length of primer oligonucleotide, as well as the size of the amplified fragments, may influence each PCR amplification (Bej *et al.*, 1991). Since the PCR method cannot be used directly for the amplification of an RNA target sequence, a complementary DNA copy (cDNA) thereof must first be synthesized. This reaction is catalysed by the enzyme, reverse transcriptase (RT), which is able to synthesize DNA from the RNA template in the presence of specific primers and DNA subunits. The single-stranded cDNA is a suitable target for PCR amplification by making use of the same or a different set of primers. RT-PCR has emerged as a sensitive and specific approach for the detection of enteric viruses containing RNA genomes (Sobsey, 1994; Schwab *et al.*, 1996; Huang *et al.*, 2000).

Although the basic PCR method is both specific and sensitive, such standard PCR reactions are not quantitative. To obtain quantitative data from PCR-based analyses, statistical methods based on most probable number (MPN) estimations have been used (Toranzos *et al.*, 1993). In MPN-PCR, DNA extracts are diluted before PCR amplification and limits are set on the number of genes in the sample by reference to known control dilutions. Another way to quantify PCR-amplified products for comparison is to include an internal control in the PCR reaction (Zachar *et al.*, 1993; Schwab *et al.*, 1998). Here, a known amount of target DNA is added to a PCR reaction containing DNA from the mixed microbial population. The known target DNA is complementary to the same primers and thus competes with the target sequences in the PCR reaction mixture. By preparing a dilution series of the known and unknown DNA species, it is possible to quantify the amount of product produced from the complementary gene in the extracted DNA. The known DNA target can be generated by cloning the gene of interest or purifying the PCR-amplified product after which a deletion is introduced to give a differently sized PCR product.

An alternative PCR assay for the direct enumeration of targeted cells was reported by Tani *et al.* (1998), who modified the standard PCR protocol so that nucleic acid sequences can be amplified *in situ*. This new PCR method was successfully applied to the direct enumeration of *E. coli* from a freshwater sample. With proper fixation and permeabilization conditions, the oligonucleotide primers and other reaction components are able to diffuse into the cells, and, upon thermal cycling, amplify a specific target sequence present in the cell. PCR products were labelled by digoxigenin during the amplification process and anti-digoxigenin antibodies conjugated with fluorescent dye were used for detection by epifluorescence microscopy. This approach allows direct visualization of the fluorescent amplification products at a single-cell level and consequently, direct enumeration of cells. Even though *in situ* PCR seems promising, it has not been used for routine detection and enumeration of microorganisms in water, as the results showed a weak fluorescence intensity signal of targeted cells.

Another promising approach for quantifying the number of cells is real-time quantitative PCR, which consists of monitoring fluorescently labelled PCR products as they are being amplified (Heid *et al.*, 1996; Mackay, 2004). The fluorescent signal can be generated by using an intercalating fluorescent dye (e.g. SYBR Green I or SYBR Gold) or a probe system (e.g. TaqMan®) (Galluzzi *et al.*, 2007). The use of intercalating dyes is the simplest and least costly approach and involves adding the fluorescent dye directly to the PCR. These dyes undergo a conformational change to become a more efficient fluorophore on binding to double stranded DNA (dsDNA). Although SYBR Green I-based assays are very sensitive, the primer's specificity for the target is crucial as any dsDNA is detected, including any primer artefacts, which can lead to false positive results. Moreover, multiplex reactions are impractical since the dye binds to all dsDNA (Galluzzi *et al.*, 2007). The TaqMan® approach depends on oligonucleotide probes complementary to a sequence located between the two primers used for PCR amplification. At one end of the probe a fluorescent reporter dye is conjugated, whilst at the other terminus there is a quencher that may be another fluorophore (also called dark-quencher). In effect the structure possesses two dyes in close proximity and in this configuration the fluorescence of one (the reporter) is quenched by the other through FRET (fluorescence resonance energy transfer). During the extension step of PCR the DNA polymerase degrades the bound TaqMan® probe, using its inherent 5′–3′ exonucleolytic activity, and thus results in the separation of reporter from quencher and an increase in fluorescence emission of the reporter molecule (Livak *et al.*, 1995). Although this approach is less prone to false-positive results compared to the use of intercalating fluorescent dyes, it is more expensive due to the requirement of the probe molecule. However, by choosing the fluorophores astutely it is possible to perform multiplex PCR (Beuret, 2004; He and Jiang, 2005). During real-time PCR, irrespective of the approach used, the accumulation of amplified product is measured automatically at each PCR cycle. The amount of target sequence is deduced from the number of PCR cycles (threshold cycle or C_t) required to cross a fixed point above a baseline, using a standard curve as reference. External quantification standards for the construction of standard curves of C_t versus copy number usually consist of the target sequence cloned into a plasmid or DNA extracted from cultured cells where the concentration or copy number of the target can be determined accurately (Galluzzi *et al.*, 2007).

Although there are numerous advantages associated with PCR as detection tool, standard PCR cannot, however, be used to detect the infectious state of an organism – only the presence or absence of pathogen-specific DNA or RNA. Yet, this viability concept is fundamental for interpreting the result in terms of public health when dealing with water samples. The PCR technique must consequently be associated with a viability test. To overcome this limitation, an indirect approach has been developed for assessing the viability of PCR-detected bacteria from water samples. This method is based on the analysis of each sample before and after a culture step in a non-selective medium: an increase in the PCR-amplified product after cultivation indicates the occurrence of bacterial multiplication and thus demonstrates the viability of the detected bacteria (Theron *et al.*, 2000, 2001). Recently, a PCR-based approach to limit detection to intact (viable) cells with an active metabolism has been reported (Nogva *et al.*, 2003; Nocker and Camper, 2009). The approach is based on the use of ethidium monoazide (EMA), which is suggested to enter only membrane-compromised cells (considered 'dead'). Once inside membrane-compromised cells, EMA intercalates into the DNA and is covalently bound to the DNA after exposure of the treated samples to bright visible light, whilst the unbound EMA, which remains free in solution, is simultaneously inactivated by reacting with water molecules. The EMA treatment is followed by extraction of genomic DNA and analysis by PCR. The result of treatment is that only unmodified DNA from intact cells whose DNA was not cross-linked with EMA can be amplified, whereas PCR amplification of modified DNA from membrane-compromised cells is efficiently suppressed. Treatment was thus suggested to lead to the exclusion of cells with damaged membranes from analysis (Nocker and Camper, 2009).

Despite its advantages, accurate characterization or identification of microbes by PCR is influenced by the same bias and variations that are inherent in many nucleic acid techniques. The main concerns are biased nucleic acid extraction (e.g. efficiency of extraction or cell lysis if using whole-cell methods), degradation of nucleic acids by nucleases and primer reactivity (*i.e.* sensitivity, specificity and accessibility) (Lim *et al.*, 2005). Additionally, a frequently encountered limitation inherent to PCR analysis of environmental samples is the inhibition of the enzymatic reaction. Whereas humic substances are known to inhibit the DNA polymerase enzyme, colloidal matter has a high affinity for DNA (Way *et al.*, 1993; Wilson, 1997). The presence of these elements in a water sample can therefore considerably decrease the amplification yield of PCR applied to the detection of greatly diluted bacteria (Straub *et al.*, 1995). Consequently, for PCR or RT-PCR, the extracted target nucleic acid is purified by protocols utilizing, for example, Sephadex (De Leon *et al.*, 1992), Chelex (Straub *et al.*, 1995) or CTAB (Jiang *et al.*, 1995).

DNA microarrays

Significant advances in the detection of sequence-specific nucleic acid hybridization have been achieved using microarrays. Microarrays are glass microslides or nylon membranes containing a high density of immobilized nucleic acids (genomic DNA, cDNA or oligonucleotides) in an ordered two-dimensional matrix (Epstein and Butow, 2000). Microarrays can be prepared by synthesizing DNA *in situ* on a glass surface using combinational chemistry or by robotic microdeposition of cDNAs (0.5- to 2-kb) amplified by PCR (Zammatteo *et al.*, 2000; Lemarchand *et al.*, 2004). The sample DNA, usually bound to a fluorescent or enzyme label, is exposed to the microarray and hybridizes with the target sequences. The detection of the probe–target hybrid at each spot on the array is achieved either by direct fluorescence scanning or by enzyme-mediated detection, yielding a semiquantitative result (de Boer and Beumer, 1999). Advantages of DNA microarray technology, compared with other techniques, include the small size of the array allowing for a higher sensitivity, the ability to simultaneously detect diverse individual sequences in complex DNA samples, and the capacity to do comparative analysis of a large number of samples (de Boer and Beumer, 1999; Van Hal *et al.*, 2000; Lemarchand *et al.*, 2004). Indeed, PCR microarrays have been used to measure relative concentrations of microbes from water (Stine *et al.*, 2003), to characterize microbial communities from environmental samples (Wu *et al.*, 2000; Call *et al.*, 2003, 2005; Franke-Whittle *et al.*, 2005) and to detect bacterial pathogens from a variety of sources, including water (Straub and Chandler, 2003; Lee *et al.*, 2008). However, obstacles such as sample size, matrix-associated inhibitors, non-specific binding and cross-hybridization must be overcome before microarrays can be used generally for the detection and differentiation of pathogens in environmental samples (Rhee *et al.*, 2004; Avarre *et al.*, 2007; Lee *et al.*, 2008). Moreover, microarrays cannot distinguish between non-viable, culturable and viable but non-culturable (VBNC) cells since DNA can persist for long periods of time after the death of cells (Masters *et al.*, 1994). The ability to discriminate between living and non-living cells is important in order to interpret the risk associated with the detection of pathogenic microbes (especially in environmental samples). The use of mRNA, which is highly labile with a short half-life, or the use of highly expressed targets may be selected to provide sensitive analyses (Keer and Birch, 2003; Zhao *et al.*, 2006).

Flow cytometry

Flow cytometry (FC) detects and quantify light scattering from fluorescent-labelled cells that have crossed a laser beam. A single sample can be analysed within 3–5 min with a quantification limit of approximately 200 cells/ml (Hammes *et al.*, 2009). FC, although an optical detection method, is used in combination with molecular techniques. Bacterial cells in water have been monitored with flow cytometry through nucleic acid staining (Berney *et al.*, 2008; Hammes *et al.*, 2008) or targeting specific cells with antibodies (Barbosa *et al.*, 2008) or FISH hybridization (Joachimsthal *et al.*, 2004; Lenaerts *et al.*, 2007).

FC is a valuable tool to differentiate between viable, intermediate and non-viable cells. Baclight™ bacterial viability kit (Live/Dead kit), widely used in flow cytometry, double stains nucleic acid with SYTO dyes (green fluorescence) and propidium iodide (PI) (red fluorescence). SYTO dyes stain the nucleic acid of all the cells, resulting in green fluorescence. The cells are afterwards stained with PI which can only move into membrane-compromised cells, staining the nucleic acid and resulting in red fluorescence. The disadvantage is that cells can be dead without showing membrane damage and hence is this rather an assay representing membrane damage than cell viability (Berney *et al.*, 2008; Hammes *et al.*, 2009). Calculating the nucleic acid content has also been used as an indicator of cell viability. The theory is that cells with higher cell viability reproduces at a higher rate and therefore will contain more copies of their genome (Mou *et al.*, 2005; Kleisteuber *et al.*, 2006). Care must be taken with the interpretation of results obtained from this approach. Bouvier *et al.* (2007) investigated the varied correlation between different nucleic acid contents and metabolic activities of subpopulations from a wide range of environmental communities.

FC combined with nucleic acid staining enable researchers to investigate the growth potential of microbial pathogens in natural waters. *Vibrio cholerae*, the causative agent of cholera, was shown through FC and SYBR Green nucleic acid staining to grow in different freshwater samples. This contradicted previous opinions that natural waters do not have sufficient nutrients to support the growth of this pathogen. Combining these experiments with assimilable organic carbon (AOC) concentrations it was concluded that *V. cholerae* would proliferate in water with a minimum AOC of 60 mg/l (Vital *et al.*, 2007). The same research group investigated the growth potential of *E. coli* O157 in freshwater samples using the same methodology. *E. coli* O157 was able to grow in freshwater samples with low carbon concentrations, once again contradicting previous opinions (Vital *et al.*, 2008).

Fluorescence-activated cell sorting (FACS) makes FC even more indispensable for detecting and differentiating between microbial pathogens in water. Cells with specific nucleic acid targets can be labelled with FISH probes, quantified and separated with FC-FACS. These cells can then be subjected for further genetic and biochemical analysis (Kalyuzhnaya *et al.*, 2006; Miyauchi *et al.*, 2007). Catalysed fluorescent reporter disposition-FISH (Sekar *et al.*, 2004) and molecular beacons (Lenaerts *et al.*, 2007) are now incorporated into FC-FACS to increase stain sensitivity and overcome the problem of sorting cells present in low numbers. FC-FACS-FISH has also been applied to sequence previously unsequenced microorganisms (Podar *et al.*, 2007) and cultured previously uncultured microorganisms (Akselband *et al.*, 2006) from environmental samples.

Molecular methodologies for microbial source tracking

Detection of pathogenic organisms provides information as to the safety and public health risks associated with a given water supply; however, it often does little to define the potential sources of the contamination. Generally, because different enteric pathogens are present in the intestines of different animals, the identification of a contamination event as being either of human or animal source would provide information as to the types of pathogens that may be expected, the risk of infection and the treatment that may be required to control transmission of disease (Scott *et al.*, 2002; Santo Domingo *et al.*, 2007). In response, molecular techniques are being developed as means to identify the source of a contaminant. DNA fingerprinting is one tool for microbial source tracking (MST) and consists of a family of techniques that are used to identify the sources of faecal contamination in various water bodies (Scott *et al.*, 2002; Meays *et al.*, 2004).

The different polymorphism-based procedures are generally coupled to a PCR reaction. In the amplified ribosomal DNA restriction analysis (ARDRA) technology, PCR-amplified 16S rRNA genes are digested with restriction endonucleases and the resulting fragments separated electrophoretically. Presence or absence of the restriction site within two strains cause differences in the length of the DNA restriction fragments, and the complexity of the pattern depends upon the number of target sequences and position of restriction sites. Comparison of the generated patterns to those obtained from a database allows assignment of isolates to species or species clusters in those cases were the banding patterns are highly similar. The separated DNA fragments may also be transferred to filters for hybridization with probes specific for an organism of interest (Parveen *et al.*, 1999; Carson *et al.*, 2001; Hartel *et al.*, 2002; Scott *et al.*, 2003). Two other protocols for generating DNA fingerprints use a single primer to amplify fragments with PCR before examination on agarose gels. PCR amplification of repetitive extragenic palindromic sequences (Rep-PCR) takes advantage of repetitive sequences found in the microbial genome (Versalovic *et al.*, 1994; Dombek *et al.*, 2000; McLellan *et al.*, 2003; Carson *et al.*, 2003). In the randomly amplified polymorphic DNA (RAPD) or arbitrarily primed PCR (Hadrys *et al.*, 1992; Breen *et al.*, 1995) technology, a short oligonucleotide primer (about 10 nucleotides), usually with random sequence that is not specific for a particular gene is used as a primer to amplify fragments. These methods yield DNA fingerprints comprised of multiple, differently sized DNA amplification products following separation by gel electrophoresis.

Detection of host-specific 16S rRNA genetic markers, using length heterogeneity PCR (LH-PCR) and terminal restriction fragment length polymorphism (T-RFLP) analysis, also holds promise as an effective method for characterizing a microbial population. The technique distinguishes members of mixtures of bacterial gene sequences by detecting differences in the number of base pairs in a particular gene fragment (Bernhard and Field, 2000a,b). Whereas LH-PCR separates PCR products for host-specific genetic markers based on the length of amplicons, T-RFLP uses restriction enzymes on amplified PCR products to determine unique size fragments. Specifically, in T-RFLP, rRNA target gene sequences are PCR-amplified using one or both of the primers with a fluorescent label. The amplification product(s) are then digested with appropriate restriction endonucleases and following electrophoresis of the resultant fragments using an automated DNA sequencer, a fluorescent electrophoretic profile of the digestion patterns is obtained (Clement *et al.*, 1998). The use of labelled primers limits the analysis (identification) to only the terminal fragments, thus allowing the study of complex microbial communities. Moreover, the possibility of discriminating fragments with differences as small as single bases gives the method a higher resolution than gel-based profiling techniques (Bernhard and Field, 2000a; Dick *et al.*, 2005; Galluzzi *et al.*, 2007).

One of the most promising technologies for microbial source tracking is, however, amplified fragment length polymorphism (AFLP) analysis. AFLP analysis appears to have the same taxonomic range as other fingerprinting techniques, but this technology combines several advantages of these different techniques, which in most cases results in the highest power of discrimination (Blears *et al.*, 1998). This technology is based on the selective amplification of a subset of genomic restriction fragments using PCR. For AFLP, purified genomic DNA is digested with two restriction endonucleases, one with an average cutting frequency and a second one with a higher cutting frequency after which oligonucleotide adapters are ligated to the genomic DNA restriction fragments. The sequence of the adapters and the adjacent restriction site serves as oligonucleotide primer binding sites for subsequent amplification of the restriction fragments by PCR. Selective nucleotides extending into the restriction fragments are added to the 3′ ends of the adapter-specific PCR primers such that only a subset of the restriction fragments are recognized and amplified. The subset of amplified fragments is then analysed by denaturing polyacrylamide gel electrophoresis to generate the fingerprint (Vos *et al.*, 1995; Savelkoul *et al.*, 1999). Since relatively small amounts of DNA are digested and detection of AFLP fragments does not depend on hybridization, the AFLP analysis method is more reproducible and robust than other fingerprinting techniques and it also displays more fragments than other fingerprinting techniques (Guan *et al.*, 2002; Leung *et al.*, 2004).

Nanotechnology materials and devices for biomolecular detection

Advances in nanotechnology and nanosciences are having a significant impact on the field of diagnostics, where a number of nanoparticle-based assays have been introduced for biomolecular detection (Fortina *et al.*, 2005; Gaspar *et al.*, 2007; Jain, 2007). The promise of increasing sensitivity and speed, and reduced cost and labour makes nanodiagnostics an appealing alternative to current molecular diagnostic techniques. New synthesis, fabrication and characterization methods for nanomaterials, which have dimensions that range from 1 to 100 nm, have evolved to a point that deliberate modulation of their size, shape and composition is possible, thereby allowing control of their properties. Along with these advances has come the ability to tailor their binding affinities for various biomolecules (proteins, nucleic acids and microbial pathogens) through surface modification and engineering (Rosi and Mirkin, 2005). Each of these capabilities allows the design of materials that can potentially be implemented into new biodiagnostic assays that can compete favourably with the current molecular diagnostic methodologies. In this section, the most promising nanomaterials (Table 2.2) and their use in the detection of nucleic acids and microbial pathogens will be highlighted.

Nanomaterials

Quantum dots

Quantum dots (QDs) are colloidal, luminescent inorganic nanocrystals with unique photochemical properties, which include high quantum yields, large extinction coefficients, pronounced photostability, as well as broad absorption spectra coupled to narrow size-tunable photoluminescent emission spectra (Bruchez *et al.*, 1998; Chan and Nie, 1998). A typical QD has a diameter of 2–8 nm and is usually composed of a core consisting of a semiconductor material, such as CdSe, enclosed in a shell of another semiconductor material with a larger spectral band-gap, such as ZnS. The shell prevents the surface quenching of excitons in the emissive core and thus increases the photostability and quantum yield for emission (Hines and Guyot-Sionnest, 1996; Dabbousi *et al.*, 1997). Since QDs are usually synthesized from organometallic precursors and salts, they have no intrinsic aqueous solubility. Consequently, the native coordinating organic ligands on the surface of the QDs must

Table 2.2 Examples of diagnostic nanotechnologies (adapted from Azzazy *et al.*, 2006)

Technology	Description	Advantages
Quantum dots (QDs)	Semiconductor nanocrystals typically composed of a core semiconductor enclosed in a shell of another semiconductor with a large spectral band-gap	Multiplexed diagnostics; immunoassays; cellular imaging
Gold nanoparticles	Gold nanoparticles in the nanometre size domain; rapid test with simple and clearer optical read-out	Immunoassays; detection of infectious agents by DNA hybridization
Magnetic nanoparticles	Label choice can be made so that its interaction with the analyte gives a magnetic signal to be used instead of an optical one; highly sensitive detection method with a high dynamic range	High sensitivity attributable to detection of subtle modifications in magnetic character; ability to detect microorganisms
Carbon nanotubes	Arrays of carbon nanotubes are used as electrodes with single-stranded DNA probes attached to their open end; hybridization of target DNA gives an electrochemical signal; they can also be adapted for analytes other than DNA	Detection sensitivity in the attomole range and reduction of signal-to-noise ratio; method also minimizes the need for amplification of target DNA
Nanowires	Lightly doped, high-aspect ratio, single-crystal nanowire with a diameter between 10 and 20 nm; allows for different interactions with different entities and more unique signals	Can be associated with almost any chemical or biological recognition system; allows real-time detection; analyte-independent; can be conjugated with almost any biomolecular recognition entity suitable for the use
Nanocantilevers	Micro-machined silicon cantilevers similar to those used in atomic force microscopy; highly sensitive mass sensors; can be functionalized with biological molecules	Protein and DNA detection and quantification
On-chip technology	Combination of PCR with nucleic acid hybridization on a single chip; less interference between parallel reactions; can be designed for use in immunoassays	Rapid and sensitive detection of DNA and proteins

either be exchanged or functionalized with a ligand that can impart both solubility and bioconjugation sites, if desired. Strategies for attaching biorecognition molecules to QDs comprise of direct chemical coupling to a functional group displayed on the QD surface (Chan and Nie, 1998; Gerion *et al.*, 2003) or by non-covalent self-assembly in which proteins are engineered to express positively charged domains that interact through electrostatic attractions with the negative surface of modified QDs (Ding *et al.*, 2004; Goldman *et al.*, 2005).

Although functionalized QDs have been used to detect complementary target nucleic acid sequences in chip-based assays and in fluorescent *in situ* hybridization (FISH) assays (Pathak *et al.*, 2001; Xiao and Barker, 2004), QDs are increasingly being used in immunoassay detection (Alivisatos *et al.*, 2005; Goldman *et al.*, 2006b). QD-based immunological assays have been applied to the detection of different bacterial and protozoan pathogens. For these immunoassays, the QDs are conjugated to organism-specific antibodies, or, alternatively, biotinylated organism-specific antibodies are used that are subsequently detected using a QD–streptavidin bioconjugate. These approaches have been used successfully for the detection of *Cryptosporidium parvum* (43 oocysts in spiked reservoir water) (Lee *et al.*, 2004), *Giardia lamblia* (1–5 × 10^3 cysts) (Zhang *et al.*, 2006a) and *Escherichia coli* O157:H7 (2 × 10^7 cells) (Su and Li, 2004; Hahn *et al.*, 2005). Moreover, QDs appear to be especially suited for multiplex immunoassays. As a demonstration of the potential of QDs in multiplexed immunoassay formats, the simultaneous detection of *E. coli* O157:H7 and *Salmonella enterica* serovar Typhimurium bacteria (10^4 cells in 2 h) (Yang and Li, 2006), and of *C. parvum* and *G. lamblia* (5 × 10^3 cysts) in environmental water samples (Zhu *et al.*, 2004a), using different coloured QDs as immunoassay labels, has been reported.

QDs have also been applied in flow cytometry because of their broad absorption spectra and narrow size-tunable photoluminescent emission spectra. The broad absorption band enables semiconductor QDs to simplify and improve the detection of multiple bacterial targets in flow cytometry samples. Each organic dye needs a separate excitation source, resulting in multiple excitation sources and complicated experimental setups when monitoring more than one organism. In comparison, only a single excitation source in the UV range is needed to excite all the visible colours of CdSe QDs. QDs have been used to simultaneously enumerate pathogenic *E. coli* O157:H7 and harmless *E. coli* DH5α in water. Cross-spectral talk of QDs are also drastically lower than the organic dyes traditionally used in flow cytometry due to their narrow emission spectra (Hahn *et al.*, 2008). Flow cytometric measurements of QDs have been compared with the widely used fluorochrome, fluorescein isothiocyanate (FITC). The minimum fluorophore concentration for detection was a 100-fold lower when paramagnetic beads were labelled with QDs (Ibáñez-Peral *et al.*, 2008).

Despite their capability for single molecule and multiplexed detection, it is, however, unlikely that QDs will completely replace traditional organic fluorophores as biological labels. Some of the challenges that have yet to be overcome include financial costs, since QDs are expensive in comparison to organic dyes and there is an initial financial investment required regarding instruments optimized for use with QDs (Azzazy *et al.*, 2006). Non-specific binding and aggregation are two factors that can lower QD fluorescence, as in the case of antibody-stained *Cryptosporidium* oocysts (Ferrari and Bergquist, 2007). Also, QDs are an order of magnitude larger than organic dyes and thus the extent to which their presence perturbs the recognition event being detected, must be determined. This is particularly important when multiplex assays are desired, since labelling several biomolecules with QDs of different sizes could result in varying degrees of perturbation due to the large differences in the QD sizes. In contrast, most organic dyes are of similar size in spite of their large differences in absorption/emission characteristics (Bailey *et al.*, 2004).

FloDots

In addition to QDs, luminescent dye-doped silica nanoparticles (FloDots), which consist of luminescent organic or inorganic dye molecules dispersed inside a silica matrix, have also been developed for ultra-sensitive bioanalysis and diagnostics (Tapec *et al.*, 2002; Yao *et al.*, 2006). Owing to the silica matrix shielding effect, the doped dye molecules are protected from environmental oxygen, enabling the fluorescence to be constant and thus providing an accurate measurement for bioanalysis (Tapec *et al.*, 2002). Moreover, the silica matrix provides a versatile substrate for surface immobilization and various biorecognition molecules, including oligonucleotides and antibodies, have been conjugated to FloDots (Santra *et al.*, 2001; Hilliard *et al.*, 2002; Tapec *et al.*, 2002).

FloDots have been used successfully for the detection of DNA hybridization and in immunoassays. Using FloDots functionalized with oligonucleotides as labels for chip-based sandwich DNA assays, Zhao *et al.* (2003) reported a detection limit of 1 fM target DNA. The high sensitivity of the assay can be ascribed to the fact that each FloDot has the fluorescence intensity of thousands of dye molecules; therefore, each gene hybridization is reported by thousands of fluorophores. FloDot-immunoassays, using FloDots conjugated with monoclonal antibodies specific for the O-antigen of *E. coli* O157:H7, have also been developed to achieve rapid *E. coli* O157:H7 detection at the single-cell level (Zhao *et al.*, 2004). The fluorescence intensity emitted by one *E. coli* O157:H7 cell was sufficient to be detected using a normal spectrofluorometer in a conventional plate-based immunological assay, or to be accurately enumerated using a flow cytometer within 1 min of sample preparation. Although the use of FloDots represents an improvement over conventional fluorophore-based assays, they remain somewhat limited by the fundamental drawbacks of conventional fluorophores, including broad adsorption and emission profiles, which ultimately limit multiplexing capabilities (Rosi and Mirkin, 2005).

Gold nanoparticles

Metal nanoparticles, such as gold, silver and iron, constitute one of the most researched branches of nanotechnology due to their electronic, optical, catalytic and thermal properties (Guo and Wang, 2007; Theron *et al.*, 2008). Among these, gold nanoparticles are intensively used in a variety of colorimetric and fluorescence biodetection assays (Penn *et al.*, 2003; Thaxton *et al.*, 2006; Gaspar *et al.*, 2007). Specific focus has been directed at colloidal gold nanoparticles ranging from 3 to 100 nm in size, since they are rather stable and their properties can be easily tailored by chemical modification of their surfaces (Mirkin and Taton, 2000; Sandström *et al.*, 2003).

Gold nanoparticle-based probes have been used in the identification of pathogenic bacteria in a chip-based system (Taton *et al.*, 2000). The assay consists of a capture DNA strand immobilized on a glass chip that recognizes the DNA of interest. A separate sequence on the captured target strand is then labelled with an oligonucleotide-functionalized gold nanoparticle probe. After catalytic reduction of silver onto the gold nanoparticle surfaces to amplify the target signal, the capture-strand/target/nanoparticle sandwich is visualized with a flatbed scanner. The target DNA was detected at concentrations of 50 fM, which represents a nearly 100-fold increase in sensitivity over traditional fluorescence-based assays. Gold nanoparticles have also been used in the detection of genomic DNA from *Staphylococcus aureus*, without enzymatic amplification, at concentrations of 33 fM (Storhoff *et al.*, 2004). The assay depends upon the target-selective binding of gold nanoparticle probes to consecutive regions of a target DNA sequence in solution, followed by transfer to an evanescently illuminated glass microscope slide for light scattering measurements. Whereas the gold particle probes scatter orange light in the presence of target, the probes scatter green light if the target is absent, whilst the target concentration is quantified by measuring the intensity of scattered light. Multiplexed detection of different viruses and bacteria, using gold nanoparticles surface functionalized with a target-specific oligonucleotide and further encoded with a Raman-active dye molecule, has also been demonstrated (Cao *et al.*, 2002). Whereas the complementary probe sequence imparts the specificity for the target sequence of interest, the presence of the target is confirmed by silver staining and the identity of the target is revealed by detecting the surface-enhanced Raman scattering (SERS) of the Raman dye near the nanoparticle surface. This assay demonstrated a sensitivity of 1 fM target concentration.

Despite its widespread use in nucleic acid hybridization assays, the use of gold nanoparticles for detection of bacterial pathogens in immunoassays has only recently been described (Lin *et al.*, 2005). The assay comprises detection of organism-specific antigens with biotinylated polyclonal antibodies after which gold particles functionalized with a secondary antibiotin antibody are added, and the particles are then visualized under a dark-field stereomicroscope. The immunoassays were demonstrated to reliably detect *Helicobacter pylori* and *E. coli* O157:H7 antigens in quantities in the order of 10 ng, which provides a sensitivity of detection comparable to those of conventional dot blot assays. A two-dot filter system has also been developed in which colloidal gold nanoparticles (2 nm) were bound onto anti-*E. coli* O157:H7 antibodies. These monoclonal antibodies were bound onto a 0.2-μm nitrocellulose filter membrane through which water was filtered. The same antibodies that captured the bacteria acted as detectors since the gold nanoparticles could be visualized under epifluorescence microscopy. This one-step detection method not only had a low detection rate (1 cfu/100 ml) and high specificity, but would be inexpensive and easy implemented in the routine testing of water samples (Kamma *et al.*, 2008).

Magnetic nanoparticles

In contrast to gold nanoparticles and QDs, magnetic nanoparticles have not been used in many biological applications. Nevertheless, advances in the synthesis of monodispersed magnetic nanoparticles, ranging in size from 2 to 20 nm (Park *et al.*, 2004; Sun *et al.*, 2004), has provided a basis from which to explore applications of magnetic nanoparticles in diagnostics. Magnetic nanoparticles are produced from materials that can be strongly attracted by magnets or be magnetized (Šafařík and Šafaříková, 2002). They can be prepared in the form of single domain or superparamagnetic (Fe_3O_4), greigite (Fe_3S_4), maghemite (γ-Fe_2O_3), and various types of ferrites ($MeO{\cdot}Fe_2O_3$, where Me = Ni, Co, Mg, Zn, Mn, etc.). Bound to biorecognition molecules, magnetic nanoparticles can be used to facilitate the separation, purification and concentration of different biomolecules (Richardson *et al.*, 2001; Yang *et al.*, 2004; Gu *et al.*, 2006). To do so, biorecognition molecules such as antibodies can be immobilized on the surface of magnetic nanoparticles through covalent or electrostatic interactions (Gupta and Gupta, 2005). After reacting these magnetic nanoparticles with sample solutions, targeted molecules can be bound by or captured on the surface of these magnetic nanoparticles. By applying a magnetic field, these nanoparticles can subsequently be concentrated and separated from the bulk solution and identified.

Biofunctional magnetic nanoparticles, in which thiolated vancomycin was attached to FePt nanoparticles, have been used to capture and detect of a wide range of bacteria at very low concentrations within 60 min (Gu *et al.*, 2003). These included capturing and detection of *Staphylococcus aureus* at 8 cfu/ml, *S. epidermidis* at 10 cfu/ml, *Enterococcus faecalis* at 26 cfu/ml, and *E. coli* at 15 cfu/ml. Although the sensitivity achieved using magnetic nanoparticles is comparable to PCR-based assays, the direct capture protocol is faster than PCR when the bacterium count is low since it obviates the need for pre-enrichment of bacteria through culturing. In an alternative approach, Ho *et al.* (2004) reported combining biofunctional magnetic nanoparticles with matrix-assisted laser desorption/ionization mass spectrometry (MALDI-MS) to detect pathogenic bacteria in water. Biofunctional nanoparticles were fabricated by attaching human immunoglobulin (IgG), which binds selectively to IgG-binding sites on the cell walls of pathogens, onto the surfaces of magnetite (Fe_3O_4) nanoparticles. Using this assay, both *S. saprophyticus* and *S. aureus* were detected at concentrations of 3×10^5 cfu/ml in aqueous sample solutions. Measuring adenosine triphosphate (ATP) bioluminescence of bacteria captured onto magnetic nanoparticles is another proposed method for detecting microorganisms. *E. coli* was detected in milk by Cheng *et al.* (2009) within a short period (1 h) and with a low detection limit (20 cfu/ml).

Biofunctional magnetic glyconanoparticles have also been engineered by covalently binding unmodified monosaccharide *d*-mannose onto iron oxide nanoparticles. These particles had the ability to recognize mannose-specific receptor sites on *E. coli* (Liu *et al.*, 2009). Magnetic nanoparticles have been developed to sequester DNA in water and capture the DNA–nanoparticles complexes by the application of high-gradient magnetic separation. Modifying magnetite clusters with poly(hexamethylene biguanide) and polyethyleneimine resulted in strong cationic nanoparticles, which enabled the binding with DNA molecules through electrostatic forces. The cationic nanoparticles can also serve as a disinfectant by binding to the negatively charged cell envelopes of bacteria. These particles were colloidally stable in fresh and ocean water for weeks at a pH ≤ 10 (Bromberg *et al.*, 2009).

Magnetic microparticle–antibody conjugates (Dynabeads®) are commercially available and kits have been developed for the detection of *Legionella* species, *Cryptosporidium* oocysts and *Giardia* cysts from concentrated water samples. Dynabeads® are also available for the detection of *E. coli*, *Salmonella* and *Listeria* species; however, the samples must be grown for 6–8 h in a pre-enrichment broth. Streptavidin-coated Dynabeads® allow researchers to design their own magnetic microparticle–antibody conjugates for specialized assays (www.invitrogen.com). Biotinylated organism-specific antibodies will bind covalently onto the streptavidin-coated Dynabeads. A wide range of biotin-labelled antibodies are available from companies such as Abcam (www.abcam.com).

Despite the promise shown by biofunctional magnetic nanoparticles, some challenges regarding their widespread use have yet to be overcome. In addition to requiring a robust surface chemistry to attach bioactive molecules onto magnetic nanoparticles without laborious synthetic efforts, more precise control of the numbers and orientations of the molecules on the surfaces of magnetic nanoparticles is also required (Gu *et al.*, 2006).

Carbon nanotubes

Since their discovery, carbon nanotubes (CNTs) have attracted great attention as nanoscale building blocks for micro- and nanodevices. CNTs can be divided essentially into single-walled carbon nanotubes (SWCNTs) and multi-walled carbon nanotubes (MWCNTs) based on the principle of hybridized carbon atom layers in the walls of CNTs. Whereas SWCNTs have diameters ranging from 0.3 to 3 nm, the MWCNTs are composed of a concentric arrangement of many cylinders and can reach diameters of up to 100 nm (Iijima and Ichihashi, 1993; Balasubramanian and Burghard, 2006). CNTs are considered to be ideal for use in biosensors for detecting individual biomolecules and other biological agents. Not only do they have high surface-to-volume and surface-to-weight ratios, but they are also conducting, act as electrodes, and can be derivatized with functional groups that allow immobilization of biomolecules (Davis *et al.*, 2003; Balasubramanian and Burghard, 2005; Sinha *et al.*, 2006).

The use of nano-size electrodes based on CNTs has the potential to greatly improve the sensitivity in recognizing DNA hybridization events. Li *et al.* (2003) developed DNA microarrays containing sensing pads constructed from MWCNTs and built into a matrix within a silicon nitride template. The upper (open) ends of the tubes act as nanoelectrodes and are functionalized with ssDNA probes. Target DNA that hybridizes to the ssDNA probe on the ends of the electrically conductive MWCNTs is detected using an electrochemical method that relies on guanine oxidation. The hybridization of less than a few attomoles of oligonucleotide targets was demonstrated. In an alternative approach, CNTs coated with alkaline phosphatase enzymes was used for the detection of amplified DNA (Wang *et al.*, 2004). This assay employs a magnetic microparticle modified with oligonucleotides that are complementary to one-half of the target DNA sequence, and alkaline phosphatase-coated carbon nanotubes that are modified with oligonucleotides that are complementary to the other half of the target DNA sequence. Binding of the target DNA promotes the formation of a magnetic microparticle/target/carbon nanotube sandwich, which is magnetically separated from the assay medium. After separation, the enzyme substrate α-naphthyl phosphate is added to the mixture, resulting in formation of α-naphthol product that is ultimately detected at a CNT-modified electrode via chronopotentiometric stripping. This method detected target DNA at concentrations as low as 54 aM (Wang *et al.*, 2004).

In addition to their potential in recognizing DNA hybridization events, the use of CNTs displaying ligands for the capturing or recognition of bacterial pathogens has recently been described. The solubilization of SWCNTs via fictionalization with derivatized galactose has been reported and it was subsequently shown that the nanotube-bound galactose could serve as polyvalent ligands that strongly interacted with receptors on *E. coli* O157:H7, resulting in significant cell agglutination (Gu *et al.*, 2005). This work has subsequently been extended to the preparation of immuno-carbon tubes (Elkin *et al.*, 2005; Lin *et al.*, 2006). For this purpose, SWCNTs and MWCNTs are functionalized with bovine serum albumin (BSA) to attain aqueous solubility and then further conjugated with an *E. coli* O157:H7-specific antibody to form immuno-carbon tubes. Limited quantitative data was provided, but the results suggest that the immuno-carbon tubes are capable of sensitively capturing the target bacteria.

While CNTs currently are not as easily functionalized as QDs or nanoparticles, they offer the distinct advantage of rapid, real-time detection and may thus become viable options as nanostructured biodiagnostic devices (Rosi and Mirkin, 2005). In addition to challenges at the fabrication level (e.g. production of pure and uncontaminated nanotubes is costly, continuous growth of defect-free CNTs to macroscopic lengths is difficult to obtain and dispersion of CNTs onto a polymer matrix is very difficult) (Sinha *et al.*, 2006), another important issue related to the use of CNTs is their toxicity (Oberdörster *et al.*, 2007). Although results suggest that chemically modifying CNTs can reduce their cytotoxicity to a certain extent (Sayes *et al.*, 2006), more research is required to address the effect of CNTs on biological systems, as well as information related to safety issues.

Nanowires

Nanowires have been explored as signal transduction motifs in the electrical detection of DNA, proteins and microbial pathogens (Lieber, 2003; Patolsky *et al.*, 2006a). Nanowire sensors operate on the basis that the change in chemical potential accompanying a target binding event can act as a field-effect gate upon the nanowire, thereby changing its conductance. The ideal nanowire sensor is a lightly doped, high-aspect ratio, single-crystal nanowire with a diameter between 10 and 20 nm (Cheng *et al.*, 2006). Recently, detailed protocols for fabrication of silicon nanowire devices, including their covalent modification with biorecognition molecules, have been described (Patolsky *et al.*, 2006b).

Real-time, label-free detection of DNA has been demonstrated by Hahm and Lieber (2004) using silicon nanowires functionalized with peptide nucleic acid (PNA). The conductance of the PNA-functionalized silicon nanowire bridging two electrodes was measured in the presence of target DNA and mutant DNA with three consecutive base deletions. Introduction of target DNA into the assay caused a rapid and immediate change in conductance, while the effect of mutant DNA was negligible. Furthermore, the conductance changes scale with target concentration, and target DNA at concentrations as low as 10 fM were detected (Hahm and Lieber, 2004).

Semiconducting silicon nanowires have also been used for the electrical detection of viruses in solutions. Patolsky *et al.* (2004) interfaced nanowires functionalized with antibodies specific for influenza A virus particles with a microfluidic sampling system. The nanowire sensing system was able to detect influenza A at the single-virus level, demonstrating that single virus/nanowire recognition events can be detected by measuring real-time changes in nanowire conductivity (Patolsky *et al.*, 2004). In addition to silicon nanowires, metallic multi-striped nanowires have also shown great promise as potential platforms for multiplex immunoassays (Nicewarner-Pena *et al.*, 2001). These nanowires are built through submicrometre layering of different metals, e.g. gold, silver and nickel, by electrodeposition within a porous alumina template. Owing to the permutations in which the metals can be deposited, a large number of unique yet easily identifiable encoded nanowires can be fabricated. The multistriped nanowires, when coated with target-specific antibodies, were reported to efficiently and accurately allow multiplex detection of *Bacillus globiggi* spores, MS2 bacteriophage and ovalbumin protein (Tok *et al.*, 2006). The sensitivity of detection for *B. globiggi* spores, MS2 bacteriophage and ova protein was estimated to be 1×10^5 cfu/ml, 1×10^5 pfu/ml and 5 ng/ml, respectively, and is comparable with results obtained using microarrays.

Nanowires are set apart from other available nanobiotechnologies due to several key features such as direct, label-free, real-time electrical signal transduction, as well as ultrahigh sensitivity, exquisite selectivity and the potential for integration of addressable arrays (Rosi and Mirkin, 2005). However, an intrinsic limitation of nanowires is that the detection sensitivity depends on solution ionic strength. Consequently, for samples with a high ionic strength, diagnostics will require a desalting step before analysis to achieve the highest sensitivity (Zheng *et al.*, 2005). Furthermore, a practical constraint might be that the synthesis and fabrication of nanowire biosensor devices require some technologies that are not common to most laboratories (Patolsky *et al.*, 2006b).

Nanocantilevers

Nanocantilevers, which are typically made of silicon, silicon nitride or silicon oxide, require only minute changes in compressive or tensile surface stress on either their upper surface or lower surface to cause measurable deflection of the cantilever, and are capable of converting biomolecular recognition reactions into micromechanical motion. Consequently, nanocantilevers offer an opportunity for the development of highly sensitive, miniature and label-free detection systems (Thundat and Majumbar, 2003; Hansen and Thundat, 2005; Carrascosa *et al.*, 2006).

Direct, label-free detection of DNA typically involves the use of silicon cantilevers with a thin gold coating on the top surface that is functionalized with thiolated capture DNA strands. Binding of the capture DNA strands with the introduced target DNA causes deflection of the cantilever, which can be measured accurately using optical detection methods. Using this approach, Fritz *et al.* (2000) demonstrated the hybridization of complementary oligonucleotides with a detection limit of 10 nM and showed that a single mismatch between two 12-mer oligonucleotides is clearly detectable. Similarly promising results were obtained by Hansen *et al.* (2001) with a 10-mer oligonucleotide.

In recent years, cantilever immunosensors have been developed to detect bacteria and viruses. Typically, the devices detect the additional mass loading that results from the interaction between specific antibodies, immobilized on the surface of the cantilever, and antigens on the cell membrane surface. In an early experiment, a cantilever biosensor was constructed and used to detect *E. coli* O157:H7, following immersion of the cantilever in a suspension containing $10^6 – 10^9$ cells/ml (Ilic *et al.*, 2000). The detection of 16 *E. coli* O157:H7 cells were demonstrated and no frequency shifts were observed when buffer alone or buffer containing *S. enterica* serovar Typhimurium was incubated with the cantilevers. A magnetoelastic cantilever immunosensor has been developed that uses a magnetic field to induce oscillation of the sensor (Ruan *et al.*, 2003). The sensor surface is coated with antibodies to permit specific capture of the desired target agent after which alkaline phosphatase-labelled antibodies to the target are added to amplify the signal, thereby increasing the total mass on the sensor. The sensor was tested with *E. coli* O157:H7 and a sensitivity of 10^2 cells/ml were reported. More recently, resonant cantilever biosensors have been developed for detection of *Listeria innocua* (Gupta *et al.*, 2004a) and vaccinia virus (Gupta *et al.*, 2004b), and the detection of a single *L. innocua* cell and vaccinia virus particle has been demonstrated.

Despite being in its early days, cantilevers provide an opportunity for label-free, real-time measurements in fluids in a single-step reaction, and can potentially serve as a powerful platform for sensitive, multiplexed detection of biomolecules. Although cantilevers can be microfabricated by standard low-cost silicon technology, leading to a decrease in production costs and allowing the possibility of integrating multiple functional devices onto the same platform (Dhayal *et al.*, 2006), some challenges must be overcome before cantilever array sensors can come into widespread use. These challenges relate especially to methods for detecting nanoscale motion and the development of immobilization techniques that can efficiently transduce the stress involved in biochemical interaction to the cantilever substrate (Cheng *et al.*, 2006).

Miniaturized technologies

Advances in microfluidics and microfabrication technologies have contributed greatly to the miniaturization of biological and chemical analytical systems, allowing the handling of low volume samples, as well as reductions in reagent consumption, waste generation, costs and assay time (Reyes *et al.*, 2002; Dittrich *et al.*, 2006). Micro-total analysis systems (μ-TAS), sometimes called 'lab-on-a-chip', are microfabricated devices capable of performing the functions of large analytical devices in small units. These devices are fabricated in glass, silicon or polymer materials, and integrate different functions and functionalities. Some sophisticated versions can perform sample introduction and handling (e.g. cell lysis, dilution and debris removal), separation (e.g. electrophoresis, chromatography) and detection, all conducted on the chip (Köhler and Henkel, 2005; Chen *et al.*, 2007; Peoples and Karnes, 2008). It is believed that μ-TAS will be particularly valuable in DNA and protein analysis (Khandurina and Guttman, 2002), genomics and proteomics (Sanders and Manz, 2000), and diagnostics (Landers, 2003).

Miniaturized immunoassays have been performed successfully with microchips (Henares *et al.*, 2008; Peoples and Karnes, 2008). These immunoaffinity microfluidic devices are considered promising platforms to achieve rapid and sensitive immunological detection of microbial cells (Rowe, 1999; Dodge *et al.*, 2001; Liu and Li, 2001). Zhu *et al.* (2004b) described a simple approach in which sample trapping and concentration steps were integrated together with whole-cell immunoassay in a silicon-based lab-on-a-chip. The immunoassay was performed by injecting the sample solution, which contained *C. parvum* and *G. lamblia*, into a microchamber. Subsequently, a solution containing fluorescently labelled target-specific antibodies was delivered and serves to simultaneously concentrate, trap and label the targeted cells at a trapping region. Following a wash step to remove unbound antibodies, the labelled parasites were detected by epifluorescence microscopy. Compared to conventional immunoassays, the total analysis time was reduced from 2–3 h to 2–5 min, and the total consumption of reagents was reduced 20-fold. In an alternative approach, Liu *et al.* (2005) injected microbeads coated with a primary antivirus antibody into a microfluidic device, which are subsequently trapped in front of a pillar-type filter region. A sample containing target virions was injected into the device and virions were captured on the surface of the microbeads. This was followed by injection of a labelling solution containing a secondary antivirus antibody labelled with QDs to allow detection by epifluorescence microscopy. In comparison to a standard ELISA performed on the same marine iridovirus, the minimal detectable concentration of the target virus was improved from 360 to 22 ng/ml, the detection time was shortened from 3 h to less than 30 min, and the amount of antibody consumed was reduced 14-fold.

Considerable effort has been directed to the development of chip-based systems for miniaturized and rapid PCR (Auroux *et al.*, 2004; Zhang *et al.*, 2006b). The devices consist of a chip containing wells, channels, electrodes, filters, pumps, valves and heating devices designed for buffer and sample storage, PCR and target DNA detection (Woolley *et al.*, 1996; Burns *et al.*, 1998; He *et al.*, 2001; Reyes *et al.*, 2002). Remarkably, a polymeric microchip with a 1.7 μl chamber containing a thermocoupler was used to successfully amplify a 500-bp DNA fragment of lambda phage in 15 cycles, in a total amplification time of 240 seconds (Giordano *et al.*, 2001). By making use of a PCR microchip coupled with a capillary electrophoresis (CE) chip, it was more recently demonstrated that bacterial targets as low as 2–3 cells could be amplified within a 200-nl PCR chamber and the PCR-amplified target DNA was subsequently resolved by CE within 10 min (Lagally *et al.*, 2004). In order to improve PCR throughput and reduce the analysis time, multi-chamber PCR microfluidics on a single chip has been reported (Zou *et al.*, 2002; Oh *et al.*, 2005). Also, chip devices with optical windows have been fabricated that allows for measurement of fluorescence intensity during the thermocycling process, thus providing a miniaturized version of real-time PCR (Northrup *et al.*, 1998; Gulliksen *et al.*, 2004; Wang *et al.*, 2006). In this regard, Cady *et al.* (2005) developed an integrated miniaturized real-time PCR detection device equipped with a microprocessor, pumps, thermocycler and light-emitting diodes (LEDs)-based fluorescence excitation/detection. Monolithic DNA purification and real-time PCR enabled fast detection of *L. monocytogenes* cells ($10^4 - 10^7$) within 45 min.

In spite of their potentially powerful application in diagnostics and environmental monitoring, the 'complete' lab-on-a-chip still requires further development. The bottlenecks blocking the realization of a truly and highly integrated chip include sample preparation and product detection (Zhang and Xing, 2007). Since the source of raw template samples is varied and the sample preparation methods are diverse, the miniaturization of conventional sample preparation and functionalities on a chip remains a challenge. As for on-chip detection, the product detection methods have not advanced as rapidly as other aspects of chip development. Consequently, miniaturized ultra-sensitive detectors are required if the sensitivity of the lab-on-a-chip devices is to be improved. Moreover, additional efforts have to be made towards the validation of the methods to demonstrate the reliability of μ-TAS systems.

Conclusions

Many years of experience have shown that monitoring microbiological water quality by standard indicator systems is and has been an important means for the supply of safe drinking water. Indicator methods are simple and easy to use, cheap and cover a wide range of pathogens. However, current indicators leave little room for improvement. These methods have served relatively well, but are less accurate and informative than what is needed. The main limitations are lack of specificity and slow response. Furthermore, current indicators are poor for addressing parasites and viruses. The detection of waterborne pathogens, as well as new and re-emerging ones posing increased risks due to newly acquired virulence properties and other traits requires examination of newer approaches to monitoring the microbiological quality of water if control and prevention of waterborne diseases are to be realized.

Developments, particularly in the fields of genomics and biotechnology in the last few years, have resulted in a wide range of molecular tools, principally based on the detection of nucleic acid material and its amplification. They offer a novel, more sensitive and specific way of detecting microorganisms. They can also identify organisms that would not be detected with current culture techniques and can be used to track new pathogenic entities, including variants of otherwise harmless microorganisms. There are probably very few groups of microorganisms that have not been detected with these amplification techniques and several test kits have already been commercialized. Despite the success of these molecular methods, several barriers must be overcome before they can be used to routinely assess water quality and the microbiological safety of drinking water. The relationship between detection by molecular approaches and the subsequent viability or infectivity of waterborne enteric pathogens remains a concern. In addition, methods used for purifying and concentrating the target microbes and their nucleic acids from water, so that they are free of contaminants that may interfere with the analysis, need further improvement, consolidation and simplification. Further research is also needed to develop and refine the prototype protocols into collaboratively tested methods that could be used routinely and expeditiously to evaluate the microbiological safety of water.

Owing to recent advances in nanoscience and nanotechnology, various different nanomaterials and devices have been developed that show great promise for diagnostic applications. Subsequently, many different nanotechnology-based diagnostic systems have been reported in the literature and many of these have the potential to become the next generation of diagnostic tools. Moreover, microfluidic chip-based systems such as the lab-on-a-chip technology should have a significant impact on environmental microbial monitoring by permitting detection and identification of targets within minutes at the sampling site with a sensitivity level of a single cell. However, some technical and practical problems need to be solved before their full potential can be realized. These include tight control over the synthesis and functionalization of nanomaterials, as small variations can change their properties and behaviour in diagnostic methods. Also, their implementation into routine functional devices remains a challenge, and note should be taken that many of the diagnostic systems must still be taken from proof-of-concept and evaluated with environmental samples. In this regard, some of the challenges that need to be resolved, in addition to those highlighted above, include sample processing, detection of multiple agents in a single sample, and improving the sensitivity and selectivity of the assays for application to complex environmental samples.

Web resources

www.abcam.com.
www.invitrogen.com.

References

Akselband, Y., Cabral, C., Castor, T.P., Chikarmane, H.M., and McGrath, P. (2006). Enrichment of slow-growing marine microorganisms from mixed cultures using gel microdrop (GMD) growth assay and fluorescence-activated cell sorting. J. Exp. Mar. Biol. Ecol. *329*, 196–205.

Alivisatos, A.P., Gu, W., and Larabell, C.A. (2005). Quantum dots as cellular probes. Ann. Rev. Biomed. Eng. *7*, 55–76.

Amann, R., Fuchs, B.M., and Behrens, S. (2001). The identification of microorganisms by fluorescence in situ hybridisation. Curr. Opin. Biotechnol. *12*, 231–236.

Amann, R., Glöckner, F.-O., and Neef, A. (1997). Modern methods in subsurface microbiology: In situ identification of microorganisms with nucleic acid probes. FEMS Microbiol. Rev. *20*, 191–200.

Amann, R., Ludwig, W., and Schleifer, K.-H. (1993). Identification and in situ detection of individual bacterial cells. FEMS Microbiol. Lett. *100*, 45–50.

Amann, R.I. (1995). In situ identification of microorganisms by whole cell hybridization with rRNA-targeted nucleic acid probes. In Molecular Microbial Ecology Manual, Akkermans, A.D.L., J.D. van Elsas, and F.J. de Bruijn, eds. (Kluwer Academic Publishers, Dordrecht, The Netherlands).

Amann, R.I., Ludwig, W., and Schleifer, K.-H. (1995). Phylogenetic identification and in situ detection of individual microbial cells without cultivation. Microbiol. Rev. *59*, 143–169.

Anderson, K.L., Whitlock, J.E., and Harwood, V.J. (2005). Persistence and differential survival of fecal indicator bacteria in subtropical waters and sediments. Appl. Environ. Microbiol. *71*, 3041–3048.

Andreescu, S., Njagi, J., Ispas, C., and Ravalli, M.T. (2009). Applications of advanced nanomaterials for environmental monitoring. J. Environ. Monit. *11*, 27–40.

Arnheim, N., and Erlich, H. (1992). Polymerase chain reaction strategy. Ann. Rev. Biochem. *61*, 131–156.

Ashbolt, N.J. (2004). Microbial contamination of drinking water and disease outcomes in developing regions. Toxicol. *198*, 229–238.

Auroux, P.-A., Koc, Y., deMello, A., Manz, A., and Day, P.J. (2004). Miniaturized nucleic acid analysis. Lab Chip *4*, 534–546.

Avarre, J.C., de Lajudie, P., and Béna, G. (2007). Hybridization of genomic DNA to microarrays: A challenge for the analysis of environmental samples. J. Microbiol. Methods *69*, 242–248.

Azzazy, H.M., Mansour, M.M., and Kazmierczak, S.C. (2006). Nanodiagnostics: A new frontier for clinical laboratory medicine. Clin. Chem. *52*, 1238–1246.

Bailey, R.E., Smith, A.M., and Nie, S. (2004). Quantum dots in biology and medicine. Physica E *25*, 1–12.

Balasubramanian, K., and Burghard, M. (2005). Chemically functionalized carbon nanotubes. Small *1*, 180–192.

Balasubramanian, K., and Burghard, M. (2006). Biosensors based on carbon nanotubes. Anal. Bioanal. Chem. *385*, 452–468.

Barbosa, J., Costa-de-Oliveira, S., Rodriques, A.G., and Pina-Vaz, C. (2008). Optimization of a flow cytometry protocol for detection and viability assessment of *Giardia lamblia*. Travel Med. Infect. Dis. *6*, 234–239.

Bej, A.K., Mahbubani, M.H., Dicesare, J.L., and Atlas, R.M. (1991). Polymerase chain reaction-gene probe detection of microorganisms by using filter-concentrated samples. Appl. Environ. Microbiol. *57*, 3529–3534.

Bernhard, A.E., and Field, K.G. (2000a). Identification of nonpoint sources of fecal pollution in coastal waters by using host-specific 16S ribosomal DNA genetic markers from fecal anaerobes. Appl. Environ. Microbiol. *66*, 1587–1594.

Bernhard, A.E., and Field, K.G. (2000b). A PCR assay to discriminate human and ruminant feces on the basis of host differences in Bacteroides-Prevotella genes encoding 16S rRNA. Appl. Environ. Microbiol. *66*, 4571–4574.

Berney, M., Vital, M., Hülshoff, I., Weilenmann, H.U., Egli, T. , and Hammes, F. (2008). Rapid, cultivation-independent assessment of microbial viability in drinking water. Water Res. *42*, 4010–4018.

Beuret, C. (2004). Simultaneous detection of enteric viruses by multiplex real-time RT-PCR. J. Virol. Methods *115*, 1–8.

Blears, M.J., De Grandis, S.A., Lee, H., and Trevors, J.T. (1998). Amplified fragment length polymorphism (AFLP): Review of the procedure and its applications. J. Ind. Microbiol. Biotechnol. *21*, 99–114.

Bottari, B., Ercolini, D., Gatti, M., and Neviani, E. (2006). Application of FISH technology for microbiological analysis: Current state and prospects. Appl. Microbiol. Biotechnol. *73*, 485–494.

Bottger, E.C., Jurs, M., Barrett, T., Wachsmuth, K., Metzger, S., and Bitter-Suermann, D. (1987). Qualitative and quantitative determination of enterobacterial common antigen (ECA) with monoclonal antibodies. Clin. Microbiol. *25*, 377–382.

Bouvier, T., Del Giorgio P.A., and Gasol, J.M. (2007). A comparative study of the cytometric characteristics of high and low nucleic-acid bacterioplankton cells from different aquatic ecosystems. Environ. Microbiol. *9*, 2050–2066.

Breen, A., Rope, A.F., Taylor, D., Loper, J.C., and Sferra, P.R. (1995). Application of DNA amplification fingerprinting (DAF) to mixed culture bioreactors. J. Ind. Microbiol. *14*, 10–16.

Bromberg, L., Raduyk, S., and Hatton, T.A. (2009). Functional magnetic nanoparticles for biodefense and biological threat monitoring and surveillance. Anal. Chem. *81*, 5637–5645.

Bruchez, M., Moronne, M., Gin, P., Weiss, S., and Alivisatos, A.P. (1998). Semiconductor nanocrystals as fluorescent biological labels. Science *281*, 2013–2016.

Burns, M.A., Johnson, B.N., Brahmasndra, S.N., Handique, K., Webster, J.R., Krishnan, M., Sammarco, T.S., Man, P.M., Jones, D., Heldsinger, D., Mastrangelo, C.H., and Burke, D.T. (1998). An integrated nanoliter DNA analysis device. Science *282*, 484–487.

Buswell, C.M., Herlihy, Y.M., Lawrence, L.M., McGuiggan, J.T.M., Marsh, P.D., Keevil, C.W., and Leach, S.A. (1998). Extended survival and persistence of Campylobacter spp. in water and aquatic biofilms and their detection by immunofluorescent-antibody and rRNA staining. Appl. Environ. Microbiol. *64*, 733–741.

Cady, N.C., Stelick, S., Kunnavakkam, M.V., and Batt, C.A. (2005). Real-time PCR detection of *Listeria monocytogenes* using an integrated microfluidics platform. Sens. Actuators A: Chem. *107*, 332–341.

Call, D.R. (2005). Challenges and opportunities for pathogen detection using DNA microarrays. Crit. Rev. Microbiol. *31*, 91–99.
Call, D.R., Borucki, M.K., and Loge, F.J. (2003). Detection of bacterial pathogens in environmental samples using DNA microarrays. J. Microbiol. Methods *53*, 235–243.
Cao, Y.C., Jin, R., and Mirkin, C.A. (2002). Nanoparticles with Raman spectroscopic fingerprints for DNA and RNA detection. Science *297*, 1536–1540.
Carrascosa, L.G., Moreno, M., Álvarez, M., and Lechuga, L.M. (2006). Nanomechanical biosensors: A new sensing tool. Trends Anal. Chem. *25*, 196–206.
Carson, C.A., Shear, B.L., Ellersieck, M.R., and Asfaw, A. (2001). Identification of fecal *Escherichia coli* from humans and animals by ribotyping. Appl. Environ. Microbiol. *67*, 1503–1507.
Carson, C.A., Shear, B.L., Ellersieck, M.R., and Schnell, J.D. (2003). Comparison of ribotyping and repetitive extragenic palindromic-PCR for identification of fecal *Escherichia coli* from humans and animals. Appl. Environ. Microbiol. *69*, 1836–1839.
Chan, W.C., and Nie, S. (1998). Quantum dot bioconjugates for ultrasensitive nonisotopic detection. Science *281*, 2016–2018.
Chen, L., Manz, A., and Day, P.J. (2007). Total nucleic acid analysis integrated on microfluidic devices. Lab Chip *7*, 1413–1423.
Cheng, M.M.-C., Cuda, G., Bunimovich, Y.L., Gaspari, M., Heath, J.R., Hill, H.D., Mirkin, C.A., Nijdam, A.J., Terracciano, R., Thundat, T., and Ferrari, M. (2006). Nanotechnologies for biomolecular detection and medical diagnostics. Curr. Opin. Chem. Biol. *10*, 11–19.
Cheng, Y., Liu, Y., Huang, J., Li, K., Zhang, W., Xian, Y., and Jin, L. (2009). Combining biofunctional magnetic nanoparticles and ATP bioluminescence for rapid detection of *Escherichia coli*. Talanta *77*, 1332–1336.
Clement, B.G., Kehl, L.E., DeBord, K.L., and Kitts, C.L. (1998). Terminal restriction fragment patterns (TRFPs), a rapid, PCR-based method for the comparison of complex bacterial communities. J. Microbiol. Methods *31*, 135–142.
Colwell, R.R., Hasan, J.A.K., Huq, A., Loomis, L., Saeibeling, R.J., Torres, M., Galvez, S., Islam, S., and Bernstein, D. (1992). Development and evaluation of a rapid, simple, sensitive, monoclonal antibody-based co-agglutination test for direct detection of *Vibrio cholerae* O1. FEMS Microbiol. Lett. *97*, 215–220.
Covert, T.C., Rodgers, M.R., Reyes, A.L., and Stelma, G.N. (1999). Occurrence of nontuberculous mycobacteria in environmental samples. Appl. Environ. Microbiol. *65*, 2492–2496.
Dabbousi, B.O., Rodriguez-Viejo, J., Mikulec, F.V., Heine, J.R., Mattoussi, H., Ober, R., Jensen, K.F., and Bawendi, M.G. (1997). (CdSe)ZnS core-shell quantum dots: Synthesis and characterization of a size series of highly luminescent nanocrystallites. J. Phys. Chem. B *101*, 9463–9475.
Davis, J.J., Coleman, K.S., Azamian, B.R., Bagshaw, C.B., and Green, M.L.H. (2003). Chemical and biochemical sensing with modified single walled carbon nanotubes. Chem. Eur. J. *9*, 3732–3739.
De Boer, E., and Beumer, R.R. (1999). Methodology for detection and typing of foodborne microorganisms. Int. J. Food Microbiol. *50*, 119–130.
De Leon, R., Matsui, S.M., Baric, R.S., Herrmann, J.E., Blacklow, N.R., Greenberg, H.B., and Sobsey, M.D. (1992). Detection of Norwalk virus in stool specimens by reverse transciptase-polymerase chain reaction and nonradioactive oligoprobes. J. Clin. Microbiol. *30*, 3151–3157.
De Silva, D.G.H., Candy, D.C.A., Mendis, L.N., Charts, H., and Rowe, B. (1992). Serological diagnosis of infection by *Shigella dysenteriae*-1 in patients with bacillary dysentery. J. Infect. Dis. *25*, 273–278.
Dhayal, B., Henne, W.A., Doorneweerd, D.D., Reifenberger, R.G., and Low, P.S. (2006). Detection of *Bacillus subtilis* spores using peptide-functionalized cantilever arrays. J. Am. Chem. Soc. *128*, 3716–3721.
Dick, L.K., Bernhard, A.E., Brodeur, T.J., Santo Domingo, J.W., Simpson, J.M., Walters, S.P., and Field, K.G. (2005). Host distributions of uncultivated fecal Bacteroidales bacteria reveal genetic markers for fecal source identification. Appl. Environ. Microbiol. *71*, 3184–3191.
Ding, S.Y., Rumbles, G., Jones, M., Tucker, M.P., Nedeljkovic, J., Simon, M.N., Wall, J.S., and Himmel, M.E. (2004). Bioconjugation of (CdSe)ZnS quantum dots using a genetically engineered multiple polyhistidine-tagged cohesin/dockerin protein polymer. Macromol. Mat. Eng. *289*, 622–628.
Dittrich, P.S., Tachikawa, K., and Manz, A. (2006). Micro total analysis systems. Latest advancements and trends. Anal. Chem. *78*, 3887–3908.
Dodge, A., Fluri, K., Verpoorte, E., and de Rooij, N.F. (2001). Electrokinetically driven microfluidic chips with surface-modified chambers for heterogeneous immunoassays. Anal. Chem. *73*, 3400–3409.
Dombek, P.E., Johnson, L.K., Zoimmerley, S.T., and Sadowsky, M.J. (2000). Use of repetitive DNA sequences and the PCR to differentiate *Escherichia coli* isolates from human and animal sources. Appl. Environ. Microbiol. *66*, 2572–2577.
Dorsch, M., Ashbolt, N.J., Cox, P.T., and Goodman, A.E. (1994). Rapid identification of *Aeromonas* species using 16S rRNA-targeted oligonucleotide primers: A molecular approach based on screening of environmental isolates. J. Appl. Bacteriol. *77*, 722–726.
Dover, J.E., Hwang, G.M., Mullen, E.H., Prorok, B.C., and Suh, S.J. (2009). Recent advances in peptide probe-based biosensors for detection of infectious agents. J. Microbiol. Methods *78*, 10–19.

Elkin, T., Jiang, X.P., Taylor, S., Lin, Y., Gu, L.R., Yang, H., Brown, J., Collins, S., and Sun, Y.P. (2005). Immuno-carbon nanotubes and recognition of pathogens. Chem. Bio. Chem. *6*, 640–643.

Epstein, C.B., and Butow, R.A. (2000). Microarray technology – enhanced versatility, persistent challenge. Curr. Opin. Biotechnol. *11*, 36–41.

Faude, U.C., and Hofle, M.G. (1997). Development and application of monoclonal antibodies for in situ detection of indigenous bacterial strains in aquatic ecosystems. Appl. Environ. Microbiol. *63*, 4534–4542.

Ferrari, B.C., and Bergquist, P.L. (2007). Quantum dots as alternatives to organic fluorophores for *Cryptosporidium* detection using conventional flow cytometry and specific monoclonal antibodies: lessons learned. Cytometry A *71*, 265–271.

Ford, T.E. (1999). Microbiological safety of drinking water: United States and global perspectives. Environ. Health Perspect. *107*, 191–206.

Ford, T.E., and MacKenzie, W.R. (2000). How safe is our drinking water? Postgrad. Med. *108*, 11–14.

Fortina, P., Kricka, L.J., Surrey, S., and Grodzinski, P. (2005). Nanobiotechnology: The promise and reality of new approaches to molecular recognition. Trends Biotechnol. *23*, 168–173.

Franke-Whittle, I.H., Klammer, S.H., and Insam, H. (2005). Design and application of an oligonucleotide microarray for the investigation of compost microbial communities. J. Microbiol. Methods *62*, 37–56.

Fritz, J., Baller, M.K., Lang, H.P., Royhuizen, H., Vettiger, P., Meyer, E., Guntherodt, H.J., Gerber, C., and Gimzewski, J.K. (2000). Translating biomolecular recognition into nanomechanics. Science *288*, 316–318.

Fuchs, B.M., Wallner, G., Beisker, W., Schwippl, I., Ludwig, W., and Amann, R. (1998). Flow cytometric analysis of the in situ accessibility of *Escherichia coli* 16S rRNA for fluorescently-labelled oligonucleotide probes. Appl. Environ. Microbiol. *64*, 4973–4982.

Galluzzi, L., Magnani, M., Saunders, N., Harms, C., and Bruce, I.J. (2007). Current molecular techniques for the detection of microbial pathogens. Sci. Prog. *90*, 29–50.

Gaspar, J.F., Baptista, P.V., and Rueff, J. (2007). Gold nanoparticle-based systems in Genetics. Curr. Pharmacogen. *5*, 39–47.

Gerion, D., Chen, F., Kannan, B., Fu, A., Parak, W.J., Chen, D.J., Majumdar, A., and Alivisatos, A.P. (2003). Room-temperature single-nucleotide polymorphism and multiallele DNA detection using fluorescent nanocrystals and microarrays. Anal. Chem. *75*, 4766–4772.

Giordano, B.C., Ferrance, J., Swedberg, S., Huhmer, A.F.R., and Landers, J.P. (2001). Polymerase chain reaction in polymeric microchips: DNA amplification in less than 240 seconds. Anal. Biochem. *291*, 124–132.

Giovannoni, S.J., De Long, E.F., Olsen, G.J., and Pace, N.R. (1988). Phylogenetic group-specific oligodeoxynucleotide probes for identification of single microbial cells. J. Bacteriol. *170*, 720–726.

Goldman, E.R., Mattoussi, H.M., Anderson, G.P., Medintz, I.L., and Mauro, J.M. (2005). Fluoroimmunoassays using antibody-conjugated quantum dots. Meth. Mol. Biol. *303*, 19–34.

Goldman, E.R., Anderson, G.P., Liu, J.L., Delehanty, J.B., Sherwood, L.J., Osborn, L.E., Cummins, L.B., and Hayhurst, A. (2006a). Facile generation of heat-stable antiviral and antitoxin single domain antibodies from a semisynthetic llama library. Anal. Chem. *78*, 8245–8255.

Goldman, E.R., Medintz, I.L., and Mattoussi, H. (2006b). Luminescent quantum dots in immunoassays. Anal. Bioanal. Chem. *384*, 560–563.

Gu, H., Xu, K., Xu, C., and Xu, B. (2006). Biofunctional magnetic nanoparticles for protein separation and pathogen detection. Chem. Commun. *9*, 941–949.

Gu, H.W., Ho, P.L., Tsang, K.W.T., Wang, L., and Xu, B. (2003). Using biofunctional magnetic nanoparticles to capture vancomycin-resistant enterococci and other gram-positive bacteria at ultralow concentration. J. Am. Chem. Soc. *125*, 15702–15703.

Gu, L., Elkin, T., Jiang, X., Li, H., Lin, Y., Qu, L., Tzeng, T.-R.J., Josepha, R., and Sun, Y.-P. (2005). Single-walled carbon nanotubes displaying multivalent ligands for capturing pathogens. Chem. Commun. *7*, 874–876.

Guan, S., Xu, R., Chen, S., Odumeru, J., and Gyles, C. (2002). Development of a procedure for discriminating among *Escherichia coli* isolates from animal and human sources. Appl. Environ. Microbiol. *68*, 2690–2698.

Gulliksen, A., Solli, L., Karlsen, F., Rogne, H., Hovig, E., Nordstrøm, T., and Sirevåg, R. (2004). Real-time nucleic acid sequence-based amplification in nanoliter volumes. Anal. Chem. *76*, 9–14.

Guo, S., and Wang, E. (2007). Synthesis and electrochemical applications of gold nanoparticles. Anal. Chim. Acta *598*, 181–192.

Gupta, A., Akin, D., and Bashir, R. (2004a). Detection of bacterial cells and antibodies using surface micromachined thin silicon cantilever resonators. J. Vac. Sci. Technol. B *22*, 2785–2791.

Gupta, A., Akin, D., and Bashir, R. (2004b). Single virus particle detection using microresonators with nanoscale thickness. Appl. Phys. Lett. *84*, 1976–1978.

Gupta, A.K., and Gupta, M. (2005). Synthesis and surface engineering of iron oxide nanoparticles for biomedical applications. Biomaterials *26*, 3995–4021.

Gutell, R.R., Larsen, N., and Woese, C.R. (1994). Lessons from an evolving rRNA: 16S and 23S rRNA structures from a comparative perspective. Microbiol. Rev. *58*, 10–26.

Hadrys, H., Balick, M., and Schierwater, B. (1992). Applications of random amplified polymorphic DNA (RAPD) in molecular ecology. Mol. Ecol. *1*, 55–63.

Hahm J.-I., and Lieber, C.M. (2004). Direct ultrasensitive electrical detection of DNA and DNA sequence variations using nanowire nanosensors. Nano Lett. *4*, 51–54.

Hahn, M.A., Keng, P.C., and Krauss, T.D. (2008). Flow cytometric analysis to detect pathogens in bacterial cell mixtures using semiconductor quantum dots. Anal. Chem. *80*, 864–872.

Hahn, M.A., Tabb, J.S., and Krauss, T.D. (2005). Detection of single bacterial pathogens with semiconductor quantum dots. Anal. Chem. *77*, 4861–4869.

Hammes, F., Berney, M., Wang, Y., Vital, M., Köster, O., and e.g.li, T. (2008). Flow-cytometric total bacterial cell counts as a descriptive microbiological parameter for drinking water treatment processes. Water Res. *42*, 269–277.

Hammes, F., Vital, M., Stucki, D., Lautenschlager, K., and e.g.li, T. (2009). Advances in microbiological methods for drinking water analysis: flow cytometry, assimilable organic carbon and pathogen growth potential. In TECHNEAU: Safe drinking water from source to tap, T. van den Hoven., and Kazner, C., eds. (London, England, IWA Publishing), pp.158–171.

Hanai, K., Satake, M., Nakanishi, H., and Venkatsewaran, K. (1997). Comparison of commercially available kits with standard methods for detection of *Salmonella* strains in foods. Appl. Environ. Microbiol. *63*, 775–778.

Hansen, K.M., and Thundat, T. (2005). Microcantilever biosensors. Methods *37*, 57–64.

Hansen, K.M., Ji, H.F., Wu, G., Datar, R., Cote, R., Majumdar, A., and Thundat, T. (2001). Cantilever-based optical deflection assay for discrimination of DNA single-nucleotide mismatches. Anal. Chem. *73*, 1567–1571.

Hartel, P.G., Summer, J.D., Hill, J.L., Collins, J.V., Entry, J.A., and Segars, W.I. (2002). Geographic variability of *Escherichia coli* ribotypes from animals in Idaho and Georgia. J. Environ. Qual. *31*, 1273–1278.

Harwood, V.J., Levine, A.D., Scott, T.M., Chivukula, V., Lukasik, J., Farrah, S.R., and Rose, J.B. (2005). Validity of the indicator organism paradigm for pathogen reduction in reclaimed water and public health protection. Appl. Environ. Microbiol. *71*, 3163–3170.

Hasan, J.A.K., Huq, A., Tamplin, R.J., Siebeling, R.J., and Colwell, R.R. (1994). A novel kit for rapid detection of *Vibrio cholerae* O1. J. Clin. Microbiol. *32*, 249–252.

He, J.-W., and Jiang, S. (2005). Quantification of enterococci and human adenoviruses in environmental samples by real-time PCR. Appl. Environ. Microbiol. *71*, 2250–2255.

He, Y., Zhang, Y.H., and Yeung, E.S. (2001). Capillary based fully integrated and automated system for nanoliter polymerase chain reaction analysis directly from cheek cells. J. Chromatogr. A *924*, 271–284.

Heid, C.A., Stevens, J., Livak, K.J., and Williams, P.M. (1996). Real-time quantitative PCR. Genome Res. *6*, 986–994.

Henares, T.G., Mizutani, F., and Hisamoto, H. (2008). Current development in microfluidic immunosensing chip. Anal. Chim. Acta *611*, 17–30.

Hicks, R., Amann, R.I., and Stahl, D.A. (1992). Dual staining of natural bacterioplankton with 4′,6-diamidino-2-phenylindole and fluorescent oligonucleotide probes targeting kingdom level 16S rRNA sequences. Appl. Environ. Microbiol. *58*, 2158–2163.

Hilliard, L.R., Zhao, X., and Tan, W. (2002). Immobilization of oligonucleotides onto silica nanoparticles for DNA hybridization studies. Anal. Chim. Acta *470*, 51–56.

Hines, M.A., and Guyot-Sionnest, P. (1996). Synthesis and characterization of strongly luminescing ZnS capped CdSe nanocrystals. J. Phys. Chem. *100*, 468–471.

Ho, K.-C., Tsai, P.-J., Lin, Y.-S., and Chen, Y.-C. (2004). Using biofunctionalized nanoparticles to probe pathogenic bacteria. Anal. Chem. *76*, 7162–7168.

Hoff, K.A. (1993). Total and specific bacterial counts by simultaneous staining with DAPI and fluorochrome-labelled antibodies. In Handbook of Methods in Aquatic Lewis Publishers Microbial Ecology, Kemp, P.F., Sherr, B.F., Sherr, E.B., and Cole, J.J., eds. (Lewis Publishers, Boca Raton, USA), pp. 149–154.

Holben, W.E., and Tiedje, J.M. (1988). Applications of nucleic acid hybridization in microbial ecology. Ecol. *69*, 561–568.

Huang, P.W., Laborde, D., Land, V.R., Matson, D.O., Smith, A.W., and Jiang, X. (2000). Concentration and detection of caliciviruses in water samples by reverse transcription-PCR. Appl. Environ. Microbiol. *66*, 4383–4388.

Ibáñez-Peral, R., Bergquist, P.L., Walter, M.R., Gibbs, M., Goldys, E.M., and Ferrari, B. (2008). Potential use of quantum dots in flow cytometry. Int. J. Mol. Sci. *9*, 2622–2638.

Iijima, S., and Ichihashi, T. (1993). Single-shell carbon nanotubes of 1-nm diameter. Nature *363*, 603–605.

Ilic, B., Czaplewski, D., Craighead, H.G., Neuzil, P., Campagnolo, C., and Batt, C. (2000). Mechanical resonant immunospecific biological detector. Appl. Phys. Lett. *77*, 450–452.

Ito, T., Kuwahara, S., and Yokota, T. (1983). Automatic and manual latex agglutination tests for measurement of cholera toxin and heat-labile enterotoxin of *Escherichia coli*. J. Clin. Microbiol. *17*, 7–12.

Jain, K.K. (2007). Applications of nanobiotechnology in clinical diagnostics. Clin. Chem. *53*, 2002–2009.

Jiang, X., Wang, J., and Estes, M.K. (1995). Characterization of SRSVs using RT-PCR and a new antigen ELISA. Arch. Virol. *140*, 363–374.

Joachimsthal, E.L., Ivanov, V., Tay, S.T.L., and Tay, J.H. (2004). Bacteriological examination of ballast water in Singapore Harbour by flow cytometry with FISH. Mar. Pollut. Bull. *49*, 334–343.

Jothikumar, N., Cliver, D.O., and Mariam, T.W. (1998). Immunomagnetic capture PCR for rapid concentration and detection of hepatitis A virus from environmental samples. Appl. Environ. Microbiol. *64*, 504–508.

Kalmbach, S., Manz, W., and Szewzyk, U. (1997). Isolation of new bacterial species from drinking water biofilms and proof of their in situ dominance with highly specific 16S rRNA probes. Appl. Environ. Microbiol. *63*, 4164–4170.

Kalyuzhnaya, M.G., Zabinsky, R., Bowerman, S., Baker, D.R., Lidstrom, M.E., and Chistoserdova, L. (2006). Fluorescence in situ hybridization-flow cytometry-cell sorting-based method for separation and enrichment of type I and type II methanotroph populations. Appl. Environ. Microbiol. *72*, 4293–4301.

Kamma, S., Tang, L., Leung, K., Ashton, E., Newman, N., and Suresh, M.R. (2008). A rapid two dot filter assay for the detection of *E. coli* 0157 in water samples. J. Immunol. Methods *336*, 159–165.

Keer, J.T., and Birch, L. (2003). Molecular methods for the assessment of bacterial viability. J. Microbiol. Methods *53*, 175–183.

Kfir, R., and Genthe, B. (1993). Advantages and disadvantages of the use of immunodetection techniques for the enumeration of microorganisms and toxins in water. Water Sci. Technol. *27*, 243–248.

Khandurina, J., and Guttman, A. (2002). Bioanalysis in microfluidic devices. J. Chromatogr. A *943*, 159–183.

Kleinsteuber, S., Riis, V., Fetzer, I., Harms, H., and Müller, S. (2006). Population dynamics within a microbial consortium during growth on diesel fuel in saline environments. Appl. Environ. Microbiol. *72*, 3531–3542.

Köhler, J.M., and Henkel, T. (2005). Chip devices for miniaturized biotechnology. Appl. Microbiol. Biotechnol. *69*, 113–125.

Lagally, E.T., Scherer, J.R., Blazej, R.G., Toriello, N.M., Diep, B.A., Ramchandani, M., Sensabaugh, G.F., Riley, L.W., and Mathies, R.A. (2004). Integrated portable genetic analysis microsystem for pathogen/infectious disease detection. Anal. Chem. *76*, 3162–3170.

Landers, J.P. (2003). Molecular diagnostics on electrophoretic microchips. Anal. Chem. *75*, 2919–2927.

Lenaerts, J., Lappin-Scott, H.M., and Porter, J. (2007). Improved fluorescent in situ hybridization method for detection of bacteria from activated sludge and river water by using DNA molecular beacons and flow cytometry. Appl. Environ. Microbiol. *73*, 2020–2023.

Lebaron, P., Catala, P., Fajon, C., Joux, F., Baudart, J., and Bernard, L. (1997). A new sensitive, whole–cell hybridization technique for detection of bacteria involving a biotinylated oligunucleotide probe targeting rRNA and tyramide signal amplification. Appl. Environ. Microbiol. *63*, 3274–3278.

Leclerc, H., Schwartzbrod, L., and Dei-Cas, E. (2002). Microbial agents associated with waterborne diseases. Crit. Rev. Microbiol. *28*, 371–409.

Lee, D.Y., Lauder, H., Cruwys, H., Falletta, P., and Beaudette, L.A. (2008). Development and application of an oligonucleotide microarray and real-time quantitative PCR for detection of wastewater bacterial pathogens. Sci. Total Environ. *398*, 203–211.

Lee, L.Y., Ong, S.L., Hu, J.Y., Ng, W.J., Feng, Y.Y., Tan, X.L., and Wong, S.W. (2004). Use of semiconductor quantum dots for photostable immunofluorescence labelling of *Cryptosporidium parvum*. Appl. Environ. Microbiol. *70*, 5732–5736.

Lee, S., Malone, C., and Kemp, P.F. (1993). Use of multiple 16S RNA-targeted fluorescent probes to increase signal strength and measure cellular RNA from natural planktonic bacteria. Marine Ecol. Prog. Ser. *101*, 193–201.

Lemarchand, K., Masson, L., and Brousseau, R. (2004). Molecular biology and DNA microarray technology for microbial quality monitoring of water. Crit. Rev. Microbiol. *30*, 145–172.

Leung, K.T., Mackereth, R., Tien, Y.-C., and Topp, E. (2004). A comparison of AFLP and ERIC-PCR analyses for discriminating *Escherichia coli* from cattle, pig and human sources. FEMS Microbiol. Ecol. *47*, 111–119.

Li, J., Ng, H.T., Cassell, A., Fan, W., Chen, H., Ye, Q., Koehne, J., Han, J., and Meyyappan, M. (2003). Carbon nanotube nanoelectrode array for ultrasensitive DNA detection. Nano. Lett. *3*, 597–602.

Lieber, C.M. (2003). Nanoscale science and technology: Building a big future from small things. MRS Bull. *28*, 486–491.

Lim, D.V., Simpson, J.M., Kearns, E.A., and Kramer, M.F. (2005). Current and developing technologies for monitoring agents of bioterrorism and biowarfare. Clin. Microbiol. Rev. *18*, 583–607.

Lin, F.Y.H., Sabri, M., Alirezaie, J., Li, D., and Sherman, P.M. (2005). Development of a nanoparticle-labeled microfluidic immunoassay for detection of pathogenic microorganisms. Clin. Diag. Lab. Immunol. *12*, 418–425.

Lin, Y., Elkin, T., Taylor, S., Gu, L., Chen, B., Veca, L.M., Zhou, B., Yang, H., Brown, J., Joseph, R., Jones, E., Jiang, X., and Sun, Y.-P. (2006). Preparation, characterization, and evaluation of immuno carbon nanotubes. Microchim. Acta *152*, 249–254.

Liu, L.H., Dietsch, H., Schurtenberger, P., and Yan, M. (2009). Photoiniated coupling of unmodified monosaccharides to iron oxide nanoparticles sensing proteins and bacteria. Bioconjug. Chem. *20*, 1349–1350.

Liu, W.T., Zhu, L., Qin, Q.W., Zhang, Q., Feng, H., and Ang, S. (2005). Microfluidic device as a new platform for immunofluorescent detection of viruses. Lab Chip *5*, 1327–1330.

Liu, Y., and Li, Y. (2001). An antibody-immobilized capillary column as a bioseparator/bioreactor for detection of *Escherichia coli* O157:H7 with absorbance measurement. Anal. Chem. *73*, 5180–5183.

Livak, K., Flood, S., Marmaro, J., Giusti, W., and Deetz, K. (1995). Oligonucleotides with fluorescent dyes at opposite ends provide a quenched probe system useful for detecting PCR product and nucleic acid hybridization. PCR Methods Appl. *4*, 357–362.

Lowery, C.J., Moore, J.E., Millar, B.C., Burke, D.P., McCorry, K.A., Crothers, E., and Dooley, S. (2000). Detection and speciation of *Cryptosporidium* spp. in environmental water samples by immunomagnetic separation, PCR and endonuclease restriction. J. Med. Microbiol. *49*, 779–785.

Mackay, I.M. (2004). Real-time PCR in the microbiology laboratory. Clin. Microbiol. Infect. *10*, 190–212.

Maidak, B.L., Larsen, N., McCaughey, M.J., Overbeek, R., Olsen, G.J., Fogel, K., Blandy, J., and Woese, C.R. (1994). The Ribosomal Database Project. Nucl. Acids Res. *22*, 3485–3487.

Maidak, B.L., Olsen, G.J., Larsen, N., Overbeek, R., McCaughey, M.J., and Woese, C.R. (1997). The RDP (Ribosomal Database Project). Nucl. Acids Res. *25*, 109–111.

Manz, W., Amann, R., Ludwig, W., Wagner, M., and Schleifer, K.-H. (1992). Phylogenetic oligodeoxynucleotide probes for the major subclasses of proteobacteria: Problems and solutions. System. Appl. Microbiol. *15*, 593–600.

Masters, C.I., Shallcross, J.A., and Mackey, B.M. (1994). Effect of stress treatments on the detection of *Listeria monocytogenes* and enterotoxigenic *Escherichia coli* by the polymerase chain reaction. J. Appl. Bacteriol. *77*, 73–79.

McKillip, J.L., Jaykus, L.-A., and Drake, M. (1998). rRNA stability in heat-killed and UV-irradiated enterotoxigenic *Staphylococcus aureus* and *Escherichia coli* O157:H7. Appl. Environ. Microbiol. *64*, 4264–4268.

McLellan, S.L., Daniels, A.D., and Salmore, A.K. (2003). Genetic characterization of *Escherichia coli* populations from host sources of fecal pollution by using DNA fingerprinting. Appl. Environ. Microbiol. *69*, 2587–2594.

Meays, C.L., Broersma, K., Nordin, R., and Mazumder, A. (2004). Source tracking fecal bacteria in water: A critical review of current methods. J. Environ. Managem. *73*, 71–79.

Meier, H., Koob, C., Ludwig, W., Amann, R.I., Frahm, E., Hoffmann, S., Obst, U., and Schleifer, K.-H. (1997). Detection of enterococci with rRNA targeted DNA probes and their use for hygienic drinking water control. Water Sci. Technol. *35*, 437–444.

Mirkin, C.A., and Taton, T.A. (2000). Semiconductors meet biology. Nature *405*, 626–627.

Miyauchi, R., Oki, K., Aoi, Y., and Tsuneda, S. (2007). Diversity of nitrate reductase genes in 'Candidatus Accumulibacter phosphatis'-dominated cultures enriched by flow-cytometric sorting. Appl. Environ. Microbiol. *73*, 5331–5337.

Moe, C.L. (1996). Waterborne transmission of infectious agents. In Manual of Environmental Microbiology, Hurst, C.J., Knudsen, G.R., McInerney, M.J., Stetzenbach, L.D., and Walter, M.V., eds. (Washington, D.C., USA, ASM Press), pp. 136–152.

Moter, A., and Gobel, U.B. (2000). Fluorescence in situ hybridization (FISH) for direct visualization of microorganisms. J. Microbiol. Methods *41*, 85–112.

Mou, X., Moran, M.A., Stepanauskas, R., González, J.M., and Hodson, R.E. (2005). Flow-cytometric cell sorting and subsequent molecular analyses for culture-independent identification of bacterioplankton involved in dimethylsulfoniopropionate transformations. Appl. Environ. Microbiol. *71*, 1405–1416.

Nicewarner-Pena, S.R., Freeman, R.G., Reiss, B.D., He, L., Pena, D.J., Walton, I.D., Cromer, R., Keating, C.D., and Natan, M.J. (2001). Submicrometer metallic barcodes. Science *294*, 137–141.

Nishimura, M., Kita-Tsukamoto, K., and Kogure, K. (1993). A new method to detect viable bacteria in natural seawater using 16S rRNA oligonucleotide probe. J. Oceanol. *49*, 51–56.

Noah, C.W., Poteet, S.S., Lister, M.M., Roderick, C.N., Smith, D.B., Colvert, R.M., Holloand, M.A., and Cerniglia, C.E. (1996). Production and characterization of monoclonal antibodies to detect *Vibrio cholerae* serogroup O1 in a rapid enzyme-linked immunosorbent assay. J. Food Protec. *59*, 1153–1157.

Nocker, A., and Camper, A.K. (2009). Novel approaches toward preferential detection of viable cells using nucleic acid amplification techniques. FEMS Microbiol. Lett. *291*, 137–142.

Nogva, H.K., Dromtorp, S.M., Nissen, H., and Rudi, K. (2003). Ethidium monoazide for DNA-based differentiation of viable and dead bacteria by 5′-nuclease. PCR BioTechniques *810*, 804–813.

Northrup, M.A., Benett, B., Hadley, D., Landre, P., Lehew, S., Richards, J., and Stratton, P. (1998). A miniature analytical instrument for nucleic acids based on micromachined silicon reaction chambers. Anal. Chem. *70*, 918–922.

Nwachcuku, N., and Gerba, C.P. (2004). Emerging waterborne pathogens: Can we kill them all? Curr. Opin. Biotechnol. *15*, 175–180.

Oberdörster, G., Stone, V., and Donaldson, K. (2007). Toxicology of nanoparticles: A historical perspective. Nanotoxicol. *1*, 2–25.

Oh, K.W., Park, C., Namkoong, K., Kim, J., Ock, K.S., Kim, S., Kim, Y.A., Cho, Y.K., and Ko, C. (2005). World-to-chip microfluidic interface with built-in valves for multichamber chip-based PCR assays. Lab Chip *5*, 845–850.

Olsen, G.J., Lane, D.J., Giovannoni, S.J., Pace, N.R., and Stahl, D.A. (1986). Microbial ecology and evolution: A ribosomal RNA approach. Annu. Rev. Microbiol. *40*, 337–365.

Olsvik, O., Popovic, T., Skjerve, E., Cudjoe, K.S., Hornes, E., Ugelstad, J., and Uhlen, M. (1994). Magnetic separation techniques in diagnostic microbiology. Clin. Microbiol. Rev. *7*, 43–54.

Pace, N.R., Stahl, D.A., Lane, D.J., and Olsen, G.J. (1986). The analysis of natural microbial populations by ribosomal RNA sequences. Adv. Microb. Ecol. *9*, 1–55.

Park, J.I., Kim, M.G., Jun, Y.W., Lee, J.S., Lee, W.R., and Cheon, J. (2004). Characterization of superparamagnetic 'core-shell' nanoparticles and monitoring their anisotropic phase transition to ferromagnetic 'solid solution' nanoalloys. J. Am. Chem. Soc. *126*, 9072–9078.

Parveen, S., Portier, K.M., Robinson, K., Edmiston, L., and Tamplin, M.L. (1999). Discriminant analysis of ribotyping profiles of *Escherichia coli* for differentiating human and nonhuman sources of fecal pollution. Appl. Environ. Microbiol. *65*, 3142–3147.

Pathak, S., Choi, S.K., Arnheim, N., and Thompson, M.E. (2001). Hydroxylated quantum dots as luminescent probes for in situ hybridization. J. Am. Chem. Soc. *123*, 4103–4104.

Patolsky, F., Zheng, G., and Lieber, C.M. (2006a). Nanowire-based biosensors. Anal. Chem. *78*, 4260–4269.

Patolsky, F., Zheng, G., and Lieber, C.M. (2006b). Fabrication of silicon nanowire devices for ultrasensitive, label-free, real-time detection of biological and chemical species. Nature Protocols *1*, 1711–1724.

Patolsky, F., Zheng, G., Hayden, O., Lakadamyali, M., Zhuang, X., and Lieber, C.M. (2004). Electrical detection of single viruses. Proc. Natl. Acad. Sci. USA *101*, 14017–14022.

Penn, S.G., He, L., and Natan, M.J. (2003). Nanoparticles for bio-analysis. Curr. Opin. Chem. Biol. *7*, 609–615.

Peoples, M.C., and Karnes, H.T. (2008). Microfluidic immunoaffinity separations for bioanalysis. J. Chromatogr. B *15*, 14–25.

Podar, M., Abulencia, C.B., Walcher, M., Hutchison, D., Zengler, K., Garcia, J.A., Holland, T., Cotton, D., Hauser, L., and Keller, M. (2007). Targeted access to the genomes of low-abundance organisms in complex microbial communities. Appl. Environ. Microbiol. *73*, 3205–3215.

Quadri, F., Chowdury, A., Hossain, J., Chodury, K., Azim, T., Shimada, T., Islam, N.K.M., Sack, R.B., and Albert, M.J. (1994). Development and evaluation of rapid monoclonal antibody based coagulation test for direct detection of *Vibrio cholerae* O139 synonym Bengal in stool samples. J. Clin. Microbiol. *32*, 1589–1590.

Reyes, D.R., Iossifidis, D., Auroux, P.-A., and Manz, A. (2002). Micro total analysis systems. 1. Introduction, theory, and technology. Anal. Chem. *74*, 2623–2636.

Rhee, S.K., Liu, X., Wu, L., Chong, S.C., Wan, X., and Zhou, J. (2004). Detection of genes involved in biodegradation and biotransformation in microbial communities by using 50-mer oligonucleotide microarrays. Appl. Environ. Microbiol. *70*, 4303–4317.

Richardson, J., Hawkins, P., and Luxton, R. (2001). The use of coated paramagnetic particles as a physical label in a magnetoimmunoassay. Biosens. Bioelectron. *16*, 989–993.

Rochelle, P.A., De Leon, R., Stewart, M.H., and Wolfe, R.L. (1997). Comparison of primers and optimization of PCR conditions for detection of *Cryptosporidium parvum* and *Giardia lamblia* in water. Appl. Environ. Microbiol. *63*, 106–114.

Rose, J.B., and Grimes, D.J. (2001). Reevaluation of microbial water safety: Powerful new tools for microbial risk assessment; Report from the American Academy of Microbiology, American Academy of Microbiology, Washington, D.C., USA.

Rosi, N.L., and Mirkin, C.A. (2005). Nanostructures in biodiagnostics. Chem. Rev. *105*, 1547–1562.

Rowe, C.A. (1999). Array biosensor for simultaneous identification of bacterial, viral, and protein analytes. Anal. Chem. *71*, 3846–3852.

Ruan, C., Zeng, K., Varghese, O.K., and Grimes, C.A. (2003). Magnetoelastic immunosensors: Amplified mass immunosorbent assay for detection of *Escherichia coli* O157:H7. Anal. Chem. *75*, 6494–6498.

Šafařík, I., and Šafaříková, M. (2002). Magnetic nanoparticles and biosciences. Monatsh. Chem. *133*, 737–759.

Saiki, R.K., Gelfand, D.H., Stoffel, S., Scharf, S.J., Higuchi, R., Horn, G.T., Mullis, K.B., and Ehrlich, H.A. (1988). Primer-directed enzymatic amplification of DNA with a thermostable DNA polymerase. Science *239*, 487–491.

Sanders, G.H.W., and Manz, A. (2000). Chip-based microsystems for genomic and proteomic analysis. Trends Anal. Chem. *19*, 364–378.

Sandström, P., Boncheva, M., and Åkerman, B. (2003). Nonspecific and thiol-specific binding of DNA to gold nanoparticles. Langmuir *19*, 7537–7543.

Santo Domingo, J.W., Bambic, D.G., Edge, T.A., and Wuertz, S. (2007). Quo vadis source tracking? Towards a strategic framework for environmental monitoring of fecal pollution. Water Res. *41*, 3539–3552.

Santra, S., Zhang, P., Wang, K., Tapec, R., and Tan, W. (2001). Conjugation of biomolecules with luminophore-doped silica nanoparticles for photostable biomarkers. Anal. Chem. *73*, 4988–4993.

Sasaki, K., Tajiri, Y., Sata, M., Fujii, Y., Matsubara, F., Zhao, M., Shimizu, S., Toyonaga, A., and Tanikawa, K. (1999). *Helicobacter pylori* in the natural environment. Scand. J. Infect. Dis. *31*, 275–279.

Savelkoul, P.M., Aarts, H.M., de Haas, J., Dijkshoorn, L., Duim, B., Otsen, M., Rademaker, J.W., Schouls, L., and Lenstra, J.A. (1999). Amplified-fragment length polymorphism analysis: The state of an art. J. Clin. Microbiol. *37*, 3083–3091.

Savichtcheva, O., and Okabe, S. (2006). Alternative indicators of fecal pollution: Relations with pathogens and conventional indicators, current methodologies for direct pathogen monitoring and future application perspectives. Water Res. *40*, 2463–2476.

Sayes, C.M., Liang, F., Hudson, J.L., Mendez, J., Guo, W., Beach, J.M., Moore, V.C., Doyle, C.D., West, J.L., Billups, W.E., Ausman, K.D., and Colvin, V.L. (2006). Functionalization density dependence of single-walled carbon nanotubes cytotoxicity *in vitro*. Toxicol. Lett. *161*, 135–142.

Schneegurt, M.A., and Kulpa, C.F. (1998). The application of molecular techniques in environmental biotechnology for monitoring microbial systems. Biotechnol. Appl. Biochem. *27*, 73–79.

Schwab, K.J., De Leon, R., and Sobsey, M.D. (1996). Immunoaffinity concentration and purification of waterborne enteric viruses for detection by reverse transcriptase PCR. Appl. Environ. Microbiol. *62*, 2086–2094.
Schwab, K.J., Neill, F.H., Estes, M.K., and Atmar, R.L. (1998). Improvements in the RT-PCR detection of enteric viruses in environmental samples. Water Sci. Technol. *38*, 83–86.
Scott, T.M., Parveen, S., Portier, K.M., Rose, J.B., Tamplin, M.L., Farrah, S.R., Koo, A., and Lukasik, J. (2003). Geographical variation in ribotype profiles of *Escherichia coli* isolates from humans, swine, poultry, beef, and dairy cattle in Florida. Appl. Environ. Microbiol. *69*, 1089–1092.
Scott, T.M., Rose, J.B., Jenkins, T.M., Farrah, S.R., and Lukasik, J. (2002). Microbial source tracking: Current methodology and future directions. Appl. Environ. Microbiol. *68*, 5796–5803.
Sekar, R., Fuchs, B.M., Amann, R., and Pernthaler, J. (2004). Flow sorting of marine bacterioplankton after fluorescence in situ hybridization. Appl. Environ. Microbiol. *70*, 6210–6219.
Sheridan, G.E., Masters, C.I., Shallcross, J.A., and MacKey, B.M. (1998). Detection of mRNA by reverse transcription-PCR as an indicator of viability in *Escherichia coli* cells. Appl. Environ. Microbiol. *64*, 1313–1318.
Sinha, N., Ma, J., and Yeow, J.T.W. (2006). Carbon nanotube-based sensors. J. Nanosci. Nanotechnol. *6*, 573–590.
Sobsey, M.D. (1994). Molecular methods to detect viruses in environmental samples. In Rapid Methods and Automation in Microbiology and Immunology, Spencer, R.C., Wright, E.P., and Newsme, S.W.B., eds. (Andover, England, Intercept Ltd), pp. 387–400.
Stahl, D.A., and Amann, R. (1991). Development and application of nucleic acid probes in bacterial systematics. In Sequencing and Hybridization Techniques in Bacterial Systematics, Stackebrandt, E., and Goodfellow, M., eds. (John Wiley and Sons, Chichester, England), pp. 205–248.
Steffan, R.J., and Atlas, R.M. (1991). Polymerase chain reaction: Application in environmental microbiology. Annu. Rev. Microbiol. *45*, 137–161.
Stine, O.C., Carnahan, A., Singh, R., Powell, J., Furuno, J.P., Dorsey, A., Silbergeld, E., Williams, H.N., and Morris, J.G. (2003). Characterization of microbial communities from coastal waters using microarrays. Environ. Monit. Assess. *81*, 327–336.
Storhoff, J.J., Lucas, A.D., Garimella, V., Bao, Y.P., and Müller, U.R. (2004). Homogeneous detection of unamplified genomic DNA sequences based on colorimetric scatter of gold nanoparticle probes. Nature Biotechnol. *22*, 883–887.
Straub, T.M., and Chandler, D.P. (2003). Towards a unified system for detecting waterborne pathogens – a review. J. Microbiol. Methods *53*, 185–197.
Straub, T.M., Pepper, I.L., and Gerba, C.P. (1995). Removal of PCR inhibiting substances in sewage sludge amended soil. Water Sci. Technol. *31*, 311–315.
Su, X., and Li, Y. (2004). Quantum dot biolabeling coupled with immunomagnetic separation for detection of *Escherichia coli* O157:H7. Anal. Chem. *76*, 4806–4810.
Sun, S.H., Zeng, H., Robinson, D.B., Raoux, S., Rice, P.M., Wang, S.X., and Li, G.X. (2004). Monodisperse MFe_2O_4 (M = Fe, Co, Mn) nanoparticles. J. Am. Chem. Soc. *126*, 273–279.
Tani, K., Kurokawa, K., and Nasu, M. (1998). Development of a direct in situ PCR method for detection of specific bacteria in natural environments. Appl. Environ. Microbiol. *64*, 1536–1540.
Tapec, R., Zhao, X.J., and Tan, W. (2002). Development of organic dye-doped silica nanoparticles for bioanalysis and biosensors. J. Nanosci. Nanotechnol. *2*, 405–409.
Taton, T.A., Mirkin, C.A., and Letsinger, R.L. (2000). Scanometric DNA array detection with nanoparticle probes. Science *289*, 1757–1760.
Thaxton, C.S., Georganopoulou, D.G., and Mirkin, C.A. (2006). Gold nanoparticle probes for the detection of nucleic acid targets. Clin. Chim. Acta *363*, 120–126.
Theron, J., and Cloete, T.E. (2002). Emerging waterborne infections: Contributing factors, agents, and detection tools. Crit. Rev. Microbiol. *28*, 1–26.
Theron, J., Cilliers, J., Du Preez, M., Brözel, V.S., and Venter, S.N. (2000). Detection of toxigenic *Vibrio cholerae* from environmental water samples by an enrichment broth cultivation-pit-stop semi-nested PCR procedure. J. Appl. Microbiol. *89*, 539–546.
Theron, J., Morar, D., Du Preez, M., Brözel, V.S., and Venter, S.N. (2001). A sensitive seminested PCR method for the detection of *Shigella* in spiked environmental water samples. Water Res. *35*, 869–874.
Theron, J., Walker, J.A., and Cloete, T.E. (2008). Nanotechnology and water treatment: Applications and emerging opportunities. Crit. Rev. Microbiol. *34*, 43–69.
Thundat, T., and Majumdar, A. (2003). Microcantilevers for physical, chemical, and biological sensing. In Sensors and Sensing in Biology and Engineering, Barth, F.G., Humphrey, J.A.C., and Secomb, T.W., eds. (Springer-Verlag, New York, USA), pp. 338–355.
Tok, J.B., Chuang, F.Y., Kao, M.C., Rose, K.A., Pannu, S.S., Sha, M.Y., Chakarova, G., Penn, S.G., and Dougherty, G.M. (2006). Metallic striped nanowires as multiplexed immunoassay platforms for pathogen detection. Angew. Chem. *45*, 6900–6904.
Toranzos, G.A., Alvarez, A.J., and Dvorsky, E.A. (1993). Application of the polymerase chain reaction technique to the detection of pathogens in water. Water Sci. Technol. *27*, 207–210.

Toranzos, G.A., and McFeters, G.A. (1996). Detection of indicator microorganisms in environmental freshwaters and drinking waters. In Manual of Environmental Microbiology, Hurst, C.J., Knudsen, G.R., McInerney, M.J., Stetzenbach, L.D., and Walter, M.V., eds. (ASM Press, Washington, D.C., USA), pp. 184–194.

Van de Peer, Y., Chapelle, S., and De Wachter, R. (1996). A quantitative map of nucleotide substitution rates in bacterial rRNA. Nucl. Acids Res. *24*, 3381–3391.

Van Hal, N.L., Vorst, O., Van Houwelingen, A.M., Kok, E.J., Peijnenburg, A., Aharoni, A., Van Tunen, A.J., and Keijer, J. (2000). The application of DNA microarrays in gene expression analysis. J. Biotechnol. *78*, 271–280.

Versalovic, J., Schneider, F., de Bruijn, F.J., and Lupski, J.R. (1994). Genomic fingerprinting of bacteria using repetitive sequence-based polymerase chain reaction. Meth. Mol. Cell. Biol. *5*, 25–40.

Vital, M., Füchslin, H.P., Hammes, F., and T. e.g.li. (2007). Growth of *Vibrio cholerae* O1 Ogawa Eltor in freshwater. Microbiology *153*, 1993–2001.

Vital, M., Hammes, F., and e.g.li, T. (2008). *Escherichia coli* O157 can grow in natural freshwater at low carbon concentrations. Environ. Microbiol. *10*, 2387–2396.

Vos, P., Hogers, R., Bleeker, M., Reijans, M., Van de Lee, T., Hornes, M., Frijters, A., Pot, J., Peleman, J., Kuiper, M., and Zabeau, M. (1995). AFLP: A new technique for DNA fingerprinting. Nucl. Acids Res. *23*, 4407–4414.

Wagner, M., Horn, M., and Daims, H. (2003). Fluorescence in situ hybridisation for the identification and characterisation of prokaryotes. Curr. Opin. Microbiol. *6*, 302–309.

Wang, J., Liu, G., and Jan, M.R. (2004). Ultrasensitive electrical biosensing of proteins and DNA: Carbon-nanotube derived amplification of the recognition and transduction events. J. Am. Chem. Soc. *126*, 3010–3011.

Wang, Z., Hansen, O., Petersen, P.K., Rogeberg, A., Kutter, J.P., Bang, D.D., and Wolff, A. (2006). Dielectrophoresis microsystem with integrated flow cytometers for on-line monitoring of sorting efficiency. Electrophoresis *27*, 5081–5092.

Ward, D.M., Weller, R., and Bateson, M.M. (1990). 16S rRNA sequences reveal numerous uncultured microorganisms in a natural community. Nature *345*, 63–65.

Way, J.S., Josephson, K.L., Pillai, S.D., Abbazadegan, M., Gerba, C.P., and Pepper, I.L. (1993). Specific detection of *Salmonella* spp. by multiplex polymerase chain reaction. Appl. Environ. Microbiol. *59*, 1473–1479.

Wilson, I.G. (1997). Inhibition and facilitation of nucleic acid amplification. Appl. Environ. Microbiol. *63*, 3741–3751.

Woese, C.R. (1987). Bacterial evolution. Microbiol. Rev. *51*, 221–271.

Woolley, A.T., Hadley, D., Landre, P., deMello, A.J., Mathies, R.A., and Northrup, M.A. (1996). Functional integration of PCR amplification and capillary electrophoresis in a microfabricated DNA analysis device. Anal. Chem. *68*, 4081–4086.

Wu, Z., Nagano, I., Matsuo, A., Uga, S., Kimata, I., Iseki, M., and Takahashi, Y. (2000). Specific PCR primers for *Cryptosporidium parvum* with extra high sensitivity. Mol. Cell. Probes *14*, 33–39.

Xiao, Y., and Barker, P.E. (2004). Semiconductor nanocrystal probes for human metaphase chromosomes. Nucl. Acids Res. *32*, e161.

Yang, H.H., Zhang, S.Q., Chen, X.L., Zhuang, Z.X., Xu, J.G., and Wang, X.R. (2004). Magnetite-containing spherical silica nanoparticles for biocatalysis and bioseparations. Anal. Chem. *76*, 1316–1321.

Yang, L., and Li, Y. (2006). Simultaneous detection of *Escherichia coli* O157:H7 and *Salmonella Typhimurium* using quantum dots as fluorescence labels. Analyst *131*, 394–401.

Yao, G., Wang, L., Wu, Y., Smith, J., Xu, J., Zhao, W., Lee, E., and Tan, W. (2006). FloDots: Luminescent nanoparticles. Anal. Bioanal. Chem. *385*, 518–524.

Zachar, V., Thomas, R.A., and Goustin, A.S. (1993). Absolute quantification of target DNA: A simple competitive PCR for efficient analysis of multiple samples. Nucl. Acids Res. *21*, 2017–2018.

Zammatteo, N., Jeanmart, L., Hamels, S., Courtois, S., Louette, P., Hevesi, L., and Remacle, J. (2000). Comparison between different strategies of covalent attachment of DNA of glass surfaces to build DNA microarrays. Anal. Biochem. *280*, 143–150.

Zarda, B., Amann, R., Wallner, G., and Schleifer, K.-H. (1991). Identification of single bacterial cells using digoxigenin-labelled, rRNA-targeted oligonucleotides. J. Gen. Microbiol. *137*, 2823–2830.

Zhang, C., and Xing, D. (2007). Miniaturized PCR chips for nucleic acid amplification and analysis: Latest advances and future trends. Nucl. Acids Res. *35*, 4223–42237.

Zhang, C., Xu, J., Ma, W., and Zheng, W. (2006b). PCR microfluidic devices for DNA amplification. Biotechnol. Adv. *24*, 243–284.

Zhang, Q., Zhu, L., Feng, H.H., Ang, S., Chau, F.S., and Liu, W.T. (2006a). Microbial detection in microfluidic devices through dual staining of quantum dots-labelled immunoassay and RNA hybridization. Anal. Chim. Acta *556*, 171–177.

Zhao, W., Yao, S., and Hsing, L.M. (2006). A microsystem-compatible strategy for viable *Escherichia coli* detection. Biosens. Bioelectron. *21*, 1163–1170.

Zhao, X., Tapec-Dytioco, R., Wang, K., and Tan, W. (2003). Efficient collection of trace amounts of DNA/mRNA molecules using genomagnetic nano-capturers. Anal. Chem. *75*, 3476–3483.

Zhao, X.J., Hilliard, L.R., Mechery, S.J., Wang, Y.P., Bagwe, R.P., Jin, S.G., and Tan, W.H. (2004). A rapid bioassay for single bacterial cell quantitation using bioconjugated nanoparticles. Proc. Nat. Acad. Sci. USA *101*, 15027–15032.

Zheng, G., Patolsky, F., Cui, Y., Wang, W.U., and Lieber, C.M. (2005). Multiplexed electrical detection of cancer markers with nanowire sensor arrays. Nature Biotechnol. *23*, 1294–1301.

Zhu, L., Ang, S., and Liu, W.T. (2004a). Quantum dots as a novel immunofluorescent detection system for *Cryptosporidium parvum* and *Giardia lamblia*. Appl. Environ. Microbiol. *70*, 597–598.

Zhu, L., Zhang, Q., Feng, H., Ang, S., Chau, F.S., and Liu, W.T. (2004b). Filter-based microfluidic device as a platform for immunofluorescent assay of microbial cells. Lab Chip *4*, 337–341.

Zou, Q., Miao, Y., Chen, Y., Sridhar, U., Chong, C.S., Chai, T., Tie, Y., The, C.H.L., Lim, T.M., and Heng, C. (2002). Micro-assembled multi-chamber thermal cycler for low-cost reaction chip thermal multiplexing. Sens. Actuators A: Chem. *102*, 114–121.

The Potential of Nanofibres and Nanobiocides in Water Purification

3

Marelize Botes and T. Eugene Cloete

Abstract

Electrospun nanofibres and nanobiocides show potential in the improvement of water filtration membranes. Biofouling of membranes caused by the bacterial load in water reduces the quality of drinking water and has become a major problem. Several studies showed inhibition of these bacteria after exposure to nanofibres with functionalized surfaces. Nanobiocides such as metal nanoparticles and engineered nanomaterials are successfully incorporated into nanofibres showing high antimicrobial activity and stability in water. Research on the applications of nanofibres and nanobiocides in water purification, the fabrication thereof and recently published patents are reviewed in this chapter.

Introduction

Drinking water contaminated with bacteria and viruses is the main cause of diseases in development countries (Jain and Pradeep, 2004). One-sixth of the world population (1.1 billion people) doesn't have access to safe water (WHO, 2004). Currently, chemical disinfectants such as chlorine and membrane-based water filtration systems are employed to control microbial pathogens. However the reliability of these methods is under question. The formation of harmful disinfection by-products from chemical disinfectants and the rise in resistant pathogens are problematic (Li *et al.*, 2008). Two significant obstacles in water treatment membrane filtration are biofouling and virus penetration (Barhate and Ramakrishna, 2007). Membrane permeability and membrane lifetime are reduced and energy costs are increased. Nanofiltration and reverse osmosis are effective in removing viruses however both are high-energy filtration schemes. Low pressure membranes with anti-biofouling and antiviral properties are therefore in urgent demand (Zodrow *et al.*, 2009).

Filtration systems may be improved by the use of nanofibrous media. The low basis weight, high permeability and small pore size make them appropriate to remove unwanted particles smaller than 0.3 μm (Barhate and Ramakrishna, 2007). Nanofibrous filtering media may differ in thickness, fibre diameter and distribution, representative pore size, porosity and tortuosity factor (indicating geometry and interconnectivity of pores) depending on the selected polymer(s) (Frenot and Chronakis, 2003). Nanobiocides such as metal nanoparticles, antibiotics and bacterial agents, can be added to the polymer solution and encapsulated in the dry nanofibres. These nanofibres are synthesized by a new and efficient fabrication process, namely electrospinning (Frenot and Chronakis, 2003).

This review will focus on the fabrication and antimicrobial activity of nanofibres and nanoparticles and possible applications in water purification.

Fabrication of nanofibres

Electrospinning is a novel technology for the fabrication of non-woven fibrous articles (nanofibres) with diameters lower than 100 nm (Renekar and Chun, 1996; Li and Zia, 2004). The first patent on the production of nanofibres was published in the USA (Cooley, 1902). Many patents followed from 1934 to 1944, which described the experimental setup using electrostatic force (Formhals, 1934). In Russia, nanofibrous filter media, known as filters of Petryanov, were kept top secret and never reached the West (Lushnikov, 1997). These filter materials were used after the Second World War to protect the environment from nuclear-active aerosol release. Five enterprises were established in Russia by the end of 1960 producing more than 20 types of fibre filter materials (Shepelev and Rykunov, 1995; Lushnikov, 1997; www.krunk.ee/eng/2/21.shtml). Production of nanofibrous materials in America gained momentum in 1980 and only started in the 1990s in Europe (www.krunk.ee/eng/2/21.shtml). Currently more than 20 enterprises are maintaining an interest in production or use of nanofibre filter media (Barhate and Ramakrishna, 2007).

Electrospinning is a simple, rapid, efficient and inexpensive method, run by only three components, namely a high voltage supplier, a capillary tube with a pipette or needle of small diameter (spinneret) and a grounded collecting plate, usually a metal screen, plate, or rotating mandrel (Liang *et al.*, 2007). The nanofibres are produced by applying a high voltage to electrically charge liquid or melt out of pipette. Taylor cones are formed when the electrical field is increased leading to the elongation of the hemispherical surface at the tip of the fluid. The electric force overcomes surface tension of the polymer solution leading to evaporation of the solvent and ejection of a thin polymer drawn towards the collector plate. Ultrafine fibres have been successfully spun from more than 50 different polymers. A polymer solution is prepared by dissolving the polymer in a specific solvent and introducing the fluid into the capillary tube for electrospinning (Huang *et al.*, 2003). Electrospun polymers with wettability and absorbency properties, applied for different types of filters are summarized in Table 3.1.

Laboratory scale electrospinning yields low quantities of nanofibres. Therefore a multi-jet electrospinning and blowing-assisted electrospinning technology were recently developed, demonstrating the production capability for fabricating nanofibrous articles on an industrially relevant scale (Chu *et al.*, 2003; Ding *et al.*, 2004a; Um *et al.*, 2004).

Researchers at Elmarco have designed a high-yield industrial application namely 'Nanospider'. This technology produces Taylor cones in close proximity to each other on a cylinder, partly immersed in a polymer solution. A thin film on the cylinder's surface is created from a controlled

Table 3.1 Polymers and solvents applied for electrospun nanofibrous material

Polymer	Solvent	Concentration	Reference
Polyurethane (PU)	Dimethyl formamide	10 wt%	Tsaia *et al.* (2002)
Polycarboate (PC)	Dichloromethane	15 wt%	Bognitzki *et al.* (2000)
	Dimethylformamide–tetrahydrofuran (1:1)	20 wt%	Tsaia *et al.* (2002)
Polylactic acid (PLA)	Dichloromethane	5 wt%	Bognitzki *et al.* (2001)
Polyethylene oxide (PEO)	Isopropyl alcohol and water	10 wt%	Morozov *et al.* (1998)
Polyvinylcarbazole	Dichlormethane	7.5 wt%	Bognitzki *et al.* (2000)
Polystyrene	Tetrahydrofuran	15 wt%	Torres (2001)
Polyamide (PA)	Dimethylacctamide	Not available	Hajra *et al.* (2003)
Cellulose acetate (CA)	Acetone, acetic acid	17%	Han *et al.* (2002)

amount of the polymer solution. Nanofibres weighing between 0.1 and $10\,g/m^2$ are synthesized and the machine produces 0.1–1 g material in less than one minute, sixty times faster than using traditional electrospinning techniques. A variety of polymers are used by Elmarco's Nanospider™ technology to produce a wide range of organic, inorganic and biodegradable nanofibres. Nanofibres may be produced by a water-based electrospinning process using poly(vinyl alcohol) (PVA), PEO and biopolymers with a mean fibre diameter of 200, 250 or 300 nm and a fibre diameter distribution of ± 30%. Solvent-based electrospinning of nylon 6, polyaramid polyurethane (PA6/12), poly(acrylic acid) (PAA), polyurethane (PUR), fluoropolymers and biopolymers (Collagen, Chitosan) produce fibres with a mean diameter of 80, 100, 150 or 250 nm, and diameter distribution of ± 30%. One of Elmarco's core principles is the production of high quality nanofibres. Elmarco's filtration material product, Nanospider AntimicrobeWeb™, has numerous applications and is currently used in the manufacture of surgical masks with almost one hundred percent effectively in the filtering of viruses and bacteria. Elmarco is also in partnership with various industries such as Cummins Filtration in USA to develop innovative solutions in motor vehicle and related filtration systems (www.elmarco.com/retrieved 24/02/2009).

Nanofibres possess an extremely high surface-to-volume ratio and a small pore size (in microns) (Huang *et al.*, 2003; Xu *et al.*, 2003; Chen and Hsieh, 2004; Gupta *et al.*, 2004; Han *et al.*, 2004; Lin *et al.*, 2004; Yang *et al.*, 2004; Zhang *et al.*, 2005). The composition of the nanofibre mat can be controlled to achieve desired properties and functionality. Parameters influencing the thickness and morphology of nanofibres include (a) molecular weight, molecular weight distribution and architecture (branched, linear, etc.) of the polymer; (b) solution properties (viscosity, conductivity, and surface tension); (c) process parameters such as electric potential, flow rate, distance between the syringe needle tip to ground collection plate; (d) ambient parameters (solution temperature, humidity, and air velocity in the electrospinning chamber); and, finally, (e) motion of target screen (Reneker *et al.*, 2000; Wang and Hsieh, 2004). In the application of water filtration membranes, nanofibres are not sufficient to withstand impacts during filtration and should be spun directly over a stronger, more rigid support, e.g. melt blown supportive fibrous media (Grafe and Graham, 2003) or should be 'sandwiched' between various layers (Wang *et al.*, 2005). To fabricate a self-supporting membrane, post-heat treatment is conducted to fuse the overlapped fibres together attaining a crystalline structure. This improves the structural integrity and mechanical strength of the fibres (Gopal *et al.*, 2006). Nanofibres with mechanical strength close to continuous film and porosity are fabricated using a 10 to 20 wt% polymer suspended in a low boiling solvent with boiling point above 100°C. Fibres adhere at large contact points without deformation (Shutov and Astakhov, 2006).

Despite the remarkable properties mentioned, nanofibres may be further functionalized by spinning blends of specific polymers, coaxial spinning of two polymer solutions, coating of fibres, or inclusion of functional components such as antimicrobial nanoparticles (Kriegel *et al.*, 2008).

Electrospun nanofibrous scaffolds have been widely investigated in the past several years with materials of different compositions applications of varying end-uses, such as filtration (Gibson *et al.*, 2001; Schreuder-Gibson *et al.*, 2002, 2004) optical and chemical sensors (Wang *et al.*, 2002; Ding *et al.*, 2004b,c; Liu and Kameoka *et al.*, 2004; Wang *et al.*, 2004; Wannatong and Sirivat, 2004), electrode materials (Kim and Yang, 2003; Kim and Park *et al.*, 2004) and biological scaffolds (Khil *et al.*, 2005; Ma *et al.*, 2005; Riboldi *et al.*, 2005; Yang and Murugan *et al.*, 2005).

Antimicrobial nanofibres

The successful preparation of antimicrobial nanofibres depends primarily on the selected surfactants and polymers. Polymer–surfactant interactions influence parameters of the electrospinning process and may alter the three critical factors of electrospun nanofibres, namely rheology, surface tension and conductivity (Yao *et al.*, 2003; Lin *et al.*, 2004). The hydrophilic and hydrophobic moieties of surfactants enable them to form association colloids and to absorb to various interfaces when dispersed in a solvent (Lange, 1999; Goodwin, 2004). Microemulsions are examples of association colloids with lipophilic compounds solubilized within their hydrophobic interior resulting in swollen micelles (Stauffer *et al.*, 1994; Weiss *et al.*, 1996, 1997, 1999, 2000; Weiss and McClements, 2000; Vandamme, 2002). The concentration of the lipophilic compound increases when solubilized in surfactant micelles leading to improved functional properties such as bioavailability, antimicrobial, or antioxidant activities (Weiss *et al.*, 2007). Gaysinksy and co-workers encapsulated lipophilic antimicrobials in microemulsions which increased their interaction with bacterial surfaces, resulting in enhanced activity against Gram-positive and Gram-negative foodborne pathogens (Gaysinsky *et al.*, 2005a,b, 2008).

Conventional membrane materials are mostly hydrophobic and should be hydrophilic for water-based filtration/purification applications. The hydrophobic polymer surface can be rendered into hydrophilic by surface modification and can hereby obtain other functionalities (Yoon *et al.*, 2009). Clean and environmentally friendly ways for surface modification of nanofibres include plasma treatment (Ramakrishana *et al.*, 2005) and surface graft polymerization (Zuwei *et al.*, 2006). Plasma treatment involves changing the surface chemical composition of fibres by treatment with oxygen, ammonia or air that generate free radicals and electrons on the surface. These functional groups change the surface adhesion and wetting properties of the membrane (Fuchs and Tiller, 2006; Fu *et al.*, 2008). Functional compounds such as poly(quaternary ammonium) are then covalently bound to the fibre surfaces to target pathogenic microorganisms (Kim and Klibanov, 2005). Surface graft polymerization is initiated by treatment with plasma and UV radiation to generate free radicals for polymerization (Yoo *et al.*, 2009). Three challenges should be kept in mind when attempting surface modification, namely the uniformity of the coating and surface modification, the filter pore size should not be affected by the coatings and industrial fabrication of these nanofibres may be complicated (Barhate and Ramakrishna, 2007). Examples of surface modification using plasma treatment, surface graft polymerization and the coupling of antimicrobial compounds are described.

Polyurethane (PU) nanofibres were treated with argon plasma, which produced surface oxide and peroxide groups. The PU fibres were then immersed in a 4-vinylpyridine monomer solution with exposure to UV irradiation, to produce poly(4-vinylpyridine) grafted PU fibres successfully. The grafted pyridine groups were functionalized through quaternization with hexylbromide to obtain antibacterial activities. The viability of Gram positive *Staphylococcus aureus* and Gram-negative *Escherichia coli* was reduced 99.9% after 4 h contact with the PU fibres (Yao *et al.*, 2008).

Yao *et al.* (2009) modified the surface of electrospun poly(vinylidene) fluorid-co-hexafluoropropylene (PVDF-HFP) fibrous membranes to achieve antimicrobial activities. The electrospun membranes were treated with poly(4-vinyl-*N*-alkylpyridinium) bromide and a combination of plasma pretreatment, UV-induced graft copolymerization and quaternization. PVDF-HFP membranes were highly active against *S. aureus* and *E. coli* and could serve as an efficient and versatile approach for the development of antimicrobial polymer fibres (Yao *et al.*, 2009).

The potential of polycarbonate (PC)/chloroform solution with quaternary ammonium salt (benzyl triethylammonium chloride, BTEAC) as antimicrobial nanofibrous membranes for ultrafiltration was investigated. The addition of small amounts of BTEAC to the PC solution decreased the average diameter and totally inhibited the growth of *S. aureus* ATCC6538, *E. coli* ATCC 25922 and *Klebsiella pneumoniae* ATCC 4352. The PC nanofibrous filter showed 99.97% filtration efficiency of 0.3 μm size particles in comparison to a polypropylene (PP) HEPA (high efficiency particulate air) filter (Kim *et al.*, 2007). A similar study was performed by Jeong *et al.* (2007). Electrospinning of polyurethane cationomers (PUCs) containing different amounts of quaternary ammonium groups were investigated for the application in antimicrobial nanofilters. Very strong antimicrobial activities against *S. aureus* and *E. coli* were found and the average diameters of the electrospun PUCs fibres decreased with increasing quaternary ammonium group content.

Mechanical properties and antimicrobial activity of nylon 6 nanofibre membranes with the addition of three different *N*-halamine additives, namely chlorinated 5,5-dimethylhydantoin (CDMH), chlorinated 2,2,5,5-tetramethyl-imidozalidin-4-one (CTMIO) and chlorinated

3-dodecyl-5,5-dimethylhydantoin (CDDMH) were determined. The growth of both *E. coli* and *S. aureus* was inhibited after a contact period of 5 to 40 min. The fastest antimicrobial action was found for CDMH. The increase of active chlorine contents resulted in a higher antimicrobial rate and larger microbial reduction. The addition of *N*-halamines had no significant effect on the mechanical properties of nylon 6 membranes and leaching of *N*-halamines was negligible (Tan and Obendorf, 2007).

A bilayer composite membrane consisting of chitosan and poly(acrylic acid) (PAA), and a bottom support layer composed of poly(ethylene glycol) diacrylate (PEGDA) and PAA was developed for antimicrobial applications (Don *et al.*, 2005). A strong and dense membrane was developed by casting PEGDA, acrylic acid and a photoinitiator and curing with UV irradiation on a glass substrate. This bottom layer was coated with a solution containing of chitosan, acrylic acid and water. Coagulation of chitosan with residual acrylic acid took place that subsequently formed a nanolayer on the top surface after application of UV irradiation. The addition of 0.5% chitosan killed 80% of *S. aureus* (CCRC 10779) and 30% of *E. coli* (CCRC 10324), and increased to 82% and 45%, respectively, with an increase of 0.5% chitosan (Don *et al.*, 2005).

Chitosan was also applied in a high-flux ultrafiltration medium by replacing flux-limiting asymmetric porous membrane with porous electrospun nanofibrous scaffolds (Yoon *et al.*, 2006). The demonstrated systems consisted of a three-tier composite structure: (1) 'non-porous' hydrophilic coating that is water permeable (chitosan), (2) polyacrylonitrile (PAN), an electrospun nanofibrous support and (3) polyethylene terephthalate (PET), a non-woven microfibrous substrate (Fig. 3.1).

PAN is resistant to most solvents and has therefore been widely used for ultrafiltration (Musale *et al.*, 1999), nanofiltration (Musale and Kumar, 2000), and reverse osmosis (Chandorilar and Bhavsar, 1983). Carbon nanofibres were also fabricated with electrospun PAN as precursor (Wang *et al.*, 2003). The PAN fibrous networking in this study was used to support a top coating layer based on chitosan (Yoon *et al.*, 2006). Chitosan has been used for anti-fouling enhancement of filtration membranes duo to its insolubility in neutral pH conditions (Gudmund *et al.*, 1989) and thus water-resistance (Wang and Spencer, 1998; Musale *et al.*, 1999; Musale and Kumar, 2000).

Three-tier composite membranes exhibited 1.0 and 1.31/m^2 h psi flux rate respectively for a 1.37 and 1.2 wt% chitosan coating comparing to a 0.61/m^2 h psi flux rate for a commercial nanofiltration filter. Mechanisms proposed to improve membrane performance (higher flux and lower fouling) include the reduction of coating layer thickness, cross-linking of chitosan layer by glutaraldehyde, and grafting of more hydrophilic polyethylene glycol molecules on the chitosan chains (Yoon *et al.*, 2006).

The polymers used in these studies are examples of material with good mechanical and physical properties. Surface modification of the nanofibres showed good filtration efficiency with a higher flux

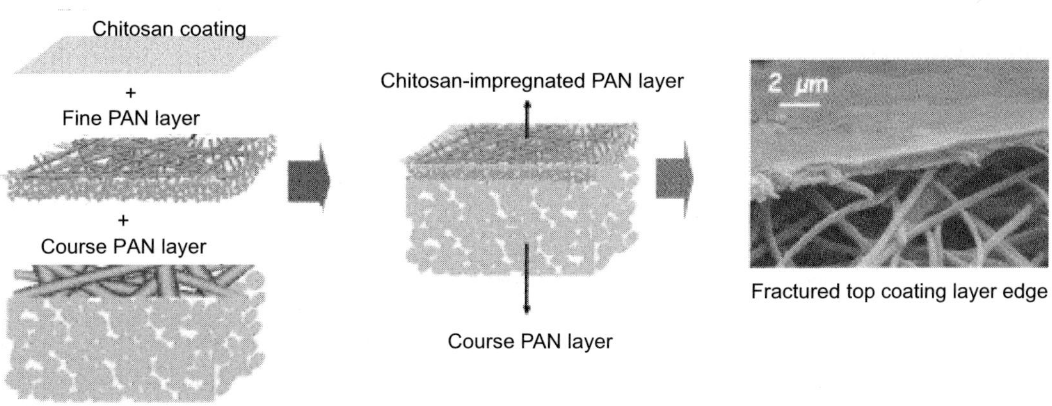

Figure 3.1 Fabrication schematics of the electrospun scaffold with a coating layer. SEM image represents the fractured composite membrane containing PAN nanofibrous scaffold (with 4 + 12 wt% sequential electrospinning) and chitosan coating (Yoon *et al.*, 2006).

rate than conventional membranes. These properties along with strong antimicrobial activity render the fibres as potential antifouling water filtration membranes.

Nanobiocides (antimcrobial nanoparticles)

Nanofibres with embedded nanobiocides are currently being used in medical treatments and air filters. Different polymers such as poly(vinyl pyrrolidone) (PVP), polyvinyl alcohol (PVA) and poly (ethylene oxide) (PEO) have been used as a template to load nanoparticles such as metal oxides by electrospinning and in polymer matrixes or films (Li *et al.*, 2006).

Manufacturing of nanoparticles

Nanoparticles are manufactured by different procedures which can be grouped into top-down and bottom-up strategies. The top-down approach involve the generation of isolated atoms from bulk materials by using various distribution techniques such as milling or attrition, repeated quenching and photolithography (Niemeyer, 2001; Gao, 2004). Bottom-up strategies are defined by the formation of more complex clusters by using molecular components as starting materials which are linked with chemical reactions and nucleation and growth process (Rotello, 2003; Gao, 2004).

Metallic nanoparticles can also be synthesized by a reliable and eco-friendly process, using bacteria and fungi (Bhainsa and Souza, 2006). *Aspergillus fumigatus* formed silver nanoparticles within 10 minutes in the presence of silver ions. The silver nanoparticles were synthesized in the extracellular cell filtrate indicated by a brown colour (Bainza and Souza, 2006). The extracellular synthesis requires no additional step of processing for release of nanoparticles from the biomass. However, this reaction mechanism is still unknown. A certain NADH dependent reductase was involved in the reduction of silver ions by *Fusarium oxysporum* (Ahmad *et al.*, 2003a). Other fungi and bacteria involved in the formation of silver or gold nanoparticles, intracellularly, include *Verticillium* spp., *Thermonospora* spp. and *Pseudomonas stutzeri* AG259 (Klaus-Jeorger *et al.*, 2001; Ahmad *et al.*, 2003b; Sastry *et al.*, 2003). This rapid synthesis of nanoparticles would be suitable for developing a biological process for mass scale production.

Nanobiocides (antimicrobial nanoparticles) are divided into three categories, namely, metal and metal oxides (nAg, ZnO, CuO, TiO_2), engineered/synthesized nanomaterials [fullerenes, e.g. nanomagnetite (nC_{60}) and carbon nanotubes] and natural antibacterial substances (antimicrobial substances, chitosan) (Li *et al.*, 2008). The antimicrobial activity of embedded metal nanoparticles in nanofibres or polymer matrices will be discussed, followed by the application of carbonaceous nanomaterial, fullerenes and natural substances as biocides.

Metal and metal oxide nanoparticles

Silver nanoparticles

Silver is the most toxic element to microorganisms and are used in various applications such as antimicrobial filters, wound dressing material, water disinfection, sensors, chemical and gas filtration, protective cloth and air filtration (Lala *et al.*, 2007). Different polymers have been used to incorporate silver nanoparticles for the production of antimicrobial nanofibres.

Polycaprolactone (PCL) and polyacrylonitrile (PAN) were synthesized by reducing silver ions photochemically whereby an electron donor (Ag nanoparticles) and electron acceptor (tetracyanoquinodimethane) were incorporated in polymer nanofibres by electrospinning (Shang *et al.* 2006). The soft segment of poly(ε-caprolactone)-based polyurethane (PCL-PU) is a biodegradable, biocompatible, and semicrystalline aliphatic polyester with excellent mechanical and thermal properties (Woo *et al.*, 2000; Heijkants *et al.*, 2005; Ping *et al.*, 2005; Hatakeyama *et al.*, 2007). Silver nanoparticles can interact with the carbonyl oxygen atoms of PCL (Son *et al.*, 2004), resulting in the even distribution of small Ag particles. PCL-PU nanofibre mats containing Ag nanoparticles may exhibit good mechanical properties with antimicrobial activity and no waste disposal problems. Jeong *et al.*

(2008) prepared PCL-PU fibres with different concentrations of $AgNO_3$ DMF/THF (7/3 w/w) for the applications of an antimicrobial nanofilter. The inclusion of $AgNO_3$ increased the tensile strength, tensile modulus, and decreased elongation of the PCL-PU nanofibre mats.

Silver bromide (AgBr) nanoparticles and poly(vinyl pyrrolidone) (PVP) were hybridized by electrospinning. Bromide ions reacted with silver ions of $AgNO_3$ in the PVP solution to form AgBr nanoparticles by a sol–gel technique. The AgBr/PVP viscous composite solutions were electrospun into PVP nanofibres containing AgBr nanoparticles. A fine dispersion of nanoparticles was observed by field-emission scanning electron microscopy (Bai *et al.*, 2008).

Nanofibres containing silver nanoparticles were electrospun with three different polymers namely cellulose acetate (CA), polyacrylonitrile (PAN) and PVP to compare antimicrobial activity against *E. coli* ATCC 25922 and *P. aeruginosa* ATCC 27853. Various amounts of $AgNO_3$ were dissolved in DMF and treated with UV irradiation to form nanoparticles. The combination of PAN in DMF containing 5 wt% of AgNO3 and UV-irradiation for 30 min was found to be most effective (Lala *et al.*, 2007).

A chelating monomer, glycidyl methacrylate–iminodiacetic acid (GMA–IDA), was grafted onto cotton fibres to form cotton fibre-graft-GMA–IDA (CFGI). CFGI–Ag^+ complexes were successfully prepared by irradiation. The cotton fibres containing Ag^+ nanoparticles with an average size of 75 nm exhibited high antibacterial activity against *E. coli* (Chen and Chiang, 2008).

Silver nanoparticles were synthesized in semi-hydrogel networks of poly(acrylamide) and starch carbohydrate polymers as an antimicrobial application. Carboxylic, amide, and hydroxyl groups of carbohydrate polymers holds large amounts of metal ions in their network, enhancing the reduction potential/anchoring ability and stabilization of nanoparticles. The silver nanoparticles were smaller and finer distributed inside the semi-IPN hydrogel network (Fig. 3.2). No particles were formed in the surrounding medium confirming the binding of networks to the silver ions and successful storing of nanoparticles (Vimala *et al.*, 2009).

The application of silver nanoparticles in water filters was also investigated. Jain and Pradeep (2005) exposed polyethane (PU) foams overnight with silver nanoparticles followed by washing and air-drying. Silver nanoparticles interacted with the nitrogen atoms of PU and could therefore be stored for extended periods without the loss of nanoparticles. Input water was infected with *E. coli* [10^5 colony-forming units (CFU) per ml] and exposed to PU foams at a flow rate of 0.5 l/min. No bacteria were detected between 1 and 4 hours of exposure which complied with the WHO requirements for drinking water. PU foams coated with silver nanoparticles may be used as a low cost and non-toxic water purification system especially in developing countries.

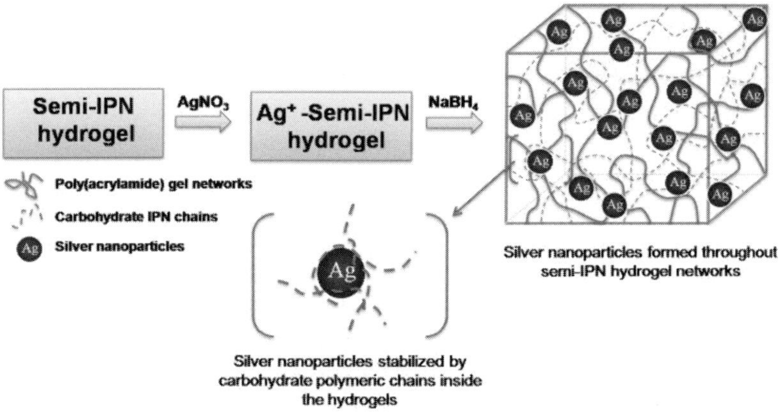

Figure 3.2 Synthesis of silver ions in a semi-IPN hydrogel network of poly(acrylamide) and starch carbohydrate polymers (Vimala *et al.*, 2009).

Researchers in Russia, investigated the biocidal activity of silver nanoparticles after application on a surface of carbonic materials, used in water filtering devices. Tap water with high concentration of *E. coli* and *Legionella pneumophila* (LP) was filtered through a carbon column with attached silver particles. The modified column reduced double the number of bacteria in comparison to the control column. The application of silver nanoparticles in water dispersion against the human immunodeficiency virus (HIV) was also tested. Up to tenfold dilution of nanoparticles was a hundred percent effective in killing the virus (www.nanomet.ru/en/appl.3.php).

Dendritic polymers are used in water filtration applications as delivery vehicles or scaffolds for antimicrobial agents such as Ag (I) and quaternary ammonium chlorides (Chen *et al.*, 2003). Poly(amidoamine) dendrimer (PAMAM)-based silver complexes and nanocomposites proved to be effective antimicrobial agents *in vitro*. The silver remained conjugated to the dendrimer in the form of ions, stable metallic silver clusters or compounds and could retain its activity. The immobilized silver diffused out into agar medium and showed high antimicrobial activity against *S. aureus, P. aeruginosa* and *E. coli* (Balogh *et al.*, 2001).

Silver nanoparticles (nAg–PSf) embedded in polysulphone ultrafiltration membranes were evaluated for biofouling reduction. Polysulphone is currently being used in water (microfiltration, ultrafiltration and nanofiltration) membranes (Zodrow *et al.*, 2009). The release of silver ions resulted in almost 100% reduction in viability of *E. coli* K12 and *Pseudomonas mendocina* KR1 and an enhanced removal of MS2 bacteriophage. The permeability of nAg-PSf was $532 \pm 117 l/m^2/h/bar$ and that of the PSf control membrane $408 \pm 180 l/m^2/h/bar$. Nanosilver incorporation also increased membrane hydrophilicity, reducing the potential for other types of membrane fouling. However a loss of silver ions from the membrane surface lead to a significant decrease in antibacterial and antiviral activity. Concentrations of 0.034 mg/l silver ions were found in the membrane permeate although no silver nanoparticles were observed in the filtrate. Future research should focus on improved fixation of Ag^+ nanoparticles on the membrane surface and encapsulating the silver to reduce release rate. Commercial membranes may be regenerated and improved by the addition of silver nanoparticles. These water filtration membranes may ensure effective control on a wide variety of bacteria and viruses and have protection against biofouling (Zodrow *et al.*, 2009).

Porous ceramics have become of great interest for drinking water treatment lately, due to the chemical, thermal, pH stability and cost-effective material (Padilla *et al.*, 1997). However, during filtration the blocked bacteria are not killed and form biofilms on the surface. Lv *et al.* (2009) attached silver nanoparticles onto porous ceramic composites by an aminosilane coupling agent, 3-aminopropyltriethoxysilane (APTES). The APTES molecule attach to silver atoms with the $-NH_2$ group on the top and to the silicon atoms in the ceramic on the bottom. The attachment of silver nanoparticles could withstand 15 min ultrasonic irradiation and several weeks' exposure to an atmospheric environment. The modified ceramic totally inhibited *E. coli* in water during on-line tests. Therefore, silver nanoparticle-decorated porous ceramic composites show potential as low cost and safe water filters (Lv *et al.*, 2009).

Zinc and copper oxide nanoparticles

Zinc oxide (ZnO) nanoparticles have strong antimicrobial activity against a broad spectrum of bacteria although the mechanism involved is still uncertain. Sawai (2003) suggested that the biocide, hydrogen peroxide is photocatalytic generated and Huang *et al.* (2008) found that ZnO nanoparticles penetrate the bacterial cell envelope and disorganize the cell membrane. ZnO is used in pharmaceutical products such as antibacterial cream, lotions and ointments and in surface coatings to prevent biofilm formation (Jones *et al.*, 2008). The application of ZnO nanoparticles in water treatment is however questioned as it is highly soluble in water increasing sensitivity of aquatic organisms towards ZnO (Franklin *et al.*, 2007).

In a study by Vicentini *et al.* (2009), zinc oxide nanoparticles were synthesized by the Pechini method by reacting citric acid with ethylene glycol. The dissolved metal ions were incorporated into blend films of chitosan and PVA with different concentrations of polyoxyethylene sorbitan monooleate and Tween 80 (T80). Film properties such as greater thermal stability, decreased tensile strength and elongation break and porosity were influenced by the addition of ZnO nanoparticles and T80. These films effectively killed *S. aureus*.

Copper oxide is not as expensive as silver and shows chemical and physical stability when mixed with polymers. The antimicrobial activity of nano CuO is not known, however properties such as extremely high surface areas and unusual crystal morphologies may render CuO as potential antimicrobial agents (Stoimenov *et al.*, 2002). Rena *et al.* (2009) showed that CuO have activity against a range of bacterial pathogens, including methicillin-resistant *S. aureus* (MRSA) and *E. coli*, with minimum bactericidal concentrations (MBCs) ranging from 100 g/ml to 5000 g/ml. A synergistic effect existed between CuO nanoparticles and sub-minimum bactericidal concentrations of silver nanoparticles.

Copper-hydrotalcite (Cu-HT) can be used as an alternative disinfectant to chlorine in a water purification system. Cu-HT has very high antibacterial activity against *Escherichia coli* and phage Q β (Sunayama *et al.*, 2002). The proposed mode of action is the direct contact between the hydroxyl layers on the surfaces of the Cu-HT particles and phage Qβ cells (Sunayama *et al.*, 2002).

Copper nanoparticles were functionalized with an acrylic group to copolymerize with other acrylic monomers. The nanoparticles became an integral part of the polymer backbone which slowed down particle release rate from the polymer matrix. The released particles were very active against marine biofouling strains, *Chlamydomonas* sp. strain PCC 6803 (freshwater) and *Synechococcus* sp. strain PCC7002 (marine) and *Phaeodactylum tricornutum* CCMP 1327. Controlled release did not affect antimicrobial activity of copper nanoparticles and correlated well with conventional copper-based biocides (Anyaogu *et al.*, 2008).

Titanium dioxide nanoparticles

Titanium dioxide (TiO_2) is a well-studied semiconductor photocatalyst (Li *et al.*, 2008). Gram-positive and Gram-negative bacteria (Wei *et al.*, 1994) and viruses (Watts *et al.*, 1995; Chou *et al.*, 2005; Hajkova *et al.*, 2007; Zan *et al.*, 2007) are sensitive to TiO_2 photocatalytic inactivation at concentrations of 100 and 1000 ppm depending on the particle size and wavelength of the light (Wei *et al.*, 1994). The mode of action includes the production of hydroxyl free radicals and peroxide formed under UV-A irradiation via oxidative and reductive pathways, respectively (Kikuchi *et al.*, 1997). TiO_2 is stable in water and cost-effective and can therefore be successfully incorporated in thin films or membrane filters for water filtration (Belhacova *et al.*, 1999; Kwak *et al.*, 2001).

Immobilised nanoparticle TiO_2 films were evaluated for photocatalytic and photolytic inactivation of *Escherichia coli* K12 in surface and distilled water. Effective photocatalytic inactivation of *E. coli* occurred both in distilled and surface water. Nitrate and sulphate anions and especially humic acid spiked into distilled water resulted in a decrease in the rate of photocatalytic disinfection. Change in the initial pH to 5.5–8.5 had no effect (Alrousan *et al.*, 2009).

Titanium dioxide nanoparticles were incorporated into an isotactic polypropylene (iPP) polymeric matrix via a cost-effective melting process using laboratory-made nanometric anatase-TiO_2 and an industrial polymer. The activity was tested against antibiotic resistant *P. aeruginosa* and *E. faecalis* bacteria. The highest biocidal activity against both bacteria was found for a sample containing a 2 wt% of anatase-TiO_2. The antimicrobial activity of the oxide alone was boosted by the nanocomposite by efficiently managing charge carrier handling through the organic–inorganic interface. This resulted in making the whole system biocide with an overall high performance comparing to current biocidal agents (Kubacka *et al.*, 2009).

Nanocomposite water membranes containing polyamide (PA) and TiO_2 nanoparticles were prepared. The TiO_2 nanoparticles enhanced the hydrophilicity of the PA-TiO_2 membranes and hereby increased the water flux. At an operating pressure of 0.1–1.0 MPa the flux rate of PA with TiO_2 nanoparticles increased from 2.5 to 25 l/m^2/h, whereas the control PA membranes exhibited a flux rate of 1 to 18 l/m^2/h. Incorporation of 5.0 wt% TiO_2 gave optimum membrane performance together with strong mechanical properties. Almost no change in TiO_2 concentration was observed after long-term filtration experiments for 2 days, implying a robust, stable nanocomposite membrane (Lee *et al.*, 2008).

TiO_2 photocatalysis is currently being used in industrial scale water purification systems. This system produce high purity water by removing micron and submicron particles, polish iron, manganese, silica, H_2O_2 and contaminants such as *Giardia, Cryptosporidium* and oocysts. The system is robust and need no chemicals for cleaning (www.purifics.com/technology/ceramic.html).

Fullerenes (C_{60}) and carbonaceous nanomaterial

Fullerenes are molecules with 60 atoms and have received much attention because of its wide application (Ju-Nam and Lead, 2008). Due to its insolubility in water, solvent exchange (Deguchi *et al.*, 2001) and ultrasonification (Deguchi *et al.*, 2006) have been applied to disperse fullerenes as nanoparticles. C_{60} nanoparticles have a broad spectrum antimicrobial activity although the mode of action is debatable (Lyon *et al.*, 2008). The toxicity of C_{60} is also under investigation. The use of C_{60} as anti-fouling agents for ceramic membranes was determined by evaluating the effect on the attachment and respiratory activity of *E. coli* K12. A decrease in bacterial numbers and increase in inactive *E. coli* cells confirm the bactericidal effect of C_{60}. The increase in membrane hydrophobicity might also play a role in antimicrobial activity (Chae *et al.*, 2009).

Carbon nanotubes are found in two main forms, namely as single-walled (SWNT) and multi-walled (MWNT). The SWNT is a single layer graphene sheet rolled-up as cylindrical shapes with diameters ranging from 0.3 to 3 nm and capped by half a fullerene whereas the MWNT is composed of many concentric layers with varying diameters of up to 100 nm (Gao, 2004; Balasubramanian and Burghard, 2006). Both SWNT and MWNT kill microorganisms when in direct contact although the rate of inactivation is slower than other conventional disinfectants (Brady-Estévez *et al.*, 2008). Carbon nanotubes have poor solubility in water and should therefore be immobilized on a reactor surface or polymer matrix (Kang *et al.*, 2007; Li *et al.*, 2008). They can therefore be applied in the prevention of biofilm formation in water filtration membranes (Li *et al.*, 2008). However the high cost involved (Theron *et al.*, 2008) and potential toxicity (Ju-Nam and Lead, 2008) may hamper the use of carbon nanotubes in water filtration.

Polyethylene mine (PEI) is a carbonaceous nanomaterial exhibiting high antimicrobial and antifungal activity (Park, 1998). The nanoparticles are formed by cross-linking and alkylation from PEI followed by methylation in order to increase degree of amino group substitution (Graveland and Kruif, 2006). The activity of PEI nanoparticles with different degree of cross-linking, particle size and zeta potential were evaluated against *Streptococcus mutants.* Only samples with long chain alkyls demonstrated high antibacterial effect for more than four weeks (Park, 1998).

Srivastava *et al.* (2004) produced carbon nanotube filters that efficiently remove *E. coli, S. aureus* and the polio-1 virus from water. The nanotube-based water filters allowed increased water flow in comparison to conventional filters and are reusable by autoclaving (Srivastava *et al.*, 2004).

Carbon nanotubes were also used to fabricate a hollow cylindrical nanofilter to remove MS2 viruses from water. A continuous spray pyrolysis method was used with *n*-hexane as a carbon source and ferrocene as a catalyst source. The nanofilter was synthesized by optimizing the catalyst concentration, temperature, gas flow rate and reaction time affecting the purity of synthesized carbon nanotubes (CNTs). The flux rate of the nanofilter at a pressure of 5 to 20 MPa was 10 to 50 m^3/m^2/s and for a conventional NF90 membrane 5 to 15 m^3/m^2/s. The MS2 viruses were efficiently removed at pressures of 8–11 bar and the fabricated nanofibre had good water permeability and filtrate flux (Mostafavia *et al.*, 2009).

A filter comprising of oxidized aluminium nanofibres on a glass fibre substrate was developed by Argonide in the USA. Bacteria and viruses are filtered from the water flow by these positively charged nanofibres. The large pores of the filter renders it more effective than traditional membranes as it manages a much higher flow rate. Up to 99.999% of viruses are retained and the filter can be used to clean water by applying muscle force with no extra energy needed, ideal for rural contexts (www.argonide.com).

NanoCeram® cartridges are synthesized by combining nano alumina into a microglass fibre matrix to form an electropositive (~+50 mV) filter medium and converted into pleated cartridges. The active ingredient, a nano-alumina fibre, only 2 nm in diameter and about 250 nm long, is grafted onto a microglass fibre backbone. The pore size is enlarged by addition of coarser cellulose or synthetic polymer fibres that also improves flexibility and strength. The media is produced by a conventional wet laying process, then pleated and end-capped to form a cartridge. These cartridges are used for drinking water purposes or as prefilters for reverse osmosis membranes. Particulate and biological contaminants are removed that subsequently reduce fouling and cleaning cycles. (www.argonide.com).

Aluminium hydroxide fibres (2 nm in diameter) with a high electropositive charge were patented in 2005. These fibres are combined into a composite filter with other fibres or particles that can filter bacteria, viruses and nano sized colloidal particles at high water flux. These alumina fibres also filter sub-micron inorganic and metallic particles to produce ultra pure water (Tepper and Kaledin, 2005).

Natural antibacterial substances

Antimicrobial peptides

Antimicrobial peptides have only recently been synthesized as nanoparticles. These peptides were engineered in the beginning according to the structures of known proteins and properties such as size, morphology and coatings (Gazit, 2007).

Antimicrobial peptide nanoparticles are designed through self-immobilization mediated by biomineralization reactions and hereby improving their effectiveness through increased stability and protection from inactivation (Eby *et al.*, 2008). A cationic decapeptide, KSL (KKVVFKVKFK)10, isolated from a combinatorial peptide array, is active against a wide range of microorganisms. Mixtures of KSL, acetic acid and tetramethyl orthosilicate (TMOS) or potassium hexafluorotitanate were prepared. The pH of the solutions was increased to 8 with 15 N NaOH and the resulting precipitate was washed and stored as a wet pellet at −20°C. Antimicrobial activity was determined as minimum inhibitory concentration (MIC) and the minimum bactericidal and fungicidal concentration (MBC and MFC) against *E. coli* (ATCC 25922), *S. aureus* (ATCC 25923), *Staphylococcus epidermidis* (ATCC 14990), and *Candida albicans* (ATCC 10231). The amorphous silica and titania did not affect bacterial growth (Eby *et al.*, 2008). Thus, the biocidal effect is explained by interaction of the KSL, attached to the particle surface, the released soluble peptide, and the bacterial cell membranes (Brogden, 2005). Cationic peptides form pores in the cell membranes of bacteria whereas surface-immobilized peptides adhere to cell surfaces and reduce bacterial viability (Etienne *et al.*, 2004; Gregory and Mello, 2005). Nisin A, a lantibiotic with a broad spectrum activity against Gram-positive bacteria, was incorporated into poly-L-lactide nano-particles (Salmaso *et al.*, 2004). A gas-compressed CO_2 process was used to yield small-sized particles with a high protein loading and efficient protein release rate. Nisin solubility and release rate and protein/polymer interactions were determined in conditions with varying pH and NaCl concentrations. An increase in pH and NaCl concentrations resulted in a decrease in nisin solubility, a slower release rate and an increased interaction between nisin and poly-L-lactide. The high antimicrobial activity against *Lactobacillus delbrueckeii* subsp. *bulgaricus* for up to 45 days indicated that nisin was efficiently encapsulated. Environmental conditions may induce the modification of the protein structure leading to altered protein/polymer interactions and affecting the protein release rate and the amount of released protein. These variables should be considered when formulating polymeric systems for the delivery of proteins (Salmaso *et al.*, 2004).

Chitosan

Chitosan is known for high antimicrobial activity towards bacteria, especially Gram-positive bacteria and effective control of fungi and viruses (Qi *et al.*, 2004). The degree of activity is determined by the organism, pH, molecular weight, degree of polymerization and the presence of lipids and proteins (No *et al.*, 2002; Rabea *et al.*, 2003). Qi and co-workers (2004) proposed that positively charged chitosan particles attacks negatively charged cell membranes leading to the formation of pores and the leakage of intercellular components. Chitosan may also penetrate the fungal cell wall and nucleus, bind to DNA and inhibit RNA synthesis (Rabea *et al.*, 2003). Activity is however lost at pH values above 6 due to the absence of pronated amino groups (Qi *et al.*, 2004). Chitosan nanoparticles were recently synthesized showing similar antimicrobial activities (Qi *et al.*, 2004; Ye *et al.*, 2006). These particles show potential in drinking water disinfection applications such as antimicrobial agents in membranes, sponges and surface coatings of water storage tanks (Li *et al.*, 2008). The high antimicrobial activity against a broad spectrum of bacteria, viruses and fungi and low toxicity render chitosan nanoparticles as an attractive alternative to other disinfectants (Li *et al.*, 2008). The method used for material preparation and presence of organics may affect antimicrobial activity (No *et al.*, 2002; Rabea *et al.*, 2003). Future research work may include the synthesis of water-soluble chitosan derivatives to improve effectiveness at alkaline pH values.

Other chitosan derivates such as chitosan–Ag^+ complexes (Yi *et al.*, 2003; Chen *et al.*, 2005) and a chitosan–Zn^{2+} complex (Wang and Du, 2004) were prepared to improve antimicrobial activity. The metal-chelating properties of chitosan have been mainly used in wastewater treatment (Jia *et al.*, 2001; Liu *et al.*, 2004; Yang and Chou *et al.*, 2005). A chitosan–Ag-nanoparticle composite (AgNP) was synthesized to test the bactericidal effect against *E. coli,* expressing recombinant green fluorescent protein (GFP). The growth of *E. coli* stopped immediately after exposure to the composite with a higher release rate of cellular GFP than with pure chitosan. Destabilization of the bacterial cell wall may be explained by the interaction of AgNPs and the sulphur-containing intracellular proteins in bacteria (Feng *et al.*, 2000; Morones *et al.*, 2005).

Chitosan triphosphate nanoparticles were individually loaded with Ag^+, Cu^{2+}, Zn^{2+}, Mn^{2+}, or Fe^{2+} and tested for antimicrobial activity against *E. coli* 25922, *Salmonella choleraesuis* ATCC 50020 and *S. aureus* 25923. The stability of aqueous nanosuspensions is measured by Zeta potential. The minimum requirement for a physically stable nanosuspension, solely stabilized by electrostatic repulsion, is ± 30 mV (Müller *et al.* 2001). The zeta potentials were enhanced significantly after loading of metal ions and were directly proportional to antimicrobial activity. Silver and zinc ions showed the highest activity and their antimicrobial effect was more pronounced against Gram-negative bacteria than Gram-positive bacteria (Du *et al.*, 2009).

Chitosan nanoparticles were modified using an O/W emulsification by grafting oleoyl onto the -NH_2 at C-2 in the chitosan molecule (OCNP). *E. coli* and *S. aureus* treated with OCNP leaded to a quick release of intracellular components. The uptake of 1-N-phenylnaphthylamine and the release of cytoplasmic β-galactosidase increased via increased permeability of inner and outer membranes. The chitosan nanoparticles adhered to the surface of *E. coli* and *S. aureus* and showed extensive cell surface alterations (Xing *et al.*, 2009).

Although the incorporation of nanobiocides are useful, two significant challenges in the long term are apparent. Firstly, the loss of antimicrobial or antiviral activity due to leaching from the membrane. The cost involved will increase as well as the impact on human health and the environment (Wiesner *et al.*, 2006). The safety of these escaped nanobiocides are not proved yet as metal and metal oxide nanoparticles have shown inflammatory and toxic effects on living cells. However in these trials much higher concentrations of nanoparticles were tested than humans would be exposed to. The nanobiocides can also be turned into harmful chemicals when exposed to chlorine and UV (Rule *et al.*, 2005; Yao *et al.*, 2009). Secondly, bacteria may become resistant towards the nanobiocides. Polysulphone ultrafiltration membranes embedded with silver nanoparticles showed no protection against silver-resistant bacteria strains such as *P. aeruginosa* (Zodrow *et al.*, 2009). The rapid depletion of nanobiocides can be addressed by improving fixation techniques by concentrating the nanobiocides closer to the surface of the membrane, through the use of cross-linkers, coupling agents or encapsulation in a polymer matrix to decrease the release rate (Zudrow *et al.*, 2009). An alternative approach for antimicrobial nanofibres is to modify the polymeric surfaces as mentioned in the previous section to achieve permanent antimicrobial surfaces.

Conclusion

In this chapter, the potential applications of antimicrobial nanofibres and nanoparticles for water purification are reviewed. The two most important requirements of water filtration membranes are total reduction of bacterial and viral numbers and prevention of biofouling. Other priorities include extended membrane life, less energy consumption, easy cleaning processes without the use of chemicals, no toxicity or production of disinfectant by-products.

Electrospun nanofibres show great potential as antibacterial coating for fibrous media or as a self-supporting membrane. Nanofibres have a much higher surface-to-volume ratio than conventional melt-blown microfibres, providing efficient separation of particulate. Electrospinning also offer opportunities to fine tune the surface functionality through polymer chemistry. Although electrospinning is a very simple method, industrial scale fabrication of nanofibres are still in the beginning phase. The design of functional nanofibres for filter membranes requires specific and controlled parameters for reproducible and controlled electrospinning. These criteria remain a challenge in the engineering of a cost-effective large-scale production system. Nanoparticles such as nAg, TiO_2, ZnO and C_{60} have been studied for antimicrobial activities in numerous articles. The high availability, low cost and high antimicrobial activity of these nanoparticles make them attractive in the use of water purification. Different fixation methods of nanobiocides to nanofibres exist which prevent leaching from the membrane and therefore reducing potential toxicity and cost. Novel biocide membranes with no toxicity may also include the incorporation of chitosan nanoparticles and antimicrobial peptides. Nanofibres modified with plasma treatment or surface graft polymerization is an alternative approach for antifouling water filtration membranes. Permanent activity against waterborne pathogens is ensured without any health risk.

With rising water-borne diseases and shrinking water sources, scientists are under high stress to improve water filtration systems. Nanofibres and nanobiocides may be the solution to ensure safe and easy access drinking water.

Acknowledgement

Reprinted with permission from *Critical Reviews in Microbiology*. M. Botes and T.E. Cloete (2010). The potential of nanofibers and nanobiocides in water purification. Crit. Rev. Microbiol. *36*, 68–81.

Web resources

- www.argonide.com accessed 10/03/2009.
- www.elmarco.com/accessed 24/02/2009.
- www.krunk.ee/eng/2/21.shtml accessed 13/03/2009.
- www.nanomet.ru/en/appl.3.php accessed 10/03/2009.
- www.purifics.com/technology/ceramic.html accessed 16/03/2009.

References

Ahmad, A., Mukherjee, P., Senapati, S., Mandal, D., Khan, M.I., Kumar, R., and Sastry, M. (2003a). Extracellular biosynthesis of silver nanoparticles using the fungus *Fusarium oxysporum*. Colloids Surf. B *28*, 313–318.

Ahmad, A., Senapati, S., Khan, M.I., Kumar, R., and Sastry, M. (2003b). Extracellular biosynthesis of monodispersed gold nanoparticles by a novel extermophillic actinomycete, *Thermonospora* sp. Langmuir *19*, 3550–3553.

Alrousan, D.M.A., Dunlop, P.S.M., McMurray, T.A., and Byrne, J.A. (2009). Photocatalytic inactivation of *E. coli* in surface water using immobilised nanoparticle TiO2 films. Water Res. *43*, 47–54.

Anyaogu, K.C., Fedorov, A.V., and Neckers, D.C. (2008). Synthesis, characterization, and antifouling potential of functionalized copper nanoparticles. Langmuir *24*, 4340–4346.

Bai, J., Li, Y., Zhang, C., Liang, X., and Yang, Q. (2008). Preparing AgBr nanoparticles in poly(vinyl pyrrolidone) (PVP) nanofibers. Colloids Surf. A *329*, 165–168.

Balasubramanian, K., and Burghard, M. (2006). Biosensors based on carbon nanotubes. Anal. Bioanal. Chem. *385*, 452–468.

Balogh, L., Swanson, D.R., Tomalia, D.A., Hagnauer, G.L., and McManus, A.T. (2001). Dendrimer silver complexes and nanocomposites as antimicrobial agents. Nano. Lett. *1*, 18–21.

Barhate, R.S., and Ramakrishna, S. (2007). Nanofibrous filtering media: filtration problems and solutions from tiny materials. J. Membr. Sci. *296*, 1–8.

Belhacova, L., Krysa, J., Geryk, J., and Jirkovsky, J. (1999). Inactivation of microorganisms in a flow-through photoreactor with an immobilized TiO_2 layer. J. Chemical Technol. Biotechnol. *74*, 149–154.

Bhainsa, K.C., and D'Souza, S.F. (2006). Extracellular biosynthesis of silver nanoparticles using the fungus *Aspergillus fumigatus*. Colloids Surf. B *47*, 160–164.

Bognitzki, M., Hou, H., Ishaque, M., Frese, D., Hellwig, M., and Schwarte, C. (2000). Polymer, metal, and hybrid nano- and mesotubes by coating degradable polymer template fibers (TUFT process). Adv. Mater. *12*, 637–640.

Bognitzki, M., Czado, W., Frese, T., Schaper, A., Hellwig, M., and Steinhart, M. (2001). Nanostructured fibers via electrospinning. Adv. Mater. *13*, 70–72.

Brady-Estévez, A.S., Kang, S., and Elimelech, M. (2008). A single-walled-carbon-nanotube filter for removal of viral and bacterial pathogens. Small *4*, 481–484.

Brogden, K.A. (2005). Antimicrobial peptides: pore formers or metabolic inhibitors in bacteria? Nat. Rev. Microbiol. *3*, 238–250.

Chae, S.-R., Wang, S., Hendren, Z.D., Wiesner, M.R., Watanabe, Y., and Gunsch, C.K. (2009). Effects of fullerene nanoparticles on *Escherichia coli* K12 respiratory activity in aqueous suspension and potential use for membrane biofouling control. J. Membr. Sci. *329*, 68–74.

Chandorilar, M.V., and Bhavsar, P.C. (1983). Evaluation of polyacrylonitrile and poly (methyl methacrylate) as membrane materials for reverse osmosis. Indian J. Technol. *2*, 124.

Chen,Y., Wang, L., Jiang, S., and Yu, H.J. (2003). Study on novel antibacterial polymer materials preparation of zeolite antibacterial agents and antibacterial polymer composite and their antibacterial properties. J. Polym. Mater. *20*, 279–284.

Chen, H., and Hsieh, Y.L. (2004). Ultrafine hydrogel fibers with dual temperature- and pH-responsive swelling behaviors. J. Pol. Sci., A Polym. Chem. *42*, 6331–6339.

Chen, C.-Y., and Chiang, C.-L. (2008). Preparation of cotton fibers with antibacterial silver nanoparticles. Mater. Lett. *62*, 3607–3609.

Chen, S.P., Wu, G.Z., and Zeng, H.Y. (2005). Preparation of high antimicrobial activity thiourea chitosan–Ag^+ complex. Carbohydr. Polym. *60*, 33–38.

Chou, W.L., Yu, D.G., and Yang, M.C. (2005). The preparation and characterization of silver-loading cellulose acetate hollow fiber membrane for water treatment. Polymer. Adv. Tech. *16*, 600–607.

Chu, B., Hsiao, B.S., Hadjiargyrou, M., Fang, D., Zong, X., and Kim, K. (2003). Application: US. New York: The Research Foundation at State University of New York.

Cooley, J.F. (1902). Apparatus for electrically dispensing fluids. US Patent No. *692,631*.

da Silva Paula, M.M., Franco, C.V., Baldin, M.C., Rodrigues, L., Barichello, T., Savi, G.D., Bellato, L.F., Fiori, M.A., and da Silva, L. (2009). Synthesis, characterization and antibacterial activity studies of poly-{styrene-acrylic acid} with silver nanoparticles. Mater. Sci. Eng. C *29*, 647–650.

Deguchi, S., Alargova, R.G., and Tsujji, K. (2001). Stable dispersions of fullerenes, C_{60} and C_{70} in water. Preparation and characterization. Langmuir *17*, 6013–6017.

Deguchi, S., Mukai, S., Tsudome, M., and Horikoshi, K. (2006). Facile generation of fullerene nanoparticles by hand-grinding. Adv. Mater. *18*, 729–732.

Ding, B., Kimura, E., Sato, T., Fujita, S., and Shiratori, S. (2004). Fabrication of blend biodegradable nanofibrous nonwoven mats via multi-jet electrospinning. Polymer *45*, 1895–1902.

Ding, B., Kim, J., Miyazaki, Y., and Shiratori, S. (2004). Electrospun nanofibrous membranes coated quartz crystal microbalance as gas sensor for NH3 detection. Sens. Actuators B B*101*, 373–380.

Ding, B., Kim, J., Fujimoto, K., and Shiratori, S. (2004). Electrospun nanofibrous polyelectrolyte membranes for advanced chemical sensors. Chem. Sensors *20*, 264–265.

Don, T.M., Chen, C.C., Lee, C.K., Cheng, W.Y., and Cheng, L.P. (2005). Preparation and antibacterial test of chitosan/PAA/PEGDA bilayer composite membranes. J. Biomater. Sci. Polym. Ed. *16*, 1503–1519.

Du, W.-L., Niu, S.-S., Xu, Y.-L., Xu, Z.-R., and Fan, C.-L. (2009). Antibacterial activity of chitosan tripolyphosphate nanoparticles loaded with various metal ions. Carbohydr. Polym. *75*, 385–389.

Eby, D.M., Farrington, K.E., and Johnson, G.R. (2008). Synthesis of bioinorganic antimicrobial peptide nanoparticles with potential therapeutic properties. Biomacromolecules *9*, 2487–2494.

Etienne, O., Picart, C., Taddei, C., Haikel, Y., Dimarcq, J.L., Schaaf, P., Voegel, J.C., Ogier, J.A., and e.g.les, C. (2004). Multilayer Polyelectrolyte Films Functionalized by Insertion of Defensin: a New Approach to Protection of Implants from Bacterial Colonization. Antimicrob. Agents Chemother. *48*, 3662–3669.

Feng, Q.L., Wu, J., Chen, G.Q., Cui, F.Z., Kim, T.N., and Kim, J.O. (2000). A mechanistic study of the antibacterial effect of silver ions on *Escherichia coli* and *Staphylococcus aureus*. J. Biomed. Mater. Res. *52*, 662–668.

Formhals, A. (1934). Process and apparatus for preparing artificial threads, US Patent No. *1,975,504*.

Franklin, N.M., Rogers, N.J., Apte, S.C., Batley, G.E., Gadd, G.E., and Casey, P.S. (2007). Comparative toxicity of nanoparticulate ZnO, bulk ZnO, and ZnCl2 to a freshwater microalga (*Pseudokirchneriella subcapitata*): the importance of particle solubility. Environ Sci. Technol. *41*, 8484–8490.

Frenot, A., and Chronakis, I.S. (2003). Polymer nanofibers assembled by electrospinning. Curr. Opin. Colloid Interface Sci. *8*, 64–75.

Fu, G.D., Yao, F., Li, Z.G., and Li, X.S. (2008). Solvent-resistant antibacterial microfibers of self-quaternized block copolymers from atom transfer radical polymerization and electrospinning. J. Mater. Chem. *18*, 859.

Fuchs, A.D., and Tiller, J.C. (2006). Contact-active antimicrobial coatings derived from aqueous suspensions. Angew. Chem. Int. Edn. *45*, 6759–6762.

Gao, G. (2004). Nanostructures and nanomaterials. Synthesis, Properties and Applications. Imperial College Press, London.

Gaysinsky, S., Davidson, P.M., Bruce, B.D., and Weiss, J. (2005a). Growth inhibition of *Escherichia coli* O157:H7 and Listeria monocytogenes by carvacrol and eugenol encapsulated in surfactant micelles. J. Food Prot. *68*, 2559–2566.

Gaysinksy, S., Davidson, P.M., Bruce, B.D., and Weiss, J. (2005b). Stability and antimicrobial efficiency of eugenol encapsulated in surfactant micelles as affected by temperature and pH. J. Food Prot. *68*, 359–1366.

Gaysinsky, S., Davidson, P.M., McClements, D.J., and Weiss, J. (2008). Formulation and characterization of phytophenol-carrying antimicrobial microemulsions. Food Biophys. *3*, 54–65.

Gazit, E. (2007). Self-assembled peptide nanostructures: the design of molecular building blocks and their technological utilization. Chem. Soc. Rev. *36*, 1263–1269.

Gibson, P., Schreuder-Gibson, H., and Rivin, D. (2001). Transport properties of porous membranes based on electrospun nanofibers. Coll. Surf. A 187*188*, 469–481.

Goodwin, J. (2004). Colloids and Interfaces with Surfactants and Polymers – An Introduction; Wiley Interscience: West Sussex, England.

Gopal, R., Kaur, S., Ma, Z., Chan, C., Ramakrishna, S., and Matsuura, T. (2006). Electrospun nanofibrous filtration membrane. J. Membr. Sci. *281*, 581–586.

Grafe, T., and Graham, K. (2003). Polymeric nanofibers and nanofiber webs: a new class of nonwovens. Nonwoven Technol Rev INJ Spring 51–55.

Graveland, B.J.F., and Kruif, C.G. (2006). Unique milk protein based nanotubes. Food Nanotech. *17*, 196–203.

Gregory, K., and Mello, C.M. (2005). Immobilization of Escherichia coli Cells by Use of the Antimicrobial Peptide Cecropin P1. Appl. Environ. Microbiol. *71*, 1130–1134.

Gudmund, S.B., Thorleif, A., and Paul, S. (1989). Chitin and Chitosan: sources, chemistry, biochemistry, physical properties and applications. New York: Elsevier.

Gupta, P., Trenor, S.R., Long, T.E., and Wilkes, G.L. (2004). In situ photo-cross-linking of cinnamate functionalized poly(methyl methacrylate-co-2-hydroxyethyl acrylate) fibers during electrospinning. Macromolecules *37*, 9211–9218.

Hajkova, P., Spatenka, P., Horsky, J., Horska, I., and Kolouch, A. (2007). Photocatalytic effect of TiO2 films on viruses and bacteria. Plasma Process. Polym. *4*, S397-S401.

Hajra, M.G., Mehta, K., and Chase, G.G. (2003). Effects of humidity, temperature, nanofibers on drop coalescence in glass fiber media. Sep. Purif. Technol. *30*, 79–88.

Han, S.O., Son, Cho, W.K., D., Youk, J.H., and Park, W.H. (2004). Preparation of porous ultra-fine fibers via selective thermal degradation of electro spun polyether imide/poly(3-hydroxybutyrate-co-3-hydroxy valerate) fibers. Polym. Degrad. Stab. *86*, 257–262.

Han, S.O., Youk, J.H., Min, K.D., Kang, Y.O., and Park, W.H. (2008). Electrospinning of cellulose acetate nanofibers using a mixed solvent of acetic acid/water: Effects of solvent composition on the fiber diameter. Mater. Lett. *62*, 759–762.

Hatakeyama, T., Izuta, Y., Hirose, S., and Hatakeyama, H. (2002). Phase transitions of lignin-based polycaprolactones and their polyurethane derivatives. Polymer *43*, 1177.

Heijkants, G.J.C., Calck, R.V., Tienen, T.G., Groot, J.H., Buma, P., Pennings, A.J., Veth, P.H., and Schouten, A.J. (2005). Biomaterials *26*, 4219.

Huang, Z.M., Zhang, Y.Z., Kotaki, M., and Ramakrishna, S. (2003). A review on polymer nanofibers by electrospinning and their application in nanocomposites. Compos. Sci.Technol. *63*, 2223–2253.

Huang, Z., Zheng, X., Yan, D., Yin, G., Liao, X., Kang, Y., Yao, Y., Huang, D., and Hao, B. (2008). Toxiciological effect of ZnO nanoparticles based on bacteria. Langmuir *24*, 4140–4144.

Jain, P., and Pradeep, T. (2005). Potential of Silver Nanoparticle-Coated Polyurethane foam as an antibacterial water filter. Biotechnol. Bioeng. *90*, 59–63.

Jeon, Kim, H.J., J.S., Kim, Kim, T.G., J.H., Yu,W.-R., and Youk, J.H. (2008). Preparation of poly(-caprolactone)-based polyurethane nanofibers containing silver nanoparticles. Appl. Surface Sci. *254*, 5886–5890.

Jeong, E.H., Yang, J., and Youk, J.H. (2007). Preparation of polyurethane cationomer nanofiber mats for use in antimicrobial nanofilter applications. Mater. Lett. *61*, 3991–3994.

Jia, Z.S., Shen, D.F., and Xu, W.L. (2001). Synthesis and antibacterial activities of quaternary ammonium salt of chitosan. Carbohydr. Res. *333*, 1–6.

Jones, N., Ray, B., Ranjit, K.T., and Manna, A.C. (2008). Antibacterial activity of ZnO nanoparticle suspensions on a broad spectrum of microorganisms. FEMS Microbiol. Lett. *279*, 71–76.

Ju-Nam, Y., and Lead, J.R., (2008). Manufactured nanoparticles: An overview of their chemistry, interactions and potential environmental implications. Sci. Tot. Environ. *400*, 396–414.

Kang, S., Pinault, M., Pfefferle, L.D., and Elimelech, M. (2007). Single-walled carbon nanotubes exhibit strong antimicrobial activity. Langmuir *23*, 8670–8673.

Klaus-Jeorger, T., Jeorger, R., Olsson, E., and Granqvist, C. (2001). Bacteria as workers in the living factory: metal accumulating bacteria and their potential for material science. Trends Biotechnol. *19*, 15–20.

Khil, M.-S., Bhattarai, S.R., Kim, H.-Y., Kim, S.-Z., and Lee, K.-H. (2005). Novel fabricated matrix via electrospinning for tissue engineering. J. Biomed. Mater. Res. 72B, 117–124.

Kikuchi, Y., Sunada, K., Iyoda, T., Hashimoto, K., and Fujishima, A. (1997). Photocatalytic bacteridal effect of TiO_2 thin films: dynamic view of the active oxygen species responsible for the effect. J. Photochem. Photobiol. A. Chem. *106*, 51–56.

Kim, C., and Yang, K.S. (2003). Electrochemical properties of carbon nanofiber web as an electrode for supercapacitor prepared by electrospinning. Appl. Phys. Lett. *83*, 1216–1218.

Kim, C., Choi, Y.O., Lee, W.J., and Yang, K.S. (2004). Supercapacitor performances of activated carbon fiber webs prepared by electrospinning of PMDA-ODA poly(amic acid) solutions. Electrochimica Acta *50*, 883–887.

Kim, C., Park, S.H., Lee, W.J., and Yang, K.S. (2004). Characteristics of supercapacitor electrodes of PBI-based carbon nanofiber web prepared by electrospinning. Electrochimica Acta *50*, 877–881.

Kim, L., and Klibanov, A.M. (2005). Surpassing nature: rational design of sterile-surface materials. Trends Biotechnol. *23*, 343.

Kim, S.J., Nam, Y.S., Rhee, D.M., Park, H.-S., and Park, W.H. (2007). Macromolecular Nanotechnology: Preparation and characterization of antimicrobial polycarbonate nanofibrous membrane. Eur. Polym. J. *43*, 3146–3152.

Kriegel, C., Arecchi, A., Kit, K., McClements, J., and Weiss, J. (2008). Fabrication, functionalization, and application of electrospun biopolymer nanofibers. Crit.Rev. Food Sci. Nutr. *48*, 775–797.

Kubacka, A., Ferrer, M., Cerrada, M.L., Serrano, C., Sánchez-Chaves, M., Fernández-García, M., de Andrés, A., Riobóo, R.J.J., Fernández-Martín, F., and Fernández-García, M. (2009). Boosting TiO2-anatase antimicrobial activity: Polymer-oxide thin films. Appl. Catal. B *89*, 441–447.

Kwak, S.Y., Kim, S.H., and Kim, S.S. (2001). Hybrid Organic/inorganic Reverse Osmosis (RO) membrane for bactericidal anti-fouling. Preparation and characterization of TiO2 nanoparticle self-assembled aromatic polyamide Thin-Film-Composite (TFC) membrane. Environ. Sci. Technol. *35*, 2388–2394.

Lala, N.L., Ramakrishnan, R., Bojun, L., Sundarrajan, S., Barhate, R.S., Ying-jun, L., and Ramakrishna, S. (2007). Fabrication of nanofibers with antimicrobial functionality used as filter: Protection against bacterial contaminants. Biotechnol. Bioeng. *97*, 1357–1365.

Lange, K.R. (1999). Surfactants a practical handbook; Hanser Publishers: Munich.

Lee, H.S., Im, S.J., Kim, J.H., and Jin, H. (2008). Polyamide thin-film nanofiltration membranes containing TiO_2 nanoparticles. Desalination *219*, 48–56.

Li, D., and Xia, Y.N. (2004). Electrospinning of nanofibers: reinventing the wheel? Adv. Mater. *16*, 1151–1170.

Li, Z.Y., Huang, H.M., and Wang, C. (2006). Electrostatic Forces Induce Poly(vinyl alcohol)-Protected Copper Nanoparticles to Form Copper/Poly(vinyl alcohol) Nanocables via Electrospinning. Macromol. Rapid Commun. *27*, 152.

Li, Q.L., Mahendra, S., Lyon, D.Y., Brunet, L., Liga, M.V., Li, D., and Alvarez, P.J.J. (2008). Antimicrobial nanomaterials for water disinfection and microbial control: Potential applications and implications. Water Res. *42*, 4591–4602.

Liang, D., Hsiao, B.J., and Chu, B. (2007). Functional electrospun nanofibrous scaffolds for biomedical applications. Adv. Drug Deliv. Rev. *59*, 1392–1412.

Lin, T., Wang, H.X., Wang, H.M., and Wang, X.G. (2004). The charge effect of cationic surfactants on the elimination of fibre beads in the electrospinning of polystyrene. Nanotechnology *15*, 1375–1381.

Liu, H., Du, Y.M., Yang, J.H., and Zhu, H.Y. (2004). Structural characterization and antimicrobial activity of chitosan/betaine derivative complex. Carbohydr. Polym. *55*, 291–297.

Liu, H.Q., Kameoka, J., Czaplewski, D.A., and Craighead, H.G. (2004). Polymeric nanowire chemical sensor. Nano Lett. *4*, 671–675.

Lushnikov, A. (1997). Obituaries:Igor Vasilievich Petryanov-Sokolov (1907–1996). J. Aerosol Sci. *28*, 545–546.

Lv, Y., Liu, H., Wang, Z., Liu, S., Hao, L., Sang, Y., Liu, D., Wang, J., and Boughton, R.I. (2009). Silver nanoparticle-decorated porous ceramic composite for water treatment. J. Membr. Sci. *331*, 50–56.

Lyon, D.Y., Brown, D.A., and Alvarez, P.J.J. (2008). Implications and potential applications of bactericidal fullerene water suspensions: effect of nC60 concentration, exposure condition and shelf life. Water Sci. Technol. *57*, 1533–1538.

Ma, Z., Kotaki, M., Inai, R., and Ramakrishna, S. (2005). Potential of nanofiber matrix as tissue-engineering scaffolds. Tissue Eng. *11*, 101–109.

Morones, J.R., Elechiguerra, J.L., Camacho, A., Holt, K., Kouri, J.B., Ramírez, J.T., and Yacaman, M.J. (2005). The bactericidal effect of silver nanoparticles. Nanotechnol. *16*, 2346–2353.

Morozov, V.N., Morozova, T.Y., and Kallenbach, N.R. (1998). Atomic force microscopy of structures produced by electrospraying polymer solutions. Int. J. Mass Spectrom. *178*, 143–159.

Mostafavia, S.T., Mehrniaa, M.R., and Rashidi, A.M. (2009). Preparation of nanofilter from carbon nanotubes for application in virus removal from water. Desalination *238*, 271–280.

Müller, R.H., Jacobs, C., and Kayser, O. (2001). Nanosuspensions as particulate drug formulations in therapy rationale for development and what we can expect for the future. Adv. Drug Deliv. Rev. *47*, 3–19.

Musale, D.A., Kumar, A., and Pleizier, G. (1999). Formation and characterization of poly(acrylonitrile)/Chitosan composite ultrafiltration membranes. J. Membr. Sci. *154*, 163.

Musale, D.A., and Kumar, A. 2000. Solvent and pH resistance of surface crosslinked chitosan/poly(acrylonitrile) composite nanofiltration membranes. J. Appl. Polym. Sci. *77*, 1782.

Niemeyer, C.M. (2001). Nanoparticles, proteins, and nucleic acids: biotechnology meets materials science. Angew. Chem. Int. Ed. *40*, 4128–4158.

No, H.K., Park, N.Y., Lee, S.H., and Meyers, S.P. (2002). Antibacterial activity of chitosans and chitosan oligomers with different molecular weights. Int. J. Food Microbiol. *74*, 65–72.

Padilla, A.P., Rodríguez, J.A., and Saitúa, H.A. (1997). Synthesis and water ultrafiltration properties of silver membrane supported on porous ceramics. Desalination *114*, 203.

Park, K.D. (1998). Bacterial adhesion on PEG modified polyurethane surfaces. Biomater. *19*, 851–859.

Ping, P., Wang, Chen, W., X., and Jing, X. (2005). Poly(ε-caprolactone) Polyurethane and Its Shape-Memory Property. Biomacromolecules. *6*, 587.

Qi, L., Xu, Z., Jiang, X., Hu, C., and Zou, X. (2004). Preparation and antibacterial activity of chitosan nanoparticles, Carbohydr. Res. *339*, 2693–2700.

Rabea, E.I., Badawy, M.E., Stevens, C.V., Smagghe, G., and Steurbaut, W. (2003). Chitosan as antimicrobial agent: applications and mode of action. Biomacromolecules *4*, 1457–1465.

Ramakrishana, S., Fujihara, K., Teo, W.-E., Lim, T.-C., and Zuwei, M. (2005). An Introduction to Electrospinning and Nanofibers, World Scientific Publishing Co. Pte. Ltd.

Ren, G., Hu, D., Cheng, E.W.C., Vargas-Reus, M.A., Reip, P., and Allaker, R.P. (2009). Characterization of copper oxide nanoparticles for antimicrobial applications. Int. J. Antimicrob. Agents 33, 587–590.

Reneker, D.H., Yarin, A.L., Fong, H., and Koombhongse, S. (2000). Bending instability of electrically charged liquid jets of polymer solutions in electrospinning. J. Appl. Phys. *87*, 4531–4547.

Reneker, D.H., and Chun, I. (1996). Nanometer diameter fibers of polymer, produced by electrospinning. Nanotechnol. *7*, 216–223.

Riboldi, S.A., Sampaolesi, M., Neuenschwander, P., Cossu, G., and Mantero, S. (2005). Electrospun degradable polyesterurethane membranes: potential scaffolds for skeletal muscle tissue engineering. Biomater. *26*, 4606–4615.

Rotello, V.M. (2003). Nanoparticles: Building Blocks for Nanotechnology, 1st ed. Springer, New York.

Rule, K.L., Ebbett, V.R., and Vikesland, P.J. (2005). Formation of chloroform and chlorinated organics by free-chlorine-mediated oxidation of Triclosan. Environ. Sci. Technol. *39*, 3176–3185.

Salmaso, S., Elvassore, N., Bertucco, A., Lante, A., and Caliceti, P. (2004). Nisin-loaded poly-L-lactide nano-particles by CO_2 anti-solvent precipitation for sustained antimicrobial activity. Int. J. Pharm. *287*, 163–173.

Sawai, J. (2003). Quantitive evaluation of antibacterial activities of metallic oxide powders (ZnO, MgO and CaO) by conductimetric assay. J. Microbiol. Methods. *54*, 177–182.

Sastry, M., Ahmad, A., Khan, M.I., and Kumar, R. (2003). Biosynthesis of metal nanoparticles using fungi and actinomycete. Curr. Sci. *85*, 162–170.

Schreuder-Gibson, H., Gibson, P., Wadsworth, L., Hemphill, S., and Vontorcik, J. (2002). Effect of filter deformation on the filtration and air flow for elastomeric nonwoven media. Adv. Filtr. Sep. Technol. *15*, 525–537.

Schreuder-Gibson, H.L., Gibson, P., Tsai, Gupta, P., P., and Wilkes, G. (2004). Cooperative charging effects of fibers from electrospinning of electrically dissimilar polymers. Int. Nonwovens J. *13*, 39–45.

Shang, T., Yang, F., Zheng, W., and Wang, C. (2006). Fabrication of electrically bistable nanofibers. Small 2, 1007–1009.

Shepelev, A.D., and Rykunov, V.A. (1995). Polymeric fiber materials for fine cleaning of gases. J. Aerosol Sci. *26*, S919–S920.

Shutov, A.A., and Astakhov, E.Y. (2006). Formation of fibrous filtering membranes by electrospinning. Techn. Phys. *51*, 1093–1096.

Sohlenius, B. (1980). Abundance, biomass and contribution to energy flow by soil nematodes in terrestrial ecosystems. Oikos *34*, 186–194.

Son, W.K., Youk, J.H., Lee, T.S., and Park, W.H. (2004). Preparation of Antimicrobial Ultrafine Cellulose Acetate Fibers with Silver Nanoparticles. Macromol. Rapid Commun. *25*, 1632.
Srivastava, A., Srivastava, O.N., Talapatra, S., Vajtai, R., and Ajayan, P.M. (2004). Carbon Nanotube Filters. Nat. Mater. *3*, 610–614.
Stauffer, D., Jan, N., He, Y., Pandey, R.B., Marangoni, D.G., and Smithpalmer, T. (1994). Micelle formation, relaxation time, and three-phase coexistence in a microemulsion model. J. Chem. Sci. *100*, 6934–6943.
Stoimenov P.K., Klinger, R.L., Marchin, G.L., and Klabunde, K.J. (2002). Metal oxide nanoparticles as bactericidal agents. Langmuir *18*, 6679–6686.
Sunayama, S., Sato, T., Kawamoto, A., Ohkubo, A., and Suzuki, T. (2002). Disinfection effect of hydrotalcite compounds containing antimicrobial metals against micro-organisms in water. Biocontrol. Sci. *7*, 75–81.
Tan, K., and Obendorf, S.K. (2007). Fabrication and evaluation of electrospun nanofibrous antimicrobial nylon 6 membranes. J. Membr. Sci. *305*, 287–298.
Tepper, F., and Kaledin, L. (2005). Nanosize electropositive fibrous adsorbent. US Patent *6,838,005*.
Theron, J., Walker, J.A., and Cloete, T.E. (2008). Nanotechnology and water treatment: Applications and emerging opportunities. Crit. Rev. Microbiol. *34*, 43–69.
Torres, B. (2001). Ultrafine fibers of polystyrene dissolved in tetrahydrofuran prepared using the electrospinning method. In: Proceeding of The National Conference On Undergraduate Research. *15–17*, 1–5.
Tsaia, P.P., Schreuder-Gibson, H., and Gibson, P. (2002). Different electrostatic methods for making electret filters. J. Electrostat. *54*, 333–341.
Um, I.C., Fang, D.F., Hsiao, B.S., Okamoto, A., and Chu, B. (2004). Electro-spinning and electro-blowing of hyaluronic acid. Biomacromolecules *5*, 1428–1436.
Vandamme, T.F. (2002). Microemulsions as ocular drug delivery systems: recent developments and future challenges. Progress in Retinal Eye Res. *21*, 15–34.
Vicentini, D.S., Smania Jr., A., and Laranjeira, M.C.M. (2009). Chitosan/poly (vinyl alcohol) films containing ZnO nanoparticles and plasticizers. Mat. Sci. Eng. C. *30*, 503–508.
Vimala, K., Sivudu, K.S., Mohan, Y.M., Sreedhar, B.K., and Raju, M. (2009). Controlled silver nanoparticles synthesis in semi-hydrogel networks of poly(acrylamide) and carbohydrates: A rational methodology for antibacterial application. Carb. Polym. *75*, 463–471.
Wang, X., and Spencer, H.G. (1998). Formation and characterization of chitosan formed-in-place ultrafiltration membranes. J. Appl. Polym Sci. *67*, 513.
Wang, X.Y., Drew, C., Lee, S.H., Senecal, K.J., Kumar, J., and Sarnuelson, L.A. (2002). Electrospun nanofibrous membranes for highly sensitive optical sensors. Nano Lett. *2*, 1273–1275.
Wang, Y., Serrano, S., and Santiago-Aviles, J.J. (2003). Raman characterization of carbon nanofibers prepared using electrospinning. Synth. Metals *138*, 423.
Wang, X.H., Du, Y.M., and Liu, H. (2004). Preparation, characterization and antimicrobial activity of chitosan–Zn complex. Carbohydr. Polym. *56*, 21–26.
Wang, Y., and Hsieh, Y.-L. (2004). Enzyme immobilization to ultra-fine cellulose fibers via amphiphilic polyethylene glycol spacers. J. Polym. Sci. A *42*, 4289–4299.
Wang, X.Y., Kim, Y.G., Drew, Ku, C., B.C., Kumar, J., and Samuelson, L.A. (2004). Electrostatic assembly of conjugated polymer thin layers on electrospun nanofibrous membranes for biosensors. Nano Lett. *4*, 331–334.
Wang, X., Chen, X., Yoon, K., Fang, D., Hsiao, B.S., and Chu, B. (2005). High flux filtration medium based on nanofibrous substrate with hydrophilic nanocomposite coating. Environ. Sci. Technol. *39*, 7684–7691.
Wannatong, L., and Sirivat, A. (2004). Electrospun fibers of polypyrrole/polystyrene blend for gas sensing applications. PMSE Preprints *91*, 692–693.
Watts, R.J., Kong, S., Orr, M.P., Miller, G.C., and Henry, B.E. (1995). Photocatalyst inactivation of coliform bacteria and viruses insecondary wastewater effluent. Water Res. *29*, 95–100.
Wei, C., Lin, W.Y., Zainal, Z., Williams, N.E., Zhu, K., Kruzic, A.P., Smith, R.L., and Rajeshwar, K. (1994). Bactericidal activity of TiO2 photocatalyst in aqueous media: toward a solar-assisted water disinfected system. Environ. Sci. Technol. *28*, 934–938.
Weiss, J., Coupland, J.N., and McClements, D.J. (1996). Solubilization of Hydrocarbon Emulsion Droplets Suspended in Nonionic Surfactant Micelle Solutions. J. Phys. Chem. A. *100*, 1066–1071.
Weiss, J., Coupland, J.N., Brathwaite, D., and McClements, D.J. (1997). Influence of molecular structure of hydrocarbon emulsion droplets on their solubilization in nonionic surfactant micelles. Colloids Surf. A *121*, 53–60.
Weiss, J., Herrmann, N., and McClements, D.J. (1999). Ostwald Ripening of Hydrocarbon Emulsion Droplets in Surfactant Solutions. Langmuir *15*, 6652–6657.
Weiss, J., and McClements, D.J. (2000). Mass Transport Phenomena in Oil-in-Water Emulsions Containing Surfactant Micelles: Solubilization. Langmuir *16*, 5879–5883.
Weiss, J., Cancelliere, C., and McClements, D.J. (2000). Mass Transport Phenomena in Oil-in-Water Emulsions Containing Surfactant Micelles: Ostwald Ripening. Langmuir *16*, 6833- 6838.
Weiss, J., McClements, J., and Takhistov, P. (2007). Growth inhibition of *Escherichia coli* O157:H7 and *Listeria monocytogenes* by carvacrol and eugenol encapsulated in surfactant micelles. Food Australia *59*, 274–275.

Wiesner, M.R., Lowry, G.V., Alverez, P., Dioysiou, D., and Biswas, P. (2006). Assessing the risks of manufactured nanomaterials. A-pages 40. Environ. Sci. Technol. 4336–4337.

Woo, G.L.Y., Mittelman, M.W., and Santerre, J.P. (2000). Synthesis and characterization of a novel biodegradable antimicrobial polymer. Biomaterials *21*, 1235.

WHO, 2004. (cited 2009 March 24); Available from: <http://www.who.int/water_sanitation_health/en/.

Xing, K., Chen, X.G., Kong, M., Liu, C.S., Cha, D.S., and Park, H.J. (2009). Effect of oleoyl-chitosan nanoparticles as a novel antibacterial dispersion system on viability, membrane permeability and cell morphology of *Escherichia coli* and *Staphylococcus aureus*. Carbohydr. Polym. *76*, 17–22.

Xu, H., Yarin, A.L., and Reneker, D.H. (2003). Characterization of fluid flow in jets during electrospinning. Abstracts of Papers of the American Chemical Society *226*, U424.

Yang, Q.B., Li, Z.Y., Hong, Y.L., Zhao, Y.Y., Qiu, S.L.C., Wang, C., and Wei, Y. (2004). Influence of solvents on the formation of ultrathin uniform poly(vinyl pyrrolidone) nanofibers with electrospinning. J. Polym. Sci. B *42*, 3721–3726.

Yang, T.C., Chou, C.C., and Li, C.F. (2005). Antibacterial activity of N-alkylated disaccharide chitosan derivatives. Int. J. Food Microbiol. *97*, 237–245.

Yang, F., Murugan, R., Wang, S., and Ramakrishna, S. (2005). Electrospinning of nano/micro scale poly(l-lactic acid) aligned fibers and their potential in neural tissue engineering. Biomater. *26*, 2603–2610.

Yao, L., Haas, T.W., Guiseppi-Elie, A., Bowlin, G.L., Simpson, D.G., and Wnek, G.E. (2003). Electrospinning and stabilization of fully hydrolyzed poly(vinyl alcohol) fibers. Chem. Mater. *15*, 1860–1864.

Yao, C., Li, X., Neoh, K.G., Shi, Z., and Kang, E.T. (2008). Surface modification and antibacterial activity of electrospun polyurethane fibrous membranes with quaternary ammonium moieties. J. Membr. Sci. *320*, 259–267.

Yao, C., Li, X., Neoh, K.G., Shi, Z., and Kang, E.T. (2009). Antimicrobial activities of surface modified electrospun poly(vinylidene fluoride-co-hexafluoropropylene) (PVDF-HFP) fibrous membranes. Appl. Surf. Sci. *255*, 3854–3858.

Ye, W.J., Xin, J.H., Li, P., Lee, K.L.D., and Kwong, T.L. (2006). Durable antibacterial finish on cotton fabric by using chitosan-based polymeric core-shell particles. J. Appl. Polym. Sci. *102*, 1787–1793.

Yi, Y., Wang, Y.T., and Liu, H. (2003). Preparation of new crosslinked chitosan with crown ether and their adsorption for silver ion for antibacterial activities. Carbohydr. Polym. *53*, 425–430.

Yoon, K., Kim, K., Wang, X., Fang, D., Hsiao, B.S., and Chu, B. (2006). High flux ultrafiltration membranes based on electrospun nanofibrous PAN scaffolds and chitosan coating. Polymer *47*, 2434–2441.

Yoon, K., Hsiao, B.S., and Chu, B. (2009). Formation of functional polyethersulfone electrospun membrane for water purification by mixed solvent and oxidation processes. Polymer *50*, 2893–2899.

Zan, L., Fa, W., Peng, T.P., and Gong, Z.K. (2007). Photocatalysis effect of nanometer TiO_2 and TiO_2-coated ceramic plate on Hepatitis B virus. J. Photochem. Photobiol. B. Biol. *86*, 165–169.

Zhang, C.X., Yuan, X.Y., Wu, L.L., Han, Y., and Sheng, J. (2005). Study on morphology of electrospun poly(vinyl alcohol) mats. Eur. Polym. J. *41*, 423–432.

Zodrow, K., Brunet, L., Mahendra, S., Li, D., Zhang, A., Li, Q., and Alvarez, P.J.J. (2009). Polysulfone ultrafiltration membranes impregnated with silver nanoparticles show improved biofouling resistance and virus removal. Water Res. *43*, 715–723.

Zuwei, M., Kotaki, M., and Ramakrishna, S. (2006). Surface modified nonwoven polysulphone (PSU) fiber mesh by electrospinning: a novel affinity membrane. J. Membr. Sci. *272*, 179–187.

Nanozymes for Biofilm Removal

4

Melanie Richards and T. Eugene Cloete

Abstract

Sessile communities of bacteria encased in extracellular polymeric substances (EPS) are known as biofilms and causes serious problems in various areas, amongst other, the medical industry, industrial water settings, paper industry and food processing industry. Although various methods of biofilm control exist, these methods are not without limitations and often fail to remove biofilms from surfaces. Biofilms often show reduced susceptibility to antimicrobials or chemicals and chemical by-products may be toxic to the environment, whereas mechanical methods may be labour intensive and expensive due to down-time required to clean the system. This has led to a great interest in the enzymatic degradation of biofilms. Enzymes are highly selective and disrupt the structural stability of the biofilm EPS matrix. Various studies have focused on the enzymatic degradation of polysaccharides and proteins for biofilm detachment since these are the two dominant components of the EPS. Due to the structural role of proteins and polysaccharides in the EPS matrix, a combination of various proteases and polysaccharases may be successful in biofilm removal. The biodegradability and low toxicity of enzymes also make them attractive biofilm control agents. Regardless of all the advantages associated with enzymes, they also suffer from various drawbacks given that they are relatively expensive, show insufficient stability or activity under certain conditions, and cannot be reused. Various approaches are being used to increase the stability of enzymes, including enzyme modification, enzyme immobilization, protein engineering and medium engineering. Although these conventional methods have been used frequently to improve the stability of enzymes, various new techniques, such as self-immobilization of enzymes, the immobilization of enzymes on nano-scale structures and the production of single-enzyme nanoparticles, have been developed. Self-immobilization of enzymes entails the cross-linking of enzyme molecules with each other and yields final preparations consisting of essentially pure proteins and high concentrations of enzyme per unit volume. The activity, stability and efficiency of immobilized enzymes can be improved by reducing the size of the enzyme-carrier. Nano-scale carrier materials allow for high enzyme loading per unit mass, catalytic recycling and a reduced loss of enzyme activity. Furthermore, enzymes can be stabilized by producing single-enzyme nanoparticles consisting of single-enzyme molecules surrounded by a porous organic–inorganic network of less than a few nanometres thick. All these new technologies of enzyme stabilization make enzymes even more attractive alternatives to other biofilm removal and control agents.

Introduction

A biofilm can be defined as a sessile community of microbes characterized by cells that are irreversibly associated with a surface, embedded in a matrix of extracellular polymeric substances (EPS), and display an altered phenotype with respect to gene expression, protein production and growth (Donlan and Costerton, 2002). Biofilms can form on a wide variety of surfaces and the thickness varies between a single-cell layer to a thick community of cells embedded in EPS (Costerton *et al.*, 1995). Materials like corrosion particles, clay, slit and mineral crystals, depending on the environment in which the biofilm has formed, may also be found in the biofilm matrix (Dolan, 2002).

Attachment and growth of microorganisms on man-made surfaces constitute a major problem in various industries. Examples of such problems include the corrosion of metal substrata, decreased heat transfer from heat exchangers, bioaccumulation of metals and toxic compounds, microbial infections (Hope and Bott, 2004), contamination of medical devices like catheters and surgical sutures (Meyer, 2003) and dental plaque formation (Xie *et al.*, 2000).

Biofilm control is generally aimed at either the removal of the extracellular polysaccharides of the biofilm or killing of its inhabiting microbes (Chen and Stewart, 2000). Mechanical cleaning, promotion of biomass detachment, removal of essential nutrients, the use of antimicrobial agents and inhibition of microbial attachment to surfaces, are all strategies employed for biofilm control (Stewart *et al.*, 2000). Although antimicrobial agents and mechanical methods such as high-pressure jetting, brushing and scrubbing (Cense *et al.*, 2006) are the most popular strategies used in the industry, it still has some disadvantages. Mechanical cleaning involves equipment down-time or significant labour expenditure and can therefore be costly. This strategy may also be ineffective due to the inaccessibility of the fouled surface (Stewart *et al.*, 2000). Biocides and disinfectants, in turn, will be insufficient when a clean surface in stead of one covered by inactive biofilm is required. Also, this strategy may be ineffective due to the reduced susceptibility of microorganisms in biofilms to antimicrobial agents (Stewart *et al.*, 2000; Thorman *et al.*, 2005; Xavier *et al.*, 2005). Some of these biocides and disinfectants have irritant properties and produce toxic by-products that can be harmful to the environment (Augustin and Ali-Vehmas, 2004; Manyak *et al.*, 2004).

There is a need to identify a biofilm control agent or method that does not rely solely on mechanical methods and antimicrobials (Manyak *et al.*, 2004). Substances capable of destroying the physical integrity of the EPS, interfering with the adhesion of bacterial cells or with the ability to initiate detachment of the cells from surfaces are good alternatives to biocides and disinfectants (Chen and Stewart, 2000).

Enzymatic control of biofilms

In general, 2–5% of the biofilm mass consists of cells, while the rest of the biofilm is made up of the EPS matrix. The EPS consists of exopolysaccharides, protein, glycoproteins, released nucleic acids, phospholipids and other surfactants. The EPS polysaccharides contain various monosaccharide units and inorganic substituents are therefore highly heterogeneous. Because the linkage types and substituents vary greatly, several bacterial species can produce exopolysaccharides of distinct chemical composition (Allison, 2003). The composition of the protein component of the biofilm is as variable as that of the polysaccharides (Lasa, 2006).

The use of enzymes may be a good alternative to other biofilm control agents. Due to the rapid biodegradability of enzymes they are expected to be harmless to the environment. Eco-toxicological tests on three different commercial enzymes concluded that they were environmentally friendly (Allermann *et al.*, 2004). The composition of the EPS is very important to understand enzyme-based control of biofilms as not only the structure but also the ratios between protein, polysaccharides and lipids can vary greatly between biofilms (Jahn *et al.*, 1999). Owing to the heterogeneity of the biofilm EPS, the use of monocompound enzymes are limited and mixture of enzymes may be required for the sufficient degradation and removal of biofilms (Johansen *et al.*, 1997). These biofilm-degrading enzyme mixtures can be employed in industrial and therapeutic settings to degrade and/or remove biofilms from the target surfaces causing a reduction of the biofilm and resulting in increased efficiency and improved hygiene (Manyak *et al.*, 2004).

Lipases hydrolyse the ester bonds between fatty acids to yield glycerol and lipids. The use of these enzymes for biofilm control has not been very abundant as lipids rarely have any structural significance in biofilms (Moss, 2006). Proteins and polysaccharides are the two most prominent components of biofilms and are therefore more popular targets for biofilm control (Kristensen *et al.*, 2008).

Polysaccharide lyases and hydrolases

Since polysaccharides are a primary component of the biofilm EPS, much attention has been given to enzymes that can destroy it (Gupta *et al.*, 2002). Enzymes capable of degrading the polysaccharides can be divided into two groups, i.e. polysaccharide lyases or – hydrolases (Sutherland, 1999; Wingender *et al.*, 1999). The substance specificity of the lyases is determined by the carbohydrate structure and, in particular, by the presence of uronic acids as they cleave the linkage between the fourth carbon and a neutral monosaccharide, and create a double bond between the fourth and fifth carbons of uronic acids (Sutherland, 1995, 1999). The activity of certain polysaccharide lyases (e.g. gellan lyases and alginate) can be reduced by the presence of acyl substituents in the carbohydrate structure (Sutherland, 1999). Polysaccharide hydrolases cleave specific bonds in the polysaccharides (Sutherland, 1995). Most of the polysaccharide-degrading enzymes are highly specific for glycosidic sugars and the anomeric sugars of the glycosidic bond (Manyak *et al.*, 2004). Glycosylases specifically hydrolyse ester-bonds in various oligo- and polysaccharides. Endoglycosylases break oligo- and polysaccharide chains at internal sites, generating oligosaccharide intermediates. This results in relatively rapid viscosity decreases of the polymer. Exoglycosylase degrades the polymer for the non-reducing termini and generates single monosaccharide end products (Manyak *et al.*, 2004; Moss, 2006). These enzymes may act on either α- or β-glycosidic linkages, and are also specific to the sites to which the glycosidic bond is attached (Moss, 2006). A limited range of linkages in polysaccharides are targeted by individual glycosylases. Owing to the great variation of linkage types in mixed biofilms, the selection of appropriate glycosylases would be complicated and a selection that can degrade one biofilm might not be able to degrade another (Leroy *et al.*, 2008). α-Amylase from various *Bacillus* strains (Gupta *et al.*, 2002) and glucoamylases from *Aspergillus niger* are among some of the enzymes with the ability to degrade polysaccharides (Orgaz *et al.*, 2006).

Proteases

Proteins also constitute an important component of the biofilm EPS matrix structure. Proteins other than the structural proteins also assist in the attachment of cells to the surface. An example of this is the enterococcal surface protein (Esp), produced by *Enterococcus faecalis,* which was shown to enhance the attachment of the bacterium to surfaces (Tendolkar *et al.*, 2004). Degradation of these proteins might aid in biofilm detachment and prevent enhanced attachment of cells to surfaces (Tendolkar *et al.*, 2004). Proteases can be used to prevent adhesion and also for the degradation of the EPS, which has been observed for a strain of *Pseudoalteromonas* (Leroy *et al.*, 2008). Recent evidence, however, suggests the presence of broadly conserved amyloid protein structures in biofilms. These structures may limit the use of proteases for EPS breakdown (Larsen *et al.*, 2007) because the structural stability of the amyloid fibrils is resistant to most proteases (Suzuki *et al.*, 2006). The two main groups of proteases, exopeptidases and endopeptidases, cleave bonds close to the polypeptide ends and bonds formed in the inner polypeptide chain, respectively. Exopeptidases are further divided into aminopeptidases and carboxypeptidases, depending on the terminus (C- or N-terminus) of the polypeptide that is cleaved. Serine proteases, cysteine proteases, aspartic proteases and metalloproteases are the four groups of endopeptidases, separated by the catalytic techniques that are used to cleave the peptides (Rao *et al.*, 1998).

Examples of enzymatic control of biofilms

Johanson *et al.* (1997) showed that a mixture of enzymes may be required to remove and kill the inhabiting cells of biofilms. The activity of Pectinex Ultra, a multicomponent enzyme preparation, was shown to remove biofilms from surfaces without any considerable bactericidal effect. DAPI-stained *P. aeruginosa* biofilms showed a reduction in cell numbers after enzyme treatment and therefore removal of the biofilm by the enzyme. CTC staining showed that the remaining cells were respiring and Pectinex Ultra is therefore not bactericidal. A combination of glucose oxidase and lactoperoxidase was shown to be bactericidal but failed to remove the biofilm from the substrata. Both the bactericidal activity and the removal of the biofilm can be obtained by combining the oxidoreductase with the polysaccharide-hydrolysing enzymes (Pectinex Ultra) (Johansen *et al.*, 1997).

Enzyme stabilization

As a result of the high specificity and targeted effectiveness of enzymes, they may offer infinite capabilities in the areas of biosensing, bioremediation and chemical conversions (Duran and Esposito, 2000). Although the practical use of enzymes has been realized in various industrial processes and products, the short catalytic lifetimes of enzymes limit their usefulness (Kim *et al.*, 2006a). Stabilization of enzymes can enable further practical applications by reducing the required amount of enzymes, prolonging the lifetime of enzyme reactors, increasing the potential of enzyme reuse and maintaining good signal of biosensors (Kim *et al.*, 2006a).

Different approaches to improve the stability of enzymes include enzyme modification, enzyme immobilization, protein engineering and medium engineering. Enzyme modification is defined by the covalent reactions to the enzyme molecule. The surface properties can be changed by the addition of functional groups or polymers on the surface of the enzyme molecule, resulting in an improvement of enzyme stability (Mozhaev, 1993; Desantis and Jones, 1999; Govardhan, 1999). Enzyme immobilization represents the incorporation or attachment of enzyme molecules into or onto large structures by covalent attachment, encapsulation or simple adsorption (Tischer and Kasche, 1999; Livage *et al.*, 2001). Immobilization provides the dual advantages on increasing enzyme stability by reducing protein unfolding, and increasing the overall size of the catalyst, making it simpler to recover (Mozhaev *et al.*, 1990). This immobilization of enzymes onto solid supports, however, reduces the specific and volumetric activity of the biocatalysts and alternative methods of self-immobilized enzyme particles have been developed. Protein engineering involves the alteration of the amino acid sequence of the enzyme using molecular biology techniques (e.g. site-specific mutagenesis), to produce an intrinsically more stable structure (Arnold *et al.*, 2001; Lehmann and Wyss, 2001; O'Fagain, 2003). The enzyme structure can also be made more stable by changing the medium around it with reaction medium engineering. This can be achieved by the use of non-aqueous organic solvents or by changing the salt composition of the aqueous solution (Mozhaev, 1993; Lee and Dordick, 2002).

Recent approaches for enzyme stabilization

Although the above-mentioned approaches have been used frequently to improve the stability and subsequent persistence of enzymes, various recent techniques have provided new methods whereby enzymes can be stabilized.

These techniques include:

- Self-immobilized enzymes. The main technologies developed in recent years include cross-linked enzyme crystals (CLECs), cross-linked enzyme aggregates (CLEAs) and self-immobilized spherical enzyme particles (SphereZymes) (Haring and Schreier, 1999; Cao *et al.*, 2000; O'Fagain, 2003; Brady *et al.*, 2008). The immobilization of enzymes by cross-linking yields final preparations consisting of basically pure proteins, with high concentrations of enzyme per unit volume (St Clair and Navia, 1992).
- Various nanostructures with large surface areas for the immobilization of enzymes molecules have been developed for the stabilization of enzymes. In addition, new enzyme composites of nanometre scale, called 'single-enzyme nanoparticles (SENs)', have also been developed for the stabilization of enzymes (Kim and Grate, 2003).

Self-immobilized enzymes

Enzymes tend to be expensive, pH and temperature sensitive, unstable and difficult to recycle (Cao *et al.*, 2003). In order to overcome these hurdles, significant research has been undertaken to develop self-immobilized enzymes.

CLECs

Preparation of CLECs includes the crystallization of the enzymes and subsequent chemical cross-linking that locks the enzymes in the crystalline form. This provides enzymes that are insoluble and stable (Khalaf *et al.*, 1996). CLECs are frequently used in X-ray studies in organic media to aid in the understanding of the molecular basis of enzyme function in organic solvents (Fitzpatrick *et al.*, 1993; Fitzpatrick *et al.*, 1994). Since CLECs are stable and can be reused, they are mainly used in biotransformation reactions (Browne *et al.*, 1998). The latest application of CLECs is in the field of microporous materials (Vilenchik *et al.*, 1998). The high stability of these enzyme crystals has led to the development of a new chemically engineered enzyme using a combination of enzyme crystal cross-linking and chemical modification of their active sites (Häring *et al.*, 1998). Although CLECs are highly active and stable biocatalysts, the synthesis of these enzyme crystals is expensive, requires highly purified enzymes and is very time consuming (Lalonde, 1997; Lopez-Serrano *et al.*, 2002). This represents a major obstacle to the wider adoption and application of the CLEC methodology in the industry (Lopez-Serrano *et al.*, 2002).

CLEAs

The addition of salts, organic solvents or non-ionic polymers to protein solutions allows physical aggregation of protein molecules into supermolecular structures without disturbing the original three-dimensional structures of the protein (Brown and Glatz, 1986). The aggregates are held together by non-covalent bonding and redissolve or collapse when dispersed in an aqueous medium. To maintain the superstructure and, hence, the activity, these physical aggregates can be chemically cross-linked to form cross-linked enzyme aggregates (Cao *et al.*, 2000). Because precipitation is often used to purify enzymes, the CLEA methodology combines purification and immobilization into a single unit operation (Sheldon, 2007). Two or more enzymes can also be used to produce combi-CLEAs, which can be used in catalytic cascade processes. The CLEA methodology is simple and amenable to rapid optimization, and it can be applied to a wide variety of enzymes. CLEAs were shown to be highly stable and recyclable catalysts with high retention of activity and tolerance to organic solvents (Sheldon, 2007).

SphereZymes

SphereZymes are formed when protein cross-linking agents are added to a water-in-oil or oil-in-water emulsion of an aqueous enzyme solution (Brady *et al.*, 2008). By cross-linking enzymes in an emulsion, hollow or solid enzyme spheres are created (Moolman *et al.*, 2005). SphereZyme particle size can be controlled by adjusting the emulsion conditions and more than one enzyme can be combined in a sphere by co-immobilizing the different enzymes. The enzymes can be orientated either towards or away from the interface by changing the continuous and dispersed phase. The size and, hence, the surface area of the sphere particles can be controlled with this technique (Pecku *et al.*, 2006). Processes utilizing either CLECs or CLEAs require excess enzyme catalysts for a particular function, since some of the enzyme active sites are not exposed in these two systems (Moolman *et al.*, 2005). SphereZymes have various other properties superior to those of both CLECs and CLEAs, including control over particle size and morphology, improved activity, recyclability and stability (Moolman *et al.*, 2005; Pecku *et al.*, 2006).

Enzymes and nanotechnology

Immobilization of enzymes has been a popular approach for large-scale applications. This is mainly due to the ease in catalyst recycling, product purification and continuous operation. Enzyme immobilization is usually carried out by either of three routes: encapsulation/entrapment, adsorption, or covalent binding (McMorn and Hutchings, 2004). The active site of the enzyme may be severely affected when covalent binding or adsorption by electrostatic interactions are used to immobilize the enzyme. This can lead to loss of enzyme activity (Bautista *et al.*, 1998). Further loss of activity may be caused by diffusional restrictions created by these procedures (Litjens *et al.*, 2001). Adsorption of the enzyme to a hydrophobic surface does not affect the active site directly. Too weak interactions between the enzyme and the surface can, however, lead to enzyme leaching, while too strong interactions may cause a gradual conformational change of the enzyme leading to a decrease in activity (Galarneau *et al.*, 2006). Encapsulation or entrapment relates to techniques of trapping the enzyme in the pore space of the carrier material (Caruso *et al.*, 2000). Enzymes immobilized in this way are usually resistant to peptidase-catalysed biodegradation and can normally be recovered and reused (Reis *et al.*, 2007).

The immobilization approach is, however, limited by the poor biocatalytic efficiency of immobilized enzymes and consequently, it is more feasible to use traditional chemical processes instead of large-scale bioprocessing (Caruana, 1997; Demirjian *et al.*, 1999). The structure of the carrier materials can be manipulated to improve the biocatalytic efficiency of the immobilized enzymes. With the use of non-porous materials, the diffusion limitation is minimal and the enzyme loading per unit mass of support is usually low, while the diffusion limitation is greater and the enzyme loading per unit mass is higher with porous materials (Wang *et al.*, 1997). The efficiency of immobilized enzymes can be improved by reducing the size of the enzyme-carrier materials (Jia *et al.*, 2003).

Nanoparticles

When surface attachment is used for the immobilization of enzymes, nanoparticles can provide a larger surface area for the attachment of enzymes, leading to higher enzyme loading per unit mass of particle (Jia *et al.*, 2003). Enzyme immobilization into smaller porous materials reduces mass-transfer resistance, as the diffusional path of substrates is much shorter compared to the diffusional path for large-sized porous materials (Govardhan, 1999). Overall, nanoparticles provide an enzyme immobilization system with minimum diffusional limitation, maximum surface area per unit mass and high enzyme loading (Kim *et al.*, 2006a).

Although most of the studies with nanoparticles have concentrated on the improvement of enzyme activity and loading, Dyal *et al.* (2003) demonstrated that the covalent attachment of lipase to magnetic γ-Fe_2O_3 nanoparticles yields good enzyme stability. The lipase immobilized on the magnetic nanoparticles exhibited high stability for a month, and was easily separated from the reaction medium with the use of a magnetic field (Dyal *et al.*, 2003).

Nanofibres

Although the use of nanoparticles has various advantages, their dispersion in reaction solutions and subsequent recovery can be a daunting task. Nanofibres can provide all the advantageous features of nanosize materials, while overcoming the limitations of nanoparticles (Kim *et al.*, 2006a). Electrospinning is a nanofibre-forming process, producing highly specific surface areas and porous structures that can be used for the immobilization of enzymes (Wu *et al.*, 2004). Electrospun nanofibres provide a large surface area for the entrapment and attachment of enzyme molecules. These nanofibre mats are very durable and separable, and by processing it in a highly porous form it can relieve mass-transfer of substrate through the mats (Kim *et al.*, 2006a). Jia *et al.* (2002) examined polystyrene nanofibres of 120nm diameter produced via electrospinning for the immobilization of chemotrypsin. It was shown that the immobilized enzymes had a much higher catalytic efficiency and a much-improved non-aqueous activity compared to other forms of immobilized enzymes. The enzyme's stability against structural denaturation was improved by the covalent bonds and the half-life of the enzyme in methanol was 18-fold longer than that of the free enzyme (Jia *et al.*, 2002). Kim *et al.* (2005) further extended this approach by employing glutaraldehyde (GA) treatment. The GA cross-links additional enzyme molecules and aggregates onto the covalently attached seed enzyme molecules. This approach produced CT coatings on the nanofibres and the activity of the CT was 9 times higher than that of nanofibres with just a layer of covalently attached CT molecules. The enzyme stability of the CT-aggregate-coated nanofibres was greatly improved as it did not show any measurable loss of activity over a month under rigorous shaking conditions. This approach creates an economically viable enzyme system with high stability and activity (Kim *et al.*, 2005).

Mesoporous silica

Mesoporous silica is an inorganic support known to be mechanically and thermally stable, non-toxic, and highly resistant against organic solvents and microbial attack (Tischer and Wedekind, 1999). The main advantages of mesoporous silica materials are the high specific surface areas, the uniform pore diameters (2–40 nm) and the high pore volume. These materials are also inert and stable at elevated temperatures. Because of these characteristics, mesoporous silica materials offer a great opportunity for immobilised biocatalytical processes (Moelans *et al.*, 2005).

For the successful immobilization of enzymes, both the proper surface characteristics of the mesoporous silica and good matching of the sizes of the enzyme molecules to the pore diameters of the mesoporous silica is necessary. If these two factors are not satisfied, the immobilization of enzymes on the mesoporous silica will barely take place (Han *et al.*, 1999; Washmon-Kriel *et al.*, 2000). MCM-41 (pore size: 4 nm) was one of the first mesoporous materials with a high surface area, high pore volume and a well-ordered pore structure (Beck *et al.*, 1992). It was shown that the MCM-type materials are successful in hosting small enzymes, but are restrictive as a host of larger enzymes due to their small pore size. This critical problem with MCM-type materials was solved with the development of SBA-15 (pore size: 5–13 nm) and mesocellular foam (MCF, pore size 15–40 nm) which made mesoporous materials more suitable for enzyme immobilization (Zhao *et al.*, 1998; Schmidt-Winkel *et al.*, 1999). Adsorption of enzymes can be accelerated by enlarging the inlet pore size or controlling the morphologies of the mesoporous silica (Jie *et al.*, 2004).

Enzyme adsorption is the most frequently used technique for enzyme immobilization into mesoporous materials. Adsorbed enzyme molecules are more stable and the enzyme activity is also more persistent (Ping *et al.*, 2003, 2004). The confinement of enzymes in mesoporous materials can restrain enzyme unfolding or denaturation when it is located in a pore of similar dimension (Minton, 2001; Ping *et al.*, 2003). Ravindra *et al.* (2004) proved that confinement has an effect on the stability of enzyme molecules. They demonstrated, with the differential scanning calorimetric (DSC) technique, that the stability of ribonuclease A (RNase A) confined in mesoporous silica is drastically increased, resulting in an increased melting temperature. They also found that the confinement inhibits the unfolding of enzymes in mesopores significantly (Ravindra *et al.*, 2004).

To achieve high enzymatic stability, it is important that the pore size of the mesoporous silica and the diameter of the enzyme are the same (Takahashi *et al.*, 2001). Mesoporous materials with large pore size usually end up with poor enzyme stability and enzyme loading as the enzymes leach out from the mesopores. The pore volume has also proven to play a role in determining the final amount of enzymes adsorbed to the mesoporous material (Takahashi *et al.*, 2000; Vinu *et al.*, 2004a). The charge interaction is important in determining the enzyme stability (Han *et al.*, 2002; Vinu *et al.*, 2004a,b). If the charge of the mesopores and the net surface charge of the enzymes are opposite, enzyme adsorption will be accelerated and a stable enzyme system will be created due to the attractive interaction between the two opposite charges. By changing the pH and buffer solution or by adding functional amino or carboxyl groups to the mesoporous material, the charge status of the enzymes and mesopores can be controlled (Vinu *et al.*, 2004a,b; Lei *et al.*, 2002). Enzyme stability is also affected by hydrophobic modification.

Although enzyme adsorption is a simple and easy approach, adsorbed enzymes are continuously leached out from the mesoporous materials, resulting in poor operational stability. To overcome this limitation, enzymes have been covalently attached into mesopores via linkers (He *et al.*, 2000; Chong and Zhao, 2004; Kim *et al.*, 2004). This approach prevents enzyme leaching and, hence, results in stabilized enzyme activity. A more stable form of immobilized enzymes will result when there are more covalent attachments per enzyme molecule (Mozhaev *et al.*, 1990). The curvature of the surface of nanopores provide an ideal configuration for multipoint covalent attachments to enzyme molecules, resulting in better stabilization of the enzymes compared to flat surfaces (Wang *et al.*, 2001).

Numerous approaches have been proposed to prevent enzyme leaching from mesoporous materials. The inlet of mesopores can be partially closed by treating the enzyme-adsorbed mesoporous silica with a silane monomer (e.g. vinyltrimehoxysilane (VTMS)). Ma *et al.* (2004) demonstrated that the treatment of enzyme-adsorbed mesoporous silica with VTMS resulted in good entrapment of enzymes within the mesopores, from which enzyme leaching is prevented, without inhibiting mass transfer through the pore inlet. Wang and Caruso (2004) coated the enzyme-adsorbed mesoporous silica with an organic/inorganic compound shell. This led to high enzyme loading and stability and also protected the entrapped enzymes from proteolysis since the proteases are not able to penetrate through the coating shell. Alginate/silica biocomposites were used in another approach to prevent enzyme leaching. Coradin and Livage (2003) saturated mesoporous silica with alginic solution containing β-galactosidase. This was followed by Ca^{2+}-induced gelation to produce the alginate/silica biocomposites.

Lee *et al.* (2005) recently developed CLEA in hierarchically ordered mesocellular mesoporous silica. The enzyme (chemotrypsin – CT) was first adsorbed onto the mesoporous silica and then cross-linked with GA treatment. CT adsorbed to mesoporous silica via the conventional approach had a half-life of 3.6 days in rigorously shaking conditions, while CLEA-CT showed excellent stability without any leaching for more than a month under the same conditions (Lee *et al.*, 2005).

Sol–gel encapsulation

The immobilization of enzymes via sol–gel encapsulation has been one of the most popular immobilization techniques since Avnir *et al.* (1994) reported that enzymes encapsulated into sol–gel matrices maintained their activities. A soluble precursor molecule [e.g. tetraethylorthosilicate (TEOS) or tetramethylorthosilicate (TMOS)] is hydrolysed to form a dispersion of colloidal particles ('sol'). A condensation reaction is then initiated when the enzyme solution is added to 'sol'. This condensation reaction caused bonds to form between the sol particles resulting in an infinite network of particles ('gel'), where enzymes are encapsulated in silicate matrices (Lakshmi *et al.*, 1997; Gill *et al.*, 1999).

Various channels and pores, ranging from 0.1 to 500nm in size, are formed in the final silicate matrices. A careful optimization process is required to prevent the leaching of encapsulated enzymes, but once it is prevented, the sol–gel approach results in a reasonably stable form of enzyme immobilization. The close fit of the enzyme molecule within the sol–gel pore prevents the unfolding and denaturation of the encapsulated enzyme (Gill *et al.*, 1999). A more tight confinement of encapsulated enzyme is achieved with this approach than with other enzyme immobilization approaches. This is because the enzymes are added from an early stage of silicate formation in the sol–gel approach while other approaches utilize pre-formed structures for immobilization (Avnir *et al.*, 1994).

Various variations of the sol–gel approach have been reported. Wei *et al.* (2000; 2001) added D-glucose into silica 'sol' as a pore forming agent to produce mesoporous silica. After gelation, the D-glucose was removed by water extraction. This formed mesopores in the final silicate gel matrices, allowing easy transfer of substrate and product and hence improved enzyme activity. Enzyme stability was also improved by preventing enzyme leaching from the gel matrices (Wei *et al.*, 2000, 2001). Luckarift *et al.* (2004) used the sol–gel approach in the presence of silaffin polypeptides from diatoms to entrap enzymes into silica nanospheres. These polypeptides catalysed the formation of silica at mild conditions (ambient temperature and pressure and neutral pH). This quick gelation process resulted in negligible loss of enzyme activity and the resulting encapsulated enzyme showed a higher stability than free enzymes. Large particles produced by general sol–gel encapsulation, usually have poor activity due to the limited mass transfer of substrate from the reaction medium to the enzyme active sites. The use of silaffin polypeptides to produce silica nanospheres can, however, improve the apparent enzyme activity by reducing the mass-transfer limitation of substrate in these nanometre-scale particles (Luckarift *et al.*, 2004).

Single-enzyme nanoparticles (SENs)

The production of single-enzyme nanoparticles is a new approach for enzyme stabilization, distinct from immobilizing enzymes into mesoporous materials or encapsulation in sol–gels, or bulk composite structures (Kim *et al.*, 2006a). SENs are enzyme composites of nanometre scale, consisting of single-enzyme molecules surrounded with a porous organic/inorganic network of less than a few nanometres thick. The synthesis of SENs involves three different steps: in the first step the enzyme surface is modified and the modified enzyme is solubilized in hexane. This is followed by vinyl polymer growth from the enzyme surface and finally an orthogonal polymerization step with silanol condensation reactions that cross-link the attached polymer chains into a network around the enzyme. High stability of the enzymes is only achieved after the second orthogonal polymerization step. Reactions are carried out so that cross-linking is largely confined to individual enzyme surfaces. This yields discrete enzyme nanoparticles rather than bulk solids (Kim and Grate, 2003).

In experiments with a protease (α-chemotrypsin, CT) the catalytic activity of the SEN-CT was greatly stabilized compared to that of the free enzymes, and the nanoscale structure of SENs did not impose a serious mass-transfer limitation of substrate. The SEN-CT showed a 5–10% decrease of CT activity after ten days incubation. This suggested that the half-life of CT activity in the form of SENs could reach up to 143 days (Kim and Grate, 2003). It was also reported that the half-life of SEN-CT was 282 times greater than free CT at 30°C in non-shaking conditions (Kim *et al.*, 2006b). When SEN-CT was stored in a buffer solution at 4°C, the sample showed negligible decrease in CT activity over 5 months, indicating the impressive storage stability of SEN-CT (Kim and Grate, 2003).

The catalytic stability of SEN-CT indicates that the activity of CT is not lost due to denaturation within the nanostructure, nor is the activity lost due to denaturation. The stabilization of SEN-CT against denaturation probably results from multiple covalent attachment points within the nanostructure (Mozhaev *et al.*, 1990; Kim and Grate, 2003).

Conclusion

The enzymatic degradation of biofilms is of high interest to the scientific community, as biofilm formation is a cause of medical, environmental and industrial problems in many areas (Chaignon *et al.*, 2007). The potential of enzymatic preparations in the control of dental biofilms (Aldridge *et al.*, 1998), pathogen bacteria (Boyd and Chakrabarty, 1994; Johansen *et al.*, 1997; Itoh *et al.*, 2005), industrial biofilms (Aldridge *et al.*, 1998), and marine biofilms (Leroy *et al.*, 2008) have already been reported. Although the use of enzymes provide several advantages, like a reduction in toxic by-products, energy consumption and effluent load, it also suffers form various drawbacks. Enzymes are relatively expensive compared to chemical catalysts and display sensitivity towards pH, temperature, solvents, oxidation and shear. Furthermore, free enzymes cannot be reused as it is difficult to recover from aqueous media (Domínguez de María *et al.*, 2006).

To overcome the problems associated with free enzymes, the enzymes are often immobilised. New methods employing self-immobilization of the enzymes or immobilization of enzymes on nano-scale enzyme-carrier materials show a number of advantages. No solid support is required for self-immobilized enzymes, final preparation consists of mainly pure protein with a high concentration of enzyme per unit volume, and the self-immobilized enzymes show an increased stability (Lopez-Serrano *et al.*, 2002; Brady *et al.*, 2008). Immobilization of enzymes on nano-scale carrier materials also allow high enzyme loading per unit mass, a reduced loss in enzyme activity, and catalyst recycling

(Jia *et al.*, 2003; Kim *et al.*, 2006a). All these advantages make self-immobilized enzymes and enzyme immobilized on nano-scale materials attractive alternatives to other biofilm control agents.

References

Aldridge, I.Y., Chmurny, A.B., Durham, D.R., and Roberts, R.L. (1998). Proteases to inhibit and remove biofilm. Patent EP0590746B1.

Allermann, K., Schneider, I., Walstrøm, E., Andersen, B.H., Andersen, T.T., and Højenvang, J. (2004). Substitution of biocides in paints for ships with enzymes. Danish Environmental protection Agency. *Miljøprojekt Nr* 910.

Allison, D.G. (2003). Molecular architecture of the biofilm matrix. In Biofilms in Medicine, Industry, and Environment Technology, Lens, P. ed. (London: IWA Publishing), pp. 81–90.

Arnold, F.H., Wintrode, P.L., Miyazaki, K., and Gershenson, A. (2001). How enzymes adapt: lessons from directed evolution. Trends Biochem. Sci. *26*, 100–106.

Augustin, M., Ali-Vehmas, T., and Atroshi, F. (2004). Assessment of enzymatic cleaning agents and disinfectants against bacterial biofilms. J. Pharm. Pharmaceut. Sci. *7*, 55–64.

Avnir, D., Braun, S., Lev, O., and Ottolenghi, M. (1994). Enzymes and other proteins entrapped in sol–gel materials. Chem. Mater. *6*, 1605–1614.

Bautista, M., Bravo, M.C., Campelo, J.M., Gracia, A., Luna, D., Marinas, J.M., and Romero, A.A. (1998). Covalent immobilization of Porcine pancreatic lipase on amorphous $AIPO_4$ and other inorganic supports. J. Chem. Technol. Biotechnol. 72, 249–254.

Beck, J.S., Vartuli, J.C., Roth, W.J., Leonowicz, M.E., Kresge, C.T., Schmitt, K.D., Chu, C.T.W., Olson, D.H., Sheppard, E.W., Mccullen, S.B., Higgins, J.B., and Schlenker, J.L. (1992). A new family of mesoporous molecular-sieves prepared with liquid-crystal templates. J. Am. Chem. Soc. *114*, 10834–19843.

Boyd, A., and Chakrabarty, A.M. (1994). Role of alginate lyase in cell detachment of *Pseudomonas aeruginosa*. Appl. Environ. Microbiol. *60*, 2355–2359.

Brady, D., Jordaan, J., Simpson, C., Chetty, A., Arumugam, C., and Moolman, F. (2008). Spherezymes: A novel structured self-immobilisation enzyme technology. BCM Biotechnol. *8*, 8.

Brown, D.L., and Glatz, C.E. (1987). Aggregate breakage in protein precipitation. Chem. Eng. Sci. *42*, 1831–1839.

Browne, J.K., McKervey, M.A., Pitarch, M., Russel, J.A., and Millership, J.S. (1998). Enzymatic synthesis of nonracemic inherently chiral calix[4]arenes by lipase-catalysed transesterification. Tetrahedron Lett. *39*, 1787–1790.

Cao, L., Van Langen, L., and Sheldon, R.A. (2003). Immobilized enzymes: carrier-bound or carrier-free? Curr. Opin. Biotechnol. *14*, 387–394.

Cao, L.Q., Van Rantwijk, F., and Sheldon, R.A. (2000). Cross-linked enzyme aggregates: a simple and effective method for the immobilization of penicillin acylase. Org. Lett. *2*, 1361–1364.

Caruana, C.M. (1997). Enzymes tackle tough processing. Chem. Eng. Prog. *93*, 13–20.

Caruso, F., Trau, D., Mohwald, H., and Renneberg, R. (2000). Enzyme encapsulation in layer-by-layer engineered polymer multilayer capsules. Langmuir. *16*, 1485–1488.

Cense, A., Peeters, E.A., Gottenbos, B., Baaijens, F.P., Nuijs, A.M., and Van Dongen, M.E. (2006). Mechanical properties and failure of *Streptococcus mutans* biofilms, studied using a microindentation device. J. Microbiol. Methods *67*, 463–472.

Chaignon, P., Sadovskaya, I., Ragunah, C., Ramasubbu, N., Kaplan, J.B., and Jabbouri, S. (2007). Susceptibility of staphylococcal biofilms to enzymatic treatments depends on their chemical composition. Appl. Microbiol. Biotechnol. 75, 125–132.

Chen, X., and Stewart, P. (2000). Biofilm removal caused by chemical treatments. Water Res. *34*, 4229–4233.

Chong, A.S.M., and Zhao, X.S. (2004). Design of large-pore mesoporous materials for immobilization of penicillin G acylase biocatalyst. Catalysis Today *93*, 293–299.

Coradin, T., and Livage, J. (2003). Mesoporous alginate/silica biocomposites for enzyme immobilisation. Comptes Rendus Chimie *6*, 147–152.

Costerton, J.W., Lewandowski, Z., Caldwell, D.E., Kober, D.R., and Lappin-Scott, H.M. (1995). Microbial bioflims. Ann. Rev. Microbiol. *49*, 711–745.

Demirjian, D., Moris-Varas, F., Gololobov, M., and Calugaru, S. (1999). Biocatalysis in chemical processing. Chemical Process. *62*, 57–58.

Desantis, G., and Jones, J.B. (1999). Chemical modification of enzymes for enhanced functionality. Curr. Opin. Biotechnol. *10*, 324–330.

Dolan, R.M., and Costerton, J.W. (2002). Biofilms: survival mechanisms of clinically relevant microorganisms. Clin. Microbiol. Rev. *15*, 167–193.

Domínguez de María, P., Sánchez-Montero, J.M., Sinisterra, J.V., and Alcántara, A.R. (2006). Understanding Candida rugosa lipases: An overview. Biotechnol. Adv. *24*, 180–196.

Donlan, R.M. (2002). Biofilms: Microbial life on surfaces. Emerg. Infect. Dis. *8*, 881–890.

Duran, N., and Esposito, E. (2000). Potential applications of oxidative enzymes and phenoloxidase-like compounds in wastewater and soil treatment: A review. Appl. Catal. B. Environ. *28*, 83–99.

Dyal, A., Loos, K., Noto, M., Chang, S.W., Spagnoli, C., Shafi, K., Ulman, A., Cowman, M., and Gross, R.A. (2003). Activity of *Candida rugosa* lipase immobilized on gamma-Fe_2O_3 magnetic nanoparticles. J. Am. Chem. Soc. *125*, 1684–1685.

Fitzpatrick, P.A., Ringe. D., and Klibanov, A.M. (1994). X-ray crystal structure of cross-linked subtilisin Carlsberg in water *vs.* acetonitrile. Biochem. Biophys. Res. Commun. *198*, 675–681.

Fitzpatrick, P.A., Steinmetz, A.C.U., Ringe, D., and Klibanov, A.M. (1993). Enzyme crystal structure in a neat organic solvent. Proc. Natl. Acad. Sci. U.S.A. *90*, 8653–8657.

Galarneau, A., Mureseanu, M., Atger, S., Renard, G., and Fajula, F. (2006). Immobilization of lipase on silicas. Relevance of textural and interfacial properties on activity and selectivity. New. J. Chem. *30*, 562–571.

Gill, I., Pastor, E., and Ballesteros, A. (1999). Lipase-silicone biocomposites: efficient and versatile immobilized biocatalysts. J. Am. Chem. Soc. *121*, 9487–9496.

Goverdhan, C.P. (1999). Crosslinking of enzymes for improved stability and performance. Curr. Opin. Biotechnol. *10*, 331–335.

Gupta, R., Beg, Q.K., and Lorenz, P. (2002). Bacterial alkaline proteases: molecular approaches and industrial applications. Appl. Microbiol. Biotechnol. *59*, 13–32.

Han, Y.-J., Stucky, G.D., and Butler, A. (1999). Mesoporous silicate sequestration and release of proteins. J. Am. Chem. Soc. *121*, 9897–9898.

Han, Y.-J., Watson, J.T.M., Stucky, G.D., and Butler, A. (2002). Catalytic activity of mesoporous silicate-immobilized chloroperoxidase. J. Mol. Catal. B: Enzym. *17*, 1–8.

Häring, D., and Schreier, P. (1999). Cross-linked enzyme crystals. Curr. Opin. Chem. Biol. *3*, 35–38.

Häring, D., Hubert, B., Schüler, E., and Schreier, P. (1998). Reasoning enantioselectivity and kinetics of selenosubtilisin from the subtilisin template. Arch. Biochem. Biophys. *354*, 263–269.

He, J., Li, X., Evans, D.G., Duan, X., and Li, C. (2000). A new support for the immobilization of penicillin acylase. J. Mol. Catal. B: Enzym. *11*, 45–53.

Hope, C.K., and Bott, T.R. (2004). Laboratory modelling of manganese biofiltration using biofilms of *Leptothrix discophora*. Water Res. *38*, 1853–1861.

Itoh, Y., Wang, X., Hinnebush, B.J., Preston, I.I.I., J.F., and Romeo, T. (2005). Depolymerization of β-*1*,6-N-acetyl-D-glucosamine disrupts the integrity of diverse bacterial biofilms. J. Bacteriol. *187*, 382–387.

Jahn, A., Griebe, T., and Nielsen, P.H. (1999). Composition of *Pseudomonas putida* biofilms: accumulation of protein in the biofilm matrix. Biofouling *14*, 49–57.

Jia, H., Zhu, G., and Wang, P. (2003). Catalytic behaviours of enzymes attached to nanoparticles: the effect of particle mobility. Biotechnol. Bioeng. *84*, 406–414.

Jia, H.F., Zhu, G.Y., Vugrinovich, B., Kataphinan, W., Reneker, D.H., and Wang, P. (2002). Enzyme-carrying polymeric nanofibers prepared via electrospinning for use as unique biocatalysts. Biotechnol. Prog. *18*, 1027–1032.

Jie, L., Jie, F., Yu, C.Z., Zhang, L.Y., Jiang, S.Y., Bo, T., and Zhao, D.Y. (2004). Immobilization of enzymes in mesoporous materials: controlling the entrance to nanospace. Micropor. Mesopor. Mater. *73*, 121–128.

Johansen, C., Falholt, P., and Gram, L. (1997). Enzymatic removal and disinfection of bacterial biofilms. Appl. Environ. Microbiol. *63*, 3724–3728.

Khalaf, N., Govardhan, C.P., Lalonde, J.J., Persichetti, R.A., Wang, Y.–F., and Margolin, A.L. (1996). Cross-linked enzyme crystals as highly active catalysts in organic solvents. J. Am. Chem. Soc. *118*, 5494–5495.

Kim, B.C., Nair, S., Kim, J., Kwak, J.H., Grate, J.W., Kim, S.H., and Gu, M.B. (2005). Preparation of biocatalytic nanofibers with high activity and stability via enzyme aggregate coating on polymer nanofibers. Nanotechnology. *16*, 382–388.

Kim, J., and Grate, J.W. (2003). Single-enzyme nanoparticles armoured by nanometer-scale organic/inorganic network. Nano Lett. *3*, 1219–1222.

Kim, J., Grate, J.W., and Wang, P. (2006a). Nanostructures for enzyme stabilization. Chemical Engineering Science *61*, 1017–1026.

Kim, J., Jia, H., Lee, C., Chung, S., Kwak, J.H., Shin, Y., Dohnalkova, A., Kim, B., Wang, P., and Grate, J.W. (2006b). Single enzyme nanoparticles in nanoporous silica: A hierarchical approach to enzyme stabilization and immobilization. Enzyme Microbial Technol. *39*, 474–480.

Kim, J.K., Park, J.K., and Kim, H.K. (2004). Synthesis and characterization of nanoporous silica support for enzyme immobilization. Colloids and Surfaces – Physicochem. Eng. Aspects *241*, 113–117.

Kristensen, J.B., Meyer, R.L., Laursen, B.S., Shipovskov, S., Besenbacher, F., and Poulsen, C.H. (2008). Antifouling enzymes and the biochemistry of marine settlement. Biotechnol. Adv. *26*, 471–481.

Lakshmi, B.B., Patrissi, C.J., and Martin, C.R. (1997). Sol–gel template synthesis of semiconductor oxide micro- and nanostructures. Chem. Mater. *9*, 2544–2550.

Lalonde, J. (1997). Practical catalysis with enzyme crystals. ChemTech. *27*, 38–45.

Larsen, P., Nielsen, J.L., Dueholm, M.S., Wetzel, R., Otzen, D., and Nielsen, P.H. (2007). Amyloid adhesions are abundant in natural biofilms. Environ. Microbiol. *9*, 3077–3090.

Lasa, I. (2006). Towards the identification of the common features of bacterial biofilm development. Int. Microbiol. *9*, 21–28.

Lee, J., Kim, J., Kim, J., Jia, H., Kim, M.-I., Kwak, J.H., Jin, S., Dohnalkova, A., Park, H.-G., Chang, H.-N., Wang, P., Grate, J.W., and Hyeon, T. (2005). Simple synthesis of hierarchically ordered mesocellular mesoporous silica materials hosting crosslinked enzyme aggregates. Small *1*, 744–753.

Lee, M.Y., and Dordick, J.S. (2002). Enzyme activation for nonaqueous media. Curr. Opin. Biotechnol. *13*, 376–384.

Lehman, M., and Wyss, M. (2001). Engineering proteins for thermostability: the use of sequence alignments versus rational desing and directed evolution. Curr. Opin. Biotechnol. *12*, 371–375.

Lei, C., Shin, Y., Liu, J., and Ackerman, E.J. (2002). Entrapping enzyme in a functionalized nanoporous support. J. Am. Chem. Soc. *124*, 11242–11243.

Leroy, C., Delbarre, C., Ghillebaert, F., Compere, C., and Combes, D. (2008). Effects of commercial enzymes on the adhesion of a marine biofilm-forming bacterium. Biofouling *24*, 11–22.

Litjens, M.J.J., Le, K.Q., Straathof, A.J.J., Jongejan, J.A., and Heijnen, J.J. (2001). Diffusion limitation causes decreased enantioselectivity of esterification of 2-butanol by immobilized *Candida Antarctica* lipase B. Biocat. Biotrans. *19*, 1–19.

Livage, J., Coradin, T., and Roux, C. (2001). Encapsulation of biomolecules in silica gels. J. Phys. Condens. Matter. *13*, 673–691.

Lopez-Serrrano, P., Cao, L., Van Rantwijk, F., and Sheldon, R.A. (2002). Cross-linked enzyme aggregates with enhanced activity: application to lipases. Biotechnology Letters. *24*, 1379–1383.

Luckarift, H.R., Spain, J.C., Naik, R.R., and Stone, M.O. (2004). Enzyme immobilization in a biomimetic silica support. Nat. Biotechnol. *22*, 211–213.

Ma, H., He, J., Evans, D.G., and Duan, X. (2004). Immobilization of lipase in a mesoporous reactor based on MCM-41. J. Mol. Catal. B: Enzym. *30*, 209–217.

Manyak, D.M., Weiner, R.M., Carlson, P.S., and Quintero, E.J. (2004). Preparation and use of biofilm-degrading multiple-specificity, hydrolytic enzyme mixtures. US patent number US2005/003503.

McMorn, P., and Hutchings, G.H. (2004). Heterogeneous enantioselective catalysts: strategies for the immobilisation of homogeneous catalysts. Chem. Soc. Rev. *33*, 108–122.

Meyer, B. (2003). Approaches to prevention, removal and killing of biofilms. Int. Biodeter. Biodegr. *51*, 249–253.

Minton, A.P. (2001). The influence of macromolecular crowding and macromolecular confinement on biochemical reactions in physiological media. J. Biol. Chem. *276*, 10577–10580.

Moelans, D., Cool, P., Baeyens, J., and Vansant, E.F. (2005). Using mesoporous silica materials to immobilise biocatalysis-enzymes. Catal. Commun. *6*, 307–311.

Moolman, F.S., Brady, D., Sewlall, A.S., Rolfes, H., and Jordaan, J. (2005). Stabilization of enzymes. Patent No. PCT/IB2005/000192.

Moss, G. (2006). Enzyme nomenclature-recommendations of the Nomenclature Committee of the International Union of Biochemistry and Molecular Biology on the nomenclature and classification of enzymes by the reactions they catalyse, Tipton, K.F. and Boyce, S., eds. (Queen Mary University of London, London).

Mozhaev, V.V. (1993). Mechanism-based strategies for protein thermostabilization. Trends Biotechnol. *11*, 88–95.

Mozhaev, V.V., Melik-Nubarov, N.S., Sergeeva, M.V., Siksnis, V., and Martinek, K. (1990). Strategy for stabilizing enzymes. Part One: increasing stability of enzymes via their multi–point interaction with a support. Biocatal. Biotransform. *3*, 179–187.

O'Fagain, C. (2003). Enzyme stabilization-recent experimental progress. Enzyme Microb. Technol. *33*, 137–149.

Orgaz, B., Kives, J., Pedregosa, A.M., Monistrol, I.F., Laborda, F., and SanJosé, C. (2006). Bacterial biofilm removal using fungal enzymes. Enzyme Microb. Technol. *40*, 51–56.

Pecku, S., Chetty, A.S., Moolman, F.S., Arumugam, C., Brady, D., and Jordaan, J. (2006). Techno-economic model for the production of self–immobilised cross-linked enzymes. SACEC 2006: South African chemical engineering congress 2006; Durban, 20–22 September *22006*, 1–10.

Ping, G., Yuan, J.M., Sun, Z.F., and Wei, Y. (2004). Studies of effects of macromolecular crowding and confinement on protein folding and protein stability. J. Mol. Recognit. *17*, 433–440.

Ping, G., Yuan, J.M., Vallieres, M., Dong, H., Sun, Z., Wei, Y., Li, F.Y., and Lin, S.H. (2003). Effects of confinement on protein folding and protein stability. J. Chem. Phys. *118*, 8042–8048.

Rao, M.B., Tanksale, A.M., Ghatge, M.S., and Deshpande, V.V. (1998). Molecular and biotechnological aspects of microbial proteases. Microbiol. Mol. Biol. Rev. *62*, 597–635.

Ravindra, R., Zhao, S., Gies, H., and Winter, R. (2004). Protein encapsulation in mesoporous silicate: the effects of confinement on protein stability, hydration, and volumetric properties. J. Am. Chem. Soc. *126*, 12224–12225.

Reis, P., Witula, T. and Holmberg, K. (2008). Mesoporous materials as host for and entrapped enzyme. Micropor. Mesopor. Mater. *110*, 355–362.

Schmidt-Winkel, P., Lukens, W.W., Zhao, D.Y., Yang, P.D., Chmelka, B.F., and Stucky, G.D. (1999). Mesocellular siliceous foams with uniformly sized cells and windows. J. Am. Chem. Soc. *121*, 254–255.

Sheldon, R.A. (2007). Cross-linked enzyme aggregates (CLEAs): stable and recyclable biocatalysts. Biochem. Soc. T. *35*, 1583–1587.

St Clair, N.L., and Navia, M.A. (1992). Cross-linked enzyme crystals as robust biocatalysts. J. Am. Chem. Soc. *114*, 7314–7316.

Stewart, P.S., McFeters, G.A., and Huang, C.T. (2000). Biofilm control by antimicrobial agents. In Biofilms II: Process Analysis and Applications, Bryers, J.D., eds. (New York: Wiley), pp. 373–405.

Sutherland, I.W. (1995). Polysaccharide lyases. FEMS Microbiol. Rev. *16*, 323–347.

Sutherland, I.W. (1999). Polysaccharases for microbial exopolysaccharides. Carbohydrate Polymers. *38*, 319–328.

Suzuki, Y., Tsujimoto, Y., Matsui, H., and Watanabe, K. (2006). Decomposition of extremely hard-to-degrade animal proteins by thermophilic bacteria. J. Biosci. Bioeng. *102*, 73–81.

Takahashi, H., Li, B., Sasaki, T., Miyazaki, C., Kajino, T., and Inagaki, S. (2000). Catalytic activity in organic solvents and stability of immobilized enzymes depend on the pore size and surface characteristics of mesoporous silica. Chem. Mater. *12*, 3301–3305.

Takahashi, H., Li, B., Sasaki, T., Miyazaki, C., Kajino, T., and Inagaki, S. (2001). Immobilized enzymes in ordered mesoporous silica materials and improvement of their stability and catalytic activity in an organic solvent. Micropor. Mesopor. Mater. *44–45*, 755–762.

Tendolkar, P.M., Baghdayan, A.S., Glimore, M.S., and Shankar, N. (2004). Enterococcal surface protein, Esp, enhances biofilm formation by *Enterococcus faecalis*. Infect. Immun. *72*, 6032–6039.

Thormann, K.M., Saville, R.M., Shukla, S., and Spormann, A.M. (2005). Induction of rapid detachment in *Shewanella oneidensis* MR-1 biofilms. J. Bacteriol. *187*, 1014–1021.

Tischer, W., and Kasche, V. (1999). Immobilized enzymes: crystals or carriers? Trends Biotechnol. *17*, 326–335.

Tischer, W., and Wedekind, F. (1999). Immobilized enzymes: methods and applications. Top. Curr. Chem. *200*, 95–126.

Vilenchik, L.A., Griffith, J.P., St Clair, N.L., Navia, M.A., and Margolin, A.L. (1998). Protein crystals as novel microporous materials. J. Amer. Chem. Soc. *120*, 4290–4294.

Vinu, A., Murugesan, V., and Hartmann, M. (2004a). Adsorption of lysozyme over mesoporous molecular sieves MCM-41 and SBA-15: influence of pH and aluminum incorporation. J. Phys. Chem. B *108*, 7323–7330.

Vinu, A., Murugesan, V., Tangermann, O., and Hartmann, M., (2004b). Adsorption of cytochrome *c* on mesoporous molecular sieves: influence of pH, pore diameter, and aluminum incorporation. Chem. Mater. *16*, 3056–3065.

Wang, P., Dai, S., Waezsada, S.D., Tsao, A.Y., and Davison, B.H. (2001). Enzyme stabilization by covalent binding in nanoporous sol–gel glass for nonaqueous biocatalysis. Biotechnol. Bioeng. *74*, 249–255.

Wang, P., Sergeeva, M.V., Lim, L., and Dordick, J.S. (1997). Biocatalytic plastics as active and stable materials for biotransformations. Nature Biotechnol. *15*, 789–793.

Wang, Y.J., and Caruso, F. (2004). Enzyme encapsulation in nanoporous silica spheres. Chem. Comm. 1528–1529.

Washmon-Kriel, L., Jimenez, V.L., and Balkus Jr, K.J. (2000). Cytochrome *c* immobilization into mesoporous molecular sieves. J. Mol. Catal. B Enzymatic *10*, 453–469.

Wei, Y., Xu, J., Feng, Q., Dong, H., and Lin, M. (2000). Encapsulation of enzymes in mesoporous host materials via the nonsurfactant-templated sol–gel process. Mater. Lett. *44*, 6–11.

Wei, Y., Xu, J., Feng, Q., Lin, M., Dong, H., Zhang, W.-J., and Wang, C. (2001). A novel method for enzyme immobilization: direct encapsulation of acid phosphatase in nanoporous silica host materials. J. Nanosci. Nanotechnol. *1*, 83–93.

Wu, L., Yuan, X., and Sheng, J. (2005). Immobilization of cellulose in nanofibrous PVA membranes by electrospinning. J. Membr. Sci. *250*, 167–173.

Xavier, J.B., Picioreanu, C., Rani, S.A., van Loosdrecht, M.C.M., and Stewart, P.S. (2005). Biofilm-control strategies based on enzymatic disruption of the extracellular polymeric substance matrix – a modelling study. Microbiology. *151*, 3817–3832.

Xie, H., Cook, G.S., Costerton, J.W., Bruce, G., Rose, T.M., and Lamont, R.J. (2000). Intergeneric communication in dental plaque biofilms. J. Bacteriol. *182*, 7067–7069.

Zhao, D.Y., Huo, Q.S., Feng, J.L., Chmelka, B.F., and Stucky, G.D. (1998). Nonionic triblock and star diblock copolymer and oligomeric surfactant syntheses of highly ordered, hydrothermally stable, mesoporous silica structures. J. Amer. Chem. Soc. *120*, 6024–6036.

Nanofiltration for Water and Wastewater Treatment

5

Ismail Koyuncu and Mehmet Cakmakci

Abstract

Nanofiltration (NF) is a new type of pressure-driven membrane process and used between reverse osmosis and ultrafiltration membranes. The most different speciality of NF membranes is the higher rejection of multivalent ions than monovalent ions. NF membranes are used in softening water, brackish water treatment, industrial wastewater treatment and reuse, product separation in the industry, salt recovery and recently desalination as two pass NF system. In this chapter, a general overview of nanofiltration membranes, membrane materials and manufacturing techniques, principles such as performance and modelling, module types, membrane characterization and applications on water and wastewater treatment were given.

Introduction

Nanofiltration (NF) is a new class of membrane processes that lie between ultrafiltration (UF) and reverse osmosis (RO) membranes (Fig. 5.1) and rejects molecules which have a size in the order of 1 nm (Van der Bruggen *et al.*, 2008). The salt separating mechanism of NF membranes are similar to non porous RO membranes where solute and solvent fluxes are uncoupled in contrast to convective solute transport in porous UF membranes (Schafer *et al.*, 2005). The molecular weight cut off value of NF membranes is in the range of 150–300 Da with the unique property of a higher rejection of multivalent ions than monovalent ions.

The high quality of RO permeate is perfect, however, it requires high feed pressures. RO permeate is sometimes mixed with feed water to balance the ions and make stable drinkable water. NF membranes have taken away all these disadvantages of RO membranes, which was a further development of RO membranes.

NF membranes were developed in the late 1970s as a loose RO membrane with reduced rejection characteristics for smaller and less charged ions such as sodium and chloride at reasonable flux and relatively low pressure (AWWA, 2007). The NF term was used as open RO, loose RO, intermediate RO/UF and tight UF until the late 1980s which was called as nanofiltration. The first applications were also reported later (Eriksson, 1988).

There were a limited number of nanofiltration papers before the 1990s which increased very rapidly during the 1990s and it became a strategic process in the water industry. The installed capacity of NF membranes increased around 10 times and reached to about 6000 Ml/day in the 2000s (Schafer *et al.*, 2005).

NF membranes are used in softening water, brackish water treatment, industrial wastewater treatment and reuse, product separation in the industry, salt recovery and recently desalination as a two pass NF system.

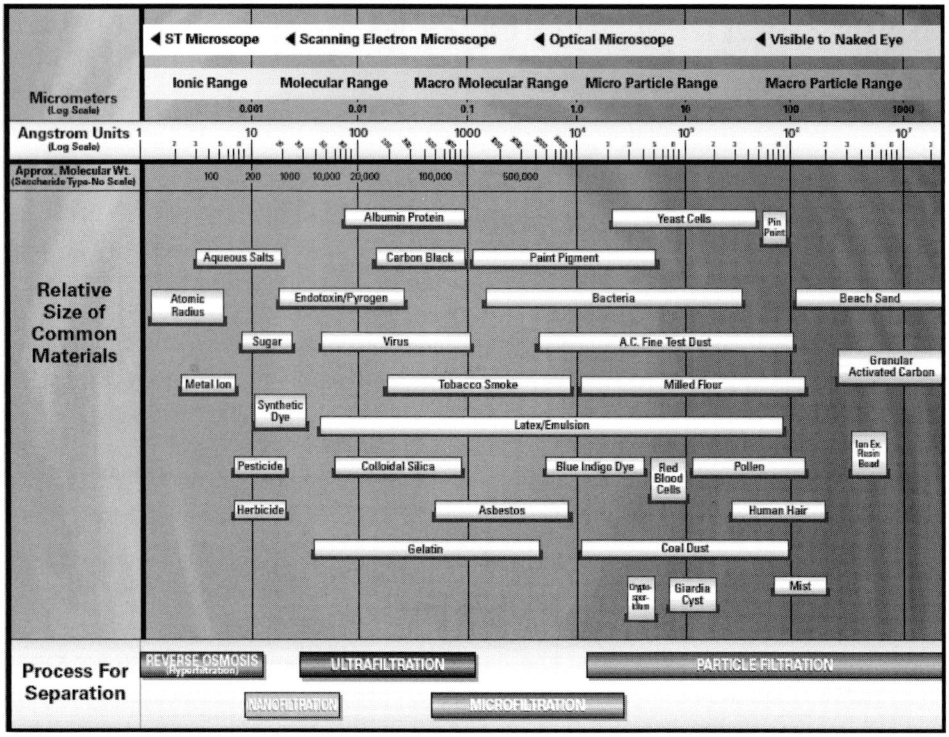

Figure 5.1 Filtration spectrum of pressure-driven membrane processes (Korean water, 2010).

The chapter gives an overview of nanofiltration membranes. Membrane materials and manufacturing techniques, principles such as performance and modelling, module types and membrane characterization will be given and finally, applications on water and wastewater treatment will be reviewed.

Nanofiltration membrane materials and manufacturing

The first NF membranes were made of cellulose acetate (CA) material in the early stage of NF membrane development. Limitations of CA in application has prevented the range of applications and impeded efforts to implement NF into new areas. The breakthrough for NF membranes was the invention of non- cellulosic polymeric membranes such as polyamide, polyethersulphone, polysulphone, chlorinated polyvinyl chloride (PVC) and polvinylidene fluoride (PVDF). UF membranes were used as a support and composite membranes were manufactured by an interfacial polymerization method. Commercial composite NF membranes were in the market after the first half of the 1980s. Ceramic and inorganic NF membranes followed the composite membranes in the 1990s (Linder and Kedem, 2005; Vankelecom *et al.*, 2005). Aqueous piperazine film was firstly interfacially polymerized on a polysulphone UF support by hydrophobic aromatic cross-linkers in the 1980s as first composite membranes and many commercial polypiperazineamide NF membranes were manufactured at low NaCl rejection (<50%) with this method. Modification of RO membranes to make NF membranes was another research area in the 1980s. Several patents were published on this subject (Strantz and Brehm, 1990). They used acidic solutions to open a selective barrier on the membrane. Polyvinyl alcohol (PVA) (Cadotte *et al.*, 1976), sulphonated polysulphone and polyethleneimine materials (Cadotte, 1977; Ikeda *et al.*, 1986) were also used to make NF composites. Recently, solvent stable NF composites were manufactured for chemical industries (Linder and Kedem, 2005; Schafer *et al.*, 2005; Vankelecom *et al.*, 2005).

Alumina, zirconia and titania are the inorganic particles used for inorganic NF membrane manufacturing. Generally, there are two layers: Al_2O_3, SiC or a metal which is made by larger particles is used for the support and a very thin layer made by small particles such as alumina, zirconia or titania is coated to this support (Vankelecom *et al.*, 2005).

Most NF membranes are negatively charged. Because of the negative charge, anions are preferentially rejected by nanofiltration membranes (Khulbe *et al.*, 2008).

Phase inversion, interfacial polymerization, coating and surface modification of membranes are the methods used for nanofiltration polymeric membrane preparation. Generally, a sol–gel method is used to make ceramic NF membranes (Schafer *et al.*, 2005; Vankelecom *et al.*, 2005).

Nanofiltration membrane performance and modelling

In comparison to ultrafiltration (UF) and reverse osmosis, nanofiltration has always been a difficult process to define and describe. Tight NF membranes are in some ways similar to RO membranes, and loose NF membranes could probably be classified as UF membranes. The specific features of NF membranes are mainly the combination of very high rejections of multivalent ions (99%) with low to moderate rejections of monovalent ions (0–70%), and the high rejection (90%) of organic compounds with a molecular weight above the molecular weight of the membrane, which is usually in the range of 150–300 Da. However, nanofiltration is still a grey zone in terms of physicochemical interactions and transport mechanisms; a transition zone with features of both UF and RO, but with its own particular characteristics as well. Understanding these phenomena is a challenge for researchers and product developers (Van der Bruggen and Geens, 2008). Technical characteristics of commercial NF membranes are given in Table 5.1.

The Hagen–Poiseuille equation is used to describe the flux value of NF membranes:

$$J_v = \frac{r_p \Delta P_e}{8\eta \Delta X} \tag{5.1}$$

where $\Delta P_e = (\Delta P - \Delta\pi)$ (Bowen and Welfort, 2008; Schafer *et al.*, 2005).

Two main approaches were used to model the transport of ionic species through NF membranes. The first is the Spiegler and Kedem model (Spiegler and Kedem, 1966; Perry and Linder, 1989; Schirg and Widmer, 1992; Levenstein *et al.*, 1996) and the second is based on the extended Nernst–Planck equations (Rautenbach and Gröschl, 1990; Tsuru *et al.*, 1991a; Machenbach *et al.*, 1997; Bowen and Mohammed, 1998). The Spiegler and Kedem model was extended to describe the salt rejection in the presence of a completely rejected organic ion (Perry and Linder, 1989; Schirg and Widmer, 1992; Levenstein *et al.*, 1996; Xu and Spencer, 1997).

The Spiegler and Kedem model uses the thermodynamics of irreversible processes which indicates the flow of each component in a solution related to the flow of other components (Bhattacharya and Ghosh, 2004). In this model the membrane is treated as a 'black box' and the nature of the membrane (i.e. charge or compactness) does not affect the transport activities through it. The model simply states that the fluxes of solute and solvent are directly related to the chemical potential differences between the two sides of the membrane.

The salt rejection of a single electrolyte has been described by Spiegler and Kedem by two coefficients: salt permeability (P_s) and the reflection coefficient (s). Assuming linear local volume and salt flows, Spiegler and Kedem derived an expression of real salt rejection R_o, as a function of J_v. The local flux equations:

$$J_v = -P_w \cdot \left[\left(\frac{dP}{dx} \right) - \sigma \left(\frac{d\pi}{dx} \right) \right] \tag{5.2}$$

$$J_s = -\overline{P_s} \left(\frac{dc_s}{dx} \right) + (1-\sigma) \cdot J_v \cdot c_s \tag{5.3}$$

Table 5.1 Technical characteristics of commercial NF membranes (Bhattacharya and Ghosh, 2004)

Membrane name	Company	Active layer	Charge	Performance			
				Eqn	R_{NaCl} (%)	R_{NaSO4} (%)	Evaluated at concentration (%)
Desal 5	Desalination Systems	Sulphonated polysulphone/ polypiperazine amide	Negative	1.9	–	–	0.1
NF-40	Film Tec	Polyamide	Negative	2.2	–	–	0.2
NF-70				1.4	–	–	0.2
XP-45				1.3	–	–	0.2
NF200				1.3	–	–	0.2
NTR-7250	Nitto Denko	Polyvinyl alcohol	Negative	2.0	–	–	0.2
NTR-7410		Sulphonated polysulphone		0.9	–	–	0.5
NTR-7450		Sulfonated polyether-sulphone		0.6	–	–	0.5
CA-30	Hoeschst	Cellulose acetate	Negative	4.3	–	–	1.2
B-TLC	Osmonics	–	Negative	0.5	–	–	0.2
AFC-30	PCI	Polyamide	Negative	2.8	–	–	0.2
SC-L100	Toray	Cellulose acetate	Neutral	1.3	–	–	0.2
UTC-20		Polyamide	Positive	1.5	–	–	0.2
UTC-60		Polyamide	Amphoteric	–	–	–	–
NP010	Nadir	Polyether sulphone	Negative	–	5–15	25–55	–
NP030		Polyether sulphone	Negative	–	25–35	80–95	–

where P_w is the specific water permeability and $\overline{P}_s$ is the local salt permeability. The integration of equation 5.3 yields:

$$\frac{J_v(1-\sigma)}{\overline{P}_s} \cdot \Delta X = \ln \frac{c_p \, \sigma}{c_p - \overline{c}_s\,(1-\sigma)} \tag{5.4}$$

with the following boundary conditions:

$c = c_{sm}$ at $x = 0$
$c = c_p$ at $x = \Delta X$

where c_{sm} and c_p are the membrane and permeate concentrations, respectively, and ΔX is the membrane thickness. Assuming single real salt rejection as $R_o = 1 - R_0 = 1 - \frac{c_p}{c_{sm}}$, equation 5.4 will be,

$$R_0 = 1 - \frac{1-\sigma}{1-\sigma F} = 1 - \frac{1-\sigma}{1-\sigma \cdot \exp\left(\frac{(1-\sigma)\,J_v}{\bar{P}_s} \cdot \Delta X\right)} \tag{5.5}$$

where

$$F = \exp\,(-J_v A),$$

$$A = \frac{1-\sigma}{\bar{P}_s} \cdot \Delta X = \frac{1-\sigma}{P_s}$$

and P_s is the overall salt permeability ($P_s = \bar{P}_s / \Delta X$).

It can be seen that σ corresponds to maximum rejection at infinite volume flux. σ and P_s can be determined from experimental data of R_o as a function of J_v by best-fit method (Spiegler and Kedem, 1966; Koyuncu and Topacik, 2002).

Perry and Linder (1989) extended the Spiegler and Kedem model to describe the salt rejection in the presence of an organic ion. The membrane phase was considered to be composed of thin differential elements separated with thin 'corresponding' aqueous layers equilibrated with the two adjacent membrane slices. In this research the concentration polarization effect was not taken into consideration. Assuming equilibrium on the membrane–solution interface, the salt concentration inside the membrane will be as follows (Koyuncu and Topacik, 2002):

$$\bar{c}_s^2 = c_{sm}\,(c_{sm} + \upsilon \cdot c_{Dm}) \tag{5.6}$$

$$\bar{c}_s = c_{sm}\left(1 + \frac{\upsilon \cdot c_{Dm}}{c_{sm}}\right)^{0.5} \tag{5.7}$$

where $\bar{c}_{bars;}$ is the salt concentration inside the membrane, c_{sm} is the salt concentration on the feed side, ν is the number of charges of organic ion and c_{Dm} is the concentration of organic ion on the feed side of the membrane. In the presence of Donnan exclusion forces, induced by the impermeable organic ions, the salt transport across the membrane proceeds as if the membrane is exposed to a salt solution having a concentration of &$\bar{c}_{bars;}$ instead of c_{sm}. Thus, the value of &$\bar{c}_{bars;}$ and not of c_{sm} must be used in equation 5.4, which yields (Perry and Linder, 1989; Koyuncu and Topacik, 2002):

$$\frac{J_v\,(1-\sigma)}{P_s} = \ln \frac{c_p\,\sigma}{c_p - \bar{c}_s\,(1-\sigma)} \tag{5.8}$$

Introducing equation 5.7 into equation 5.8 and assuming constant σ and P_s, the real salt rejection in the presence of organic ion can now be written as (Perry and Linder, 1989; Koyuncu and Topacik, 2002):

$$R_s = 1 - \frac{(1-\sigma)\left(1 + \frac{\upsilon\, c_{Dm}}{c_{sm}}\right)^{0.5}}{1-\sigma \exp\left(\frac{(1-\sigma)\cdot J_v}{P_s}\right)} \tag{5.9}$$

Levenstein *et al.* (1996) modified the Perry–Linder model by neglecting the convective term in the Spiegler and Kedem model. The new model was presented in two forms: differential and integral form. If the integral and Perry–Linder models are combined, the real salt rejection will be as follows (Perry and Linder, 1989; Koyuncu and Topacik, 2002):

$$R_s = 1 - \left(\frac{B}{B+J_v}\right)\cdot\left(1 + \upsilon\frac{c_{Dm}}{c_{sm}}\right)^{0.5} \tag{5.10}$$

where B is the diffusivity parameter ($B = B_0 c_m{}^n$). In the case of quaternary system with two non-permeating ions salt rejection is as follows:

$$R_s = 1 - \left(\frac{B}{B+J_v}\right)\cdot\left(1 + \frac{\upsilon_1\cdot c_{D1m} + \upsilon_2\cdot c_{D2m}}{c_{sm}}\right)^{0.5} \tag{5.11}$$

where ν_1 is the valency of c_{D1m} and ν_2 is the valency of c_{D2m} (Levenstein *et al.*, 1996; Koyuncu and Topacik, 2002).

Concentration and gel polarization effects of organic ion were considered and gel layer-controlled mass transport was taken into account by Koyuncu and Topacik (2002). Concentration and gel polarization layer effects of organic dye ions on salt (NaCl) and dye concentrations at the membrane–feed interface were incorporated into the previous models and theoretical and experimental results fitted very well (Fig. 5.2). According to Koyuncu and Topacik, the observed salt rejection was written as follows:

$$R_{sOBS} = 1 - (1 - R_o)\cdot(\alpha)\cdot\left(1 + \frac{\upsilon\cdot c_{Db}}{\alpha\cdot c_{sb}}\exp\left(\frac{J_v}{k_D}\right)\right)^{0.5} \tag{5.12}$$

where R_{sOBS} is the observed salt rejection in the presence of an organic ion, R_{o} is the real single salt rejection, υ is the number of charges of the organic ion, c_{sm} and c_{Dm} are the feed salt and organic ion concentrations, α is the gel polarization effect on salt removal and k_{D} is the mass transport coefficient of dye. Model parameters (α and k_{D}) were calculated from the statistical regression of the observed NaCl rejection and flux data. α is defined as

$$\alpha = \left(\frac{c_p}{c_{sb}} + \exp\left(\frac{J_v}{k_{avg}}\right) - \frac{c_p}{c_{sb}}\cdot\exp\left(\frac{J_v}{k_{avg}}\right)\right) = 1 - R_{sOBS} + R_{sOBS}\cdot\exp\left(\frac{J_v}{k_{avg}}\right) \tag{5.13}$$

Even if α is a function of R_{sOBS} and J_{v} theoretically, α might be considered as a constant with increasing pressure (Koyuncu and Topacik, 2002, 2003a). Therefore, if the model parameter, α is accepted as a constant, this approach will provide an easy method to determine the observed salt rejection

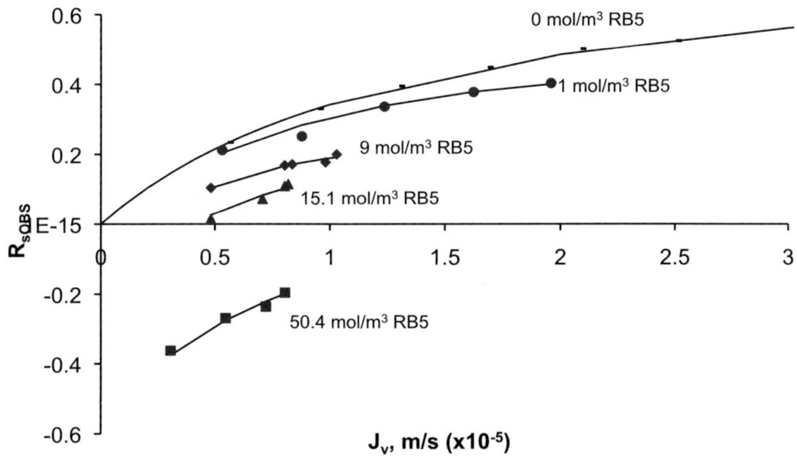

Figure 5.2 Observed NaCl retention in the presence of organic ion (Reactive Black 5) for 340 mol/m^3 NaCl solution (Koyuncu and Topacik, 2002).

rate as a function of permeate flux, which is affected by an increase in pressure. The value of α is constant for the same salt and dye concentrations in the model. However, if one of the salt or dye concentrations was changed, the α value was also influenced. According to Koyuncu and Topacik the coefficient of the gel polarization effect on salt removal (α) was influenced by the ratio of NaCl and dye concentrations, based on experimental results. An increase in dye/salt concentration ratios increased the coefficient of the gel polarization effect on salt removal (α) linearly. The relationship between α and c_D/c_S for the investigated NaCl concentrations was written as (Koyuncu and Topacik, 2002, 2003)

$$\alpha = \gamma + \beta \cdot \frac{c_{Db}}{c_{sb}} \tag{5.14}$$

The slope (β) and intercept (γ) of lines were determined from the linear regression of α and dye/salt concentration ratios (c_D/c_S). β was a fitting parameter different from the β symbol used in Perry and Linder. β increased linearly with increasing NaCl concentrations. The value of γ was almost one for all investigated NaCl concentrations according to the experimental results. Thus, equation 5.14 was written as (Koyuncu and Topacik, 2002, 2003)

$$\alpha = 1 + \beta \cdot \frac{c_{Db}}{c_{sb}} \tag{5.15}$$

In this formula, if the dye concentration is zero, the value of α will be equal to 1 and equation 5.12 is converted into the Spiegler and Kedem model. This supports the assumption of γ in equation 5.14 as 1 (Koyuncu and Topacik, 2002; 2003).

Another model for transport definition through the charged nanofiltration membrane is the extended Nernst–Planck (ENP) model (Dresner, 1972; Tsuru *et al.*, 1991a–c; Bowen and Mohammed, 1998; Bhattacharya and Ghosh, 2004). This equation was first used by Lakshminarayanaiah (1969) and Dresner (1972) and is given by

$$J_s = J_v c_i - D_i \frac{dc_i}{dx} - \frac{F\, z_i c_i D_i}{R_g T} \frac{d\psi}{dx} \tag{5.16}$$

The ENP equation has three terms that are attributed to convection, diffusion and migration in the electric field. Sieving and electrical mechanisms are combined in this model. Rejection data is analysed as a function of volumetric flux and is dependent on three parameters: effective pore radius (r_p), effective ratio of membrane thickness to porosity ($\Delta x/A_k$) and effective membrane charge density (x_d) (Schafer *et al.*, 2005; Bowen and Welfort, 2008). Tsuru *et al.* (1991a) separated single and mixed electrolyte solutions by low pressure RO membranes, which had negative charges. In this research, the rejections of the divalent ions were much larger than those of monovalent ions for mixed electrolyte solutions having the same counter ion as the membrane fixed charge. The negative rejection was dependent on the mole fraction of the feed solution and on the applied pressure. The extended Nernst–Planck equations were applied to the analysis of experimental data by considering the effective charge density (Koyuncu and Topacik, 2002).

Nanofiltration membrane characterization

Nanofiltration membranes can be either porous or non-porous. Most polymeric NF membranes are non-porous while inorganic NF membranes can be non-porous with pore sizes of 0.5–2 nm. As the separation performance of these non-porous and porous NF membranes are almost the same, separation mechanisms can be different (Mulder *et al.*, 2005; Schafer *et al.*, 2005). Characterization methods for NF membranes are classified into three groups: performance parameters, morphology parameters and charge parameters (Table 5.2). Rejection and flux measurements are considered as performance parameters (Mulder *et al.*, 2005; Schafer *et al.*, 2005).

There are different methods to describe the mechanisms of rejection and permeability which was discussed in the previous section. Morphology related parameters can be membrane pore structure, hydrophobicity and chemical structure. Membrane porosity can be investigated by thermoporometry, scanning electron microscopy (SEM), atomic force microscopy (AFM), transmission electron

Table 5.2 Characterization methods for NF membranes (Mulder *et al.*, 2005)

Parameters	Method	Characteristics
Performance parameters	Retention measurements with charged solutes	Surface charge, pore size
	Retention measurements with uncharged solutes	Pore size
	Water permeability measurements	Membrane resistance
	Solvent permeability measurements	Membrane resistance
Morphology parameters	Gas adsorption/desorption	Pore size/surface area
	Permporometry	Pore size/porosity
	Microscopy (SEM-EDAX, AFM, TEM, VSI)	Pore size/porosity
	Spectroscopy (ATR-FTIR, NMR, Raman spectroscopy, XPS)	Surface roughness, pore size, porosity, chemical composition
	Contact angle	Hydrophobicity
Charge parameters	Electrokinetic measurement	Zeta potential, surface charge
	Titration	Ion exchange capacity
	Impedance spectroscopy	Ionic conductivity

microscopy (TEM), gas adsorption–desorption and bubble permeability methods. Characterization for inorganic membranes can be different than polymeric membranes such as gas adsorption-desorption method used generally for inorganic membranes. ATR-FTIR is employed to characterize the chemical structure of the membrane surface. Contact angle measurement is used to characterize the membrane surface hydrophobicity. SEM, AFM and TEM are also new microscopic methods to describe the membrane surface characteristics (Mulder *et al.*, 2005; Schafer *et al.*, 2005). Recently, VSI (vertical scanning interferometry) was also used to describe the membrane surface at high scan areas and comparisons with AFM, which was performed at low scan areas, were done (Fig. 5.3) (Koyuncu *et al.*, 2006a). It was observed that membrane surface roughness increases with scan-size until a critical scan-size is reached. This critical scan-size appears to be similar for all the polymeric membrane surfaces studied. Because this value is significantly larger than that which can be obtained with an AFM, suggests that AFM derived roughness statistics provide an incomplete assessment of membrane morphology. Conversely, interferometry allows for rapid and accurate characterization of membrane roughness, at substantially larger scan-sizes. Finally, for a complete evaluation of membrane surface roughness scan-sizes of at least 250,000 μm^2 should be used (Fig. 5.4) (Koyuncu *et al.*, 2006a).

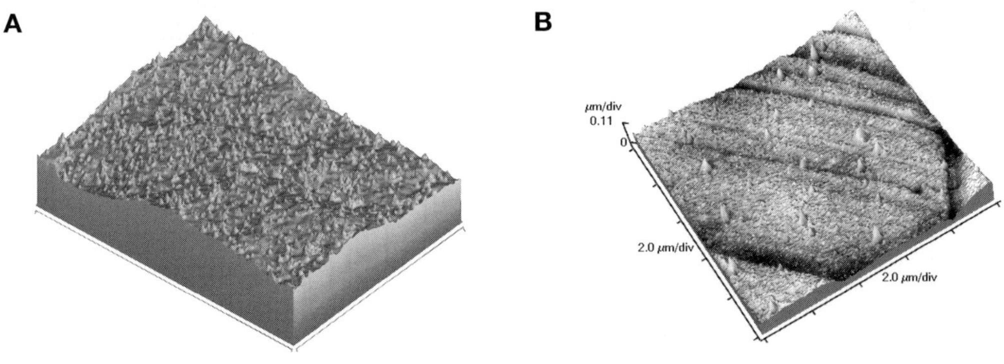

Figure 5.3 Comparison of HL NF membranes with (A) VSI and (B) AFM at different scan sizes (Koyuncu *et al.*, 2006a).

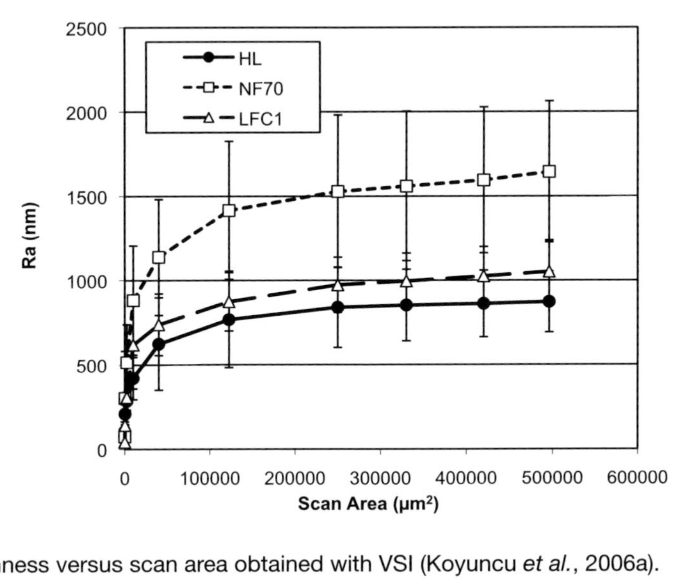

Figure 5.4 Roughness versus scan area obtained with VSI (Koyuncu *et al.*, 2006a).

Nanofiltration membrane modules

The common membrane modules used for nanofiltration membranes are spiral wound, plate and frame, tubular, hollow fibre and capillary modules (Fig. 5.5). Packing density, membrane change (replacement), energy requirement, fouling control and cleaning are the important module characteristics that should be important for NF membranes. The characteristics of various modules are compared in Table 5.3. Spiral wound and tubular modules are the most suitable types for nanofiltration membranes. While spiral wound modules are most common and standardized, there may be some limitations for highly polluted waters. Tubular modules can be used if the feed contains suspended solids (Fane, 2005; Schafer *et al.*, 2005).

Applications

Water treatment applications

Nanofiltration membranes have been applied in several areas such as water treatment. Many researchers have used NF membranes to obtain the best treated water at low cost. Detailed information about NF for water treatment is summarized in Table 5.3.

Drinking water treatment

NF membranes have been used for treating both surface water such as lakes, rivers, streams or other fresh water sources and ground water for drinking water purposes and to remove hardness, natural organic matter, micropollutants, etc. (Reiss *et al.*, 1999; Park *et al.*, 2004; Orecki *et al*; 2004; Wang *et al.*, 2005; Kim *et al.*, 2008; Low *et al.*, 2008; Marconnet *et al.*, 2009).

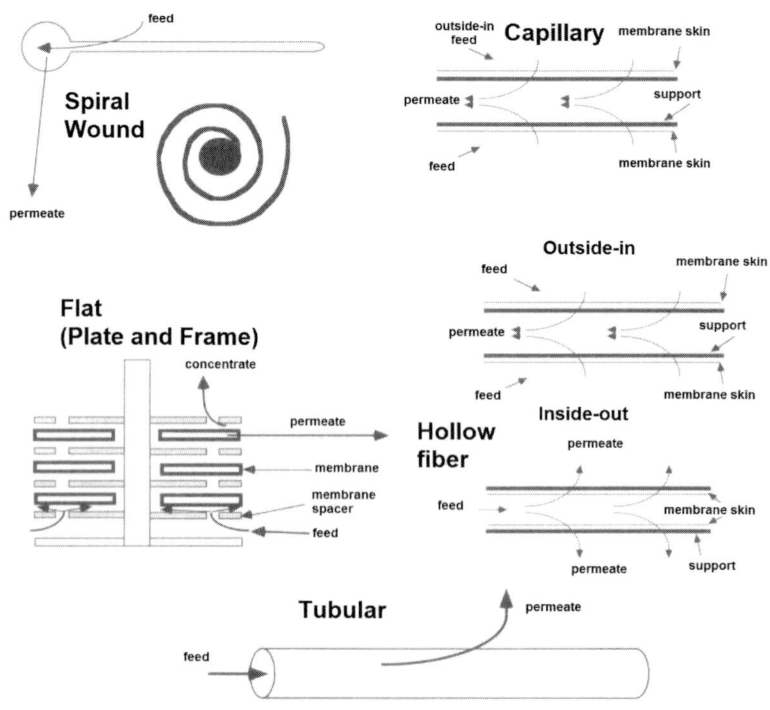

Figure 5.5 Nanofiltration membrane modules (Wiesner *et al.*, 2008).

Table 5.3 Comparison of different NF membrane modules (Fane, 2005; Schafer *et al.*, 2005)

Characteristics	Plate and frame	Spiral wound	Tubular	Hollow fibre	Submerged	High shear
Packing density (m^2/m^3)	Moderate (200–500)	High (500–1000)	Low–moderate (70–400)	High (500–5000)	Moderate	Low–moderate
Membrane replacement	Sheet	Element	Tubes	Element	Element	Sheet
Energy Requirement	Low–moderate	Moderate	High	Low	Low	Moderate
Fouling control	Moderate	Good (no solids) Poor (solids)	Good	Moderate–good	Moderate–poor	Very good
Cleaning	Moderate	Poor	Good	Good (backflush possible)	Good (backflush possible)	Good
Standardization	No	Yes	No	No	No	No
Ease of manufacture	Simple	Complex	Simple	Moderate	Moderate	Complex
Limitations for NF	Pressure containment	None	None	Burst pressure of fibres	Low driving force and flux	Pressure containment

Water softening

NF membranes are utilized in water softening for industrial process water or drinking water supplies from hard water source including calcium and magnesium. NF is operated typically in the range of 5–30 bar and more effective while total dissolved solids are more than 500 mg/l, and lower than 2000 mg/l. Scaling is a more serious problem in NF systems than water softening. $CaCO_3$, $CaSO_4$ are the main constituents of scale and rarely $BaSO_4$, $SrSO_4$, $Ca(PO_4)_2$, ferric and aluminium hydroxides. Acidification, antiscalant addition and changes on operation parameters can be used for scale control (Thorsen and Fløgstad, 2006).

NF membrane type, coated material, pH, ionic concentration and pressure are important parameters for water softening (Schaep *et al.*, 1998; Mika *et al.*, 1999; Van der Bruggen and Vandecasteele, 2003; Hilal *et al.*, 2004; Mazzoni and Bandini, 2006; Walha *et al.*, 2007; Diawara, 2008). Schaep *et al.* (1998) used three NF membranes including NF70 (Dow-Filmtec), UTC20 (Toray Industries) and UTC60 (Toray Industries) with rejection rates of calcium at 99%, 96% and 72%, respectively. Osmonics NF membranes (Desal 5 and BQ01) and a Hydranautics membrane (TFV-7450) were used in a cross-flow cell system at 500 kPa for water softening (Mika *et al.*, 1999). Rejection rates for Desal 5, BQ01 and TFV-7450 membranes were 80.7%, 31.4% and 68.1% for calcium and 87%, 26.7% and 65.3% for magnesium, respectively. Desal-5 DK membranes were used to treat water–$CaCl_2$ mixtures by Mazzoni and Bandini (2006). Effects of feed pH and salt on removal efficiency were also investigated. Although calcium rejection reaches maximum value as the ionic concentration increases, rejection rate decreased by elevated pH feed. Van der Bruggen *et al.* (2004a) applied 0.01 M solutions of NaCl, Na_2SO_4, $MgCl_2$, $MgSO_4$ and $NaNO_3$ on electrodialysis (AMV/CMV Selemion membranes and ACS/CMS Tokuyama membranes) and NF membranes (NTR 7450 and UTC-60). Both NF membranes had good separation efficiency for the separation of monovalent and divalent anions at 60%, while performances of electrodialysis membranes were low. The best separation was observed with a UTC-60 membrane due to the size exclusion effects for the larger divalent ions. The $MgCl_2$ rejection for a NF270 membrane was about 95% at a flux value of 0.85 $m^3/(m^2\ day)$ at 4.8 bar

(Ouyang *et al.* 2008). A tubular NF membrane from gamma-filtration (MPT 03) was used by Walha *et al.* (2007), which has a material of polyamide and polyethersulphone and molecular weight cut off of 200 Da. Rejection rates with the MPT 03 NF membrane were 42.5% for calcium and 38.3% for magnesium at pressures lower than 20 bar and water temperature at 25 ^{0}C. The cost of a water softening plant with a NF membrane (Fig. 5.6) was investigated by Bergman (1996). Drinking water was supplied from the groundwater including high concentration of calcium. Spiral wound NF membranes having molecular weight cut off values in the range of 200–500 Da were used. Operation and maintenance cost was found at \$0.53–0.88 l/day. It was also reported as lower than US\$0.20/m^3 by several researchers.

Natural organic matter (NOM) removal

Natural organic matter (NOM) is found both in surface and ground waters. The removal of natural organic matter prior to disinfection is important for the formation of disinfection by-products (DBPs) control strategy. DBPs in drinking water are suspected to be toxic, carcinogenic and mutagenic to humans. New DBPs regulations such as US Environmental Protection Agency Disinfectant/Disinfection By-Products Rule (USEPA, 1998) require higher removal rates of NOM. An important alternative to NOM treatment is the membrane process.

Recently, NOM removal studies by NF membranes increased due to the stringent DBPs rules. Membrane fouling is one of the important issues of NOM removal. In particular, dissolved and colloidal organic matter can be a major contributor to membrane fouling. Although rejection of NOM with a reverse osmosis (RO) membrane is higher than the NF membrane, a NF membrane is more likely to remove organic matter because of the low pressure requirement (Yoon and Lueptow, 2005). It has potential for removing organic and inorganic pollutants for drinking water and it applies to the new DBPs rules (Lee and Lee, 2007).

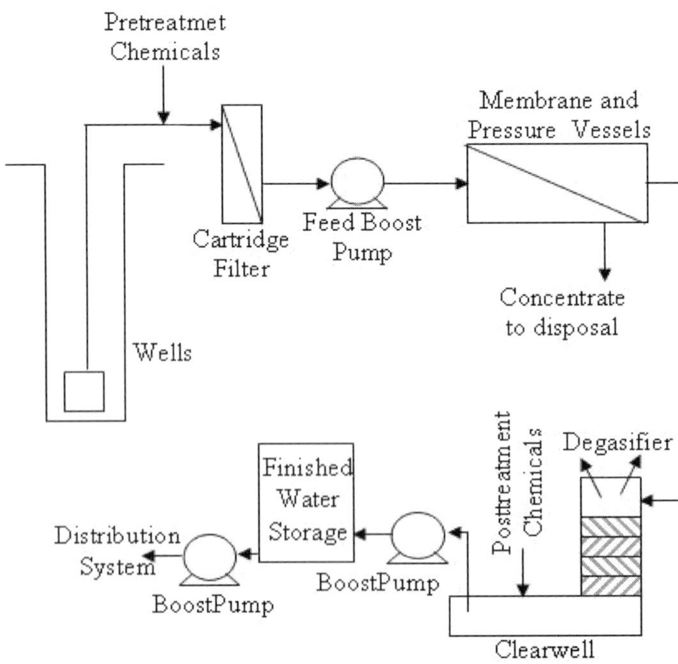

Figure 5.6 Schematic diagram of membrane softening plant in Florida (Bergman, 1996).

The charge and hydrophobicity of NF membranes have an effect on NOM removal performance. Lee and Lee, (2007) used four different NF membranes for NOM removal. These membranes were NF-45 (negatively charged), MPF-34 (positively charged), CA865 (hydrophilic) and GR90 (hydrophobic). They found that NF fouling by NOM adsorption was an important factor in hydrophobic and positively charged membranes, whereas, it was negligible for hydrophilic and negatively charged membranes.

Suspended solids, colloidal matter and other pollutants have to be reduced for minimizing membrane fouling potential. Some researchers used different pretreatment methods in order to reduce the NF membrane fouling caused from NOM. High TOC river water was treated by NF membranes (Reis *et al.*, 1999) and microfiltration (Memcor/Zenon), in-line coagulation-microfiltration and coagulation-sedimentation-filtration (CSF) were used as pretreatment methods (Fig. 5.7). The pretreatment performance was ordered as Memcor MF > Zenon MF > CSF. Van der Bruggen *et al.* (2004b) used microfiltration (MF) as a pretreatment process for NF membranes. They used 0.1-μm MF and five NF membranes. These NF membranes were UTC-20 (Toray Ind. Inc.), Desal 5 DL and Desal 51 HL (Osmonics), NF-PES-10 and N30F (Nadir). Higher flux decline in MF was observed because smaller particles led to more compact cake formation and higher specific cake resistance in dead-end mode. Internal fouling was also observed in MF due to 0.1-μm particles. A more severe flux decline in NF was observed without MF pretreatment. This situation led to slower back-diffusion from the membrane surface to the bulk solution. It is concluded that MF pretreatment is effective in avoiding a large fraction of the fouling in NF. Ozon, powdered activated carbon (PAC) and ultrafiltration was also used as pretreatment before NF. PAC or UF showed a remarkable improvement in flux, whereas ozone pretreatment was ineffective in controlling NF fouling. These results showed that the removal of colloids is crucial for preventing NF fouling (Lee and Lee, 2007). Dissolved air flotation (DAF) was used as pretreatment for Tagus river surface water (Valadas, Portugal) as a pretreatment before the spiral-wound module nanofiltration (SWNF) by Geraldesa *et al.* (2008). It was determined that pretreatment as coagulation/flocculation and the DAF process was necessary to provide a good pretreatment for the SWNF of Tagus river surface water. Microfiltration (MF), coagulation-sand filtration (CS) and ozonation-activated carbon adsorption after CS treatment (CSOB) were used as pretreatments before the UTC 60, UTC 70 and UTC 270 NF membranes by Kim *et al.* (2007). CSOB pretreatment was the most effective in removing organic foulants and

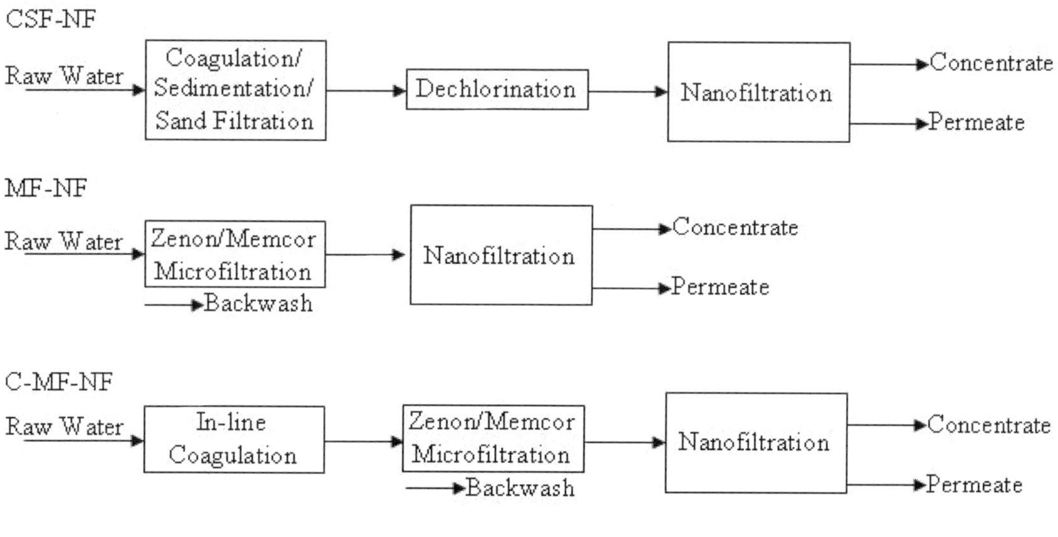

Figure 5.7 Treatment process trains for NF membrane (Reis *et al.*, 1999).

very high initial flux of NF-UTC 270 caused a rapid build-up of a cake layer. Koyuncu *et al.* (2006b) reported that GAC pretreatment, combined with ozone, reduced fouling of the tested NF membranes to the greatest extent. However, UV irradiation and pre-filtration through a 5 μm membrane did not measurably reduce fouling. The effect of GAC age on fouling was also investigated and GAC columns that had been in service for longer periods of time provided the same benefit in reducing fouling as GAC columns that had been generated recently. This suggests that the primary benefit of GAC pretreatment may be related to biological activity on the GAC filters rather than adsorption of organic matter. UF and NF membranes were compared for NOM removal by Costa and Pinho (2006). It was observed that NF presents a much better performance in terms of organics removal than UF and assures high quality of treated water.

Micropollutants removal

The presence of small concentrations of pollutants has an impact on human health and biotic environment and these pollutants are generally defined as micropollutants. More stringent regulations, increasing demand for water and market self-regulation requires the use of new technologies for the removal of micropollutants. Recently, researchers have investigated the separation of organic and a few inorganic micropollutants such as DBPs, pesticides, endocrine-disrupting chemicals (EDCs) and arsenic with NF membranes.

NOM is not harmful to human health; however, by-products from chlorine reactions can have an impact. The main DBPs are trihalomethanes (THMs) and halogenated acetic acids (HAAs) (Singer, 1999; Uyak *et al.*, 2008; Ratasuk *et al.*, 2009). The disinfectant/disinfection by-product (D/DBP) regulation in the USA has set maximum contaminant levels for total THM (TTHM) species and five HAA species at 80 and 60 μg/l, respectively (USEPA, 2003). The European Union (EU) regulated TTHM limit at 100 μg/l (EECD, 1998) and the WHO guidelines are 200 μg/l for chloroform, 60 μg/l for dichlorobromomethane, 100 μg/l for dibromochloromethane, 90 μg/l for dichloroacetonitrile and 100 μg/l for dibromoacetonitrile (Singer, 1999).

Membrane filtration, particularly nanofiltration (NF), is one of the most effective methods for THM, HAA and DBP precursor control (Allgeier and Summers, 1995; Koffskey and Lykins, 1999; Chellam, 2000; Uyak *et al.*, 2008; Ratasuk *et al.*, 2009). According to these studies, DBPs removal rate was generally higher than 80% depending on the contaminant type, membrane used, operating conditions and chemistry of the feed solution.

Uyak *et al.* (2008) used a NF200 and a DS5 NF membrane to remove THM from surface water. Removal rate of THM species such as chloroform (CFM), bromodichloromethane (BDCM), and dibromochloromethane (DBCM) were investigated and results are shown in Fig. 5.8. Rejections decreased with increasing pressure and high rejections were obtained at low pressures. The molecular weight (MW) of THM species affected removal efficiency because a higher rejection rate was obtained for DBCM which has a higher MW than other species.

Ratasuk *et al.* (2009) investigated HAAs removal by NF membranes. The five regulated haloacetic acids (HAA5) include chloroacetic acid (CAA), dichloroacetic acid (DCAA), trichloroacetic acid (TCAA), bromoacetic acid (BAA) and dibromoacetic acid (DBAA). A dense negatively charged NF membrane (ES10) was more efficient for the removal of HAA5 than the loose negatively charged NF membrane (NTR 7410) and the neutral NF membrane (NTR 729HF). The removal efficiency of HAA5 was about 90%–100% at low pressure of 1×10^5 Pa with the ES10 membrane.

The use of persistent pesticides may result in extremely environmental hazards. Trace elements of these substances can be detected in many regions of the world due to widespread and long term applications. Thus, pesticides are common in both surface and ground waters. Recently, investigations were focused on pesticides removal by NF and low-pressure RO membranes. The molecular structure and functional groups of pesticides affects the removal efficiency of the membranes (Montovay *et al.*, 1996; Berg *et al.*, 1997: Boussahel *et al.*, 2000; Kiso *et al.*, 2000, 2001; Van der Bruggen *et al.*, 2001; Zhang *et al.*, 2004; Kosutic *et al.*, 2005; Sarkar *et al.*, 2007; Ahmad *et al.*, 2009).

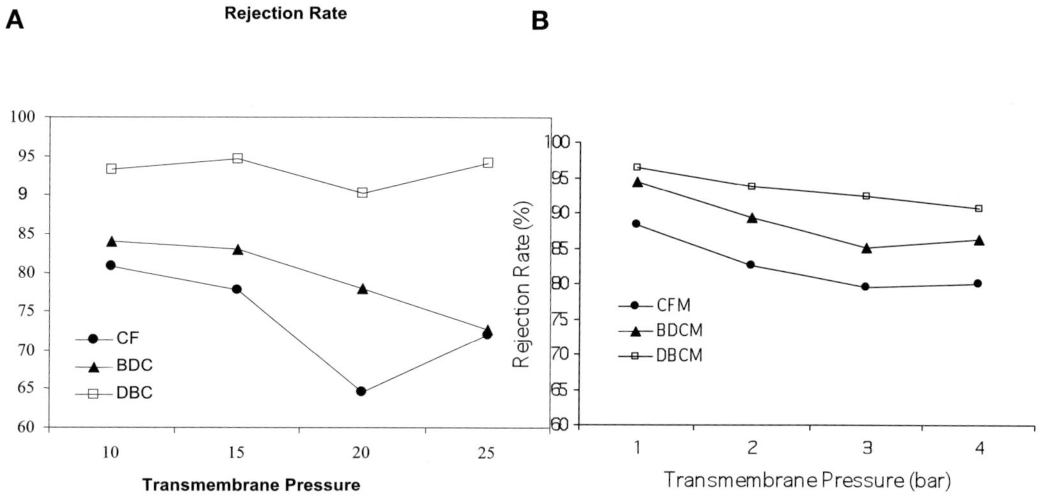

Figure 5.8 Influence of transmembrane pressure on permeate concentration of THM species with 200 μg/l feed concentration for (A) DS5 membrane and (B) NF200 membrane (Uyak *et al.*, 2008).

The treatment of atrazine and simazine in distilled water, tap water and river water were investigated by Zhang *et al.* (2004). Five NF membranes (DESAL 51 HL, DESAL 5 DL, UTC-20, UTC-60) were investigated during this study. It was found that removal efficiency was affected at different pH levels. The highest removal rate was obtained at pH of 8. On the other hand, the most suitable membrane for the removal of pesticides was found as UTC-20. Kosutic *et al.* (2005) found that the removal efficiency of pesticides increased by using a negatively charged NF270 membrane. The rejections of uncharged organic molecules were however influenced by the membrane material and pore size distribution. A NF48 membrane was used to remove isoproturon as target pesticides from contaminated surface water (Sarkar *et al.*, 2007). Coagulation was used as pretreatment and the concentration of isoproturon was decreased from 1 mg/L to 3–4 μg/L by using NF48 membrane.

Endocrine-disrupting chemicals (EDCs) consist of a vast number of natural and synthetic organics and inorganics. EDCs may interfere with the normal function of the hormonal systems of humans and animals, including abnormal sexual differentiation and development of the nervous and immune systems (Royalsoc., 2000). Recently a few researchers used a NF membrane for EDCs removal (Bodzek *et al.*, 2006; Kim and Kim, 2006; Yoon *et al.*, 2006, 2007; Kim *et al.*, 2008; Kazner *et al.*, 2008; Yangali-Quintanilla *et al.*, 2009; Comerton *et al.*, 2009a,b; Braeken and Van der Bruggen, 2009; Koyuncu *et al.*, 2008a). Clean and fouled nanofiltration membranes (NF200 and NF90) were used by Yangali-Quintanilla *et al.* (2009) for the removal of nine pharmaceuticals and five endocrine disruptors. Removal efficiencies of hydrophilic neutral compounds with a clean NF200 membrane varied from 35% to 70% while that of a NF90 membrane was in the range of 62–96%. Additionally, hydrophobic neutral compounds were removed up to 95–98% by a clean NF90. However, hydrophobic neutral compounds were removed by a fouled NF200 membrane up to 5–38% while a NF90 membrane did not remove these compounds. UTC-20 and NF270 membranes removed oestradiol and oestrone at rejection rates of 75– 85% and 65%–83%, respectively (Braeken and Van der Bruggen, 2009). Yoon *et al.* (2007) used thin film composites with a aromatic polyamide NF membrane (ESNA, Hydranautics, USA) for EDCs and personal care products removal. The NF membrane removal rates varied between 30%–90%. The solution chemistry, organic matter and salinity affected the rejection of tetracyclines and sulphonamides and selected hormones by NF membranes (Koyuncu *et al.*, 2008a). Tetracyclines have a high adsorptive affinity for the membrane while the adsorption rates for hormones are lower. An addition of antibiotics to the hormone solution increases the hormone rejections while almost complete rejections were observed for tetracyclines. Humics, other organics and also antibiotics influence the hormone rejection. Removal of pharmaceuticals from groundwater was studied by Radjenovic *et al.* (2008). Uncharged solutes with higher MWCO such as carbamazepine, hydrochlorothiazide, propyphenazone and glibenclamide were removed at 85% while lower MWCO such as acetaminophen was removed at 44.8–73%. More than 95% negatively charged pharmaceuticals such as ketoprofen, diclofenac and sulphamethoxazole were removed while 90% positively charged pharmaceuticals such as charged sotalol and metoprolol were removed. Some negatively charged pharmaceuticals such as gemfibrozil and mefenamic acid were rejected at 50–70% and 30–50%, respectively. The potential of nanofiltration for the separation of pharmaceutical and estrogenic compounds from salts in urine was investigated by Pronk *et al.* (2006). All micropollutants in non-hydrolysed urine were rejected above 92% by NF270 membranes, while synthetic urine rejection was above 73%.

Arsenic occurs naturally in the ground and surface waters and is not desired in drinking water due to carcinogenic effects on the human body. The common types of arsenic are arsenate (As(V)) and arsenite (As(III)) (Cakmakci *et al.*, 2008). On the other hand, the lethal range of inorganic arsenic for humans is estimated at a dose of 1–3 mg arsenic/kg (Shih, 2005). Recently, arsenic in drinking water was taken into considerable account because some of the drinking water resources contain high concentrations of arsenic which cause acute and chronic symptoms in many countries, especially in Bangladesh, China, Mongolia and Taiwan (Shih, 2005). In order to minimize the negatively effects, arsenic concentration in drinking water is promulgated and maximum contaminant level (MCL) of 10 µg/l arsenic in drinking water is recommended by WHO, European Union (EU) and U.S. Environmental Protection Agency (USEPA, 2000).

Recently, limited studies were applied with NF membranes to remove arsenic from water due to lower pressure requirements than for the RO membranes (Urase *et al.*, 1998; Kang *et al.*, 2000; Seidel *et al.*, 2001; Sato *et al.*, 2002; Saitua *et al.*, 2005; Shih, 2005; Uddin *et al.*, 2007; Xia *et al.*, 2007). Ionic species removal by a NF membrane strongly depended on the properties of the membrane. Membrane pore size and surface charge influenced ionic species retention. Sato *et al.* (2002) used negatively charged membranes such as ES-10, NTR-7250 and NTR-729HF (Nitto Electric Industrial Co., Japan). Arsenate was removed over 85% by these membranes whereas arsenite removal efficiency was very low (lower than 22%) by NTR-7250 and NTR-729HF. A ES-10 membrane removed arsenite more than 75%. A polyamide thin-film composite NF membrane was studied and 45–90% arsenate and 10–20% arsenite removal efficiencies were observed (Seidel *et al.*, 2001). Aromatic polyamide NF membranes were also studied to remove arsenite and 50 and 89% of removal efficiencies were obtained at a pH of 3 and 10, respectively (Urase *et al.*, 1998). Uddin *et al.* (2007) used a NF200 membrane and found the removal efficiency of arsenite under 10 bar pressure as 23% and 25% for the feed arsenite concentrations of 0.05 and 0.25 mg/L. NF200 and DS5 membranes were used by Cakmakci *et al.* (2008) at higher arsenite concentrations. They have found that arsenite removal efficiency decreased with increasing arsenite concentration and removal rate reached to 90% for the concentrations of 0.5 and 1.0 mg/l.

Dual-stage desalination

Fresh water sources are not adequate in the world and water demand increases every day. Thus, treatment of seawater and non-conventional water sources becomes more important to supply enough potable water. Although RO membranes have been reliably used in many seawater plants, the energy requirement is high when compared with the surface and ground water treatment by conventional systems. A dual-staged nanofiltration process is being evaluated as an alternative to reverse osmosis for seawater desalination. The primary goal of this system is to reduce energy consumption while producing potable water at an acceptable recovery rate. Dual-staged NF (NF^2) membranes developed by Long Beach Water Department (Gouellec *et al.*, 2006) are shown in Fig. 5.9 (lbwater, 2009). Seawater is pumped from the sea and then filtered to remove suspended materials. Filtered

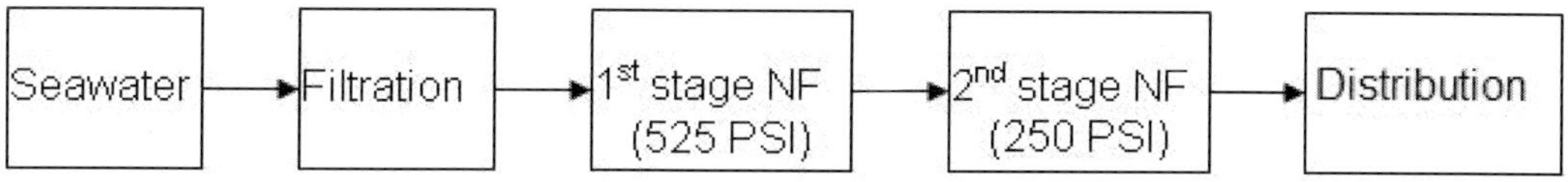

Figure 5.9 Seawater desalination by The Long Beach NF^2 (lbwater, 2009).

seawater is pumped under 525 psi (36.2 bar) pressure through the first NF membrane. This first NF membrane allows water molecules to pass but blocks all but the smallest 12% of salt molecules. The permeate of the first NF including the smallest 12% of salt molecules is pumped under 250 psi (17.2 bar) through the second NF membrane. This membrane blocks passage of almost all remaining salts. Long Beach Water Department found that NF^2 is the most-cost effective manner for seawater desalination as energy cost is reduced by 20–30% when compared with the traditional desalination method (lbwater, 2009). Harrison *et al.* (2007) used two commercial NF membranes for seawater desalination at bench scale study. They found that fouling is apparent and flux decline was about 20% after 3 days' operation.

NF membranes have also been used as a pretreatment before RO for seawater desalination. Thus, seawater RO (SWRO) is protected from membrane fouling by the removal of turbidity, bacteria and hardness such as Ca^{2+}, Mg^{2+}, SO_4^{2-} and HCO_3^- (Hassan *et al.*, 1999; Anne *et al.*, 2001). Hassan *et al.* (1999) used a pilot scale seawater treatment plant (Fig. 5.10) that included NF and RO membranes. The system consists of a dual media filter followed by a fine sand filter. They observed that SO_4^{2-}, total hardness and total dissolved solids were removed at 98% and 93%, respectively. On the other hand, NF as a pretreatment provides a lower operating pressure for SWRO. NF90, NF270 and N30F were used by Hilal *et al.* (2005) to treat highly concentrated (NaCl) salt solutions of up to 25,000 mg/l. An NF90 membrane rejected 95% of salt for salinity of 5000 mg/l and at pressure of 9 bars while its rejection declined to 41% at salinity of 25.000 mg/l at the same pressure. Rejection levels for NF270 and NF30F varied between 11–29% and 3–6%, respectively. Similar results were also observed with real seawater studies (Hilal *et al.*, 2007). NF90 and NF270 membranes were also used by Al-Zoubi and Omar (2009) for seawater treatment. They found that a NF90 membrane have a distinct ability

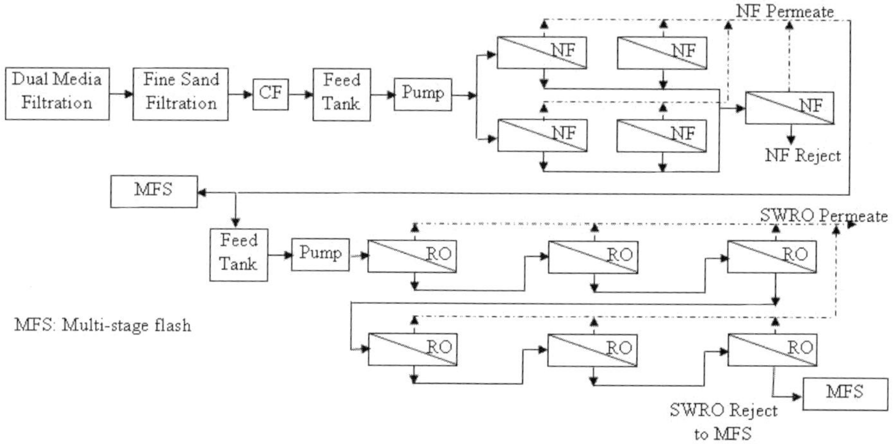

Figure 5.10 Schematic flow diagram of NF-SWRO Desalination Pilot Plant (Hassan *et al.*, 1999).

to reject both monovalent and divalent ions but with a relatively low flux. However, low values of monovalent ions were rejected by NF270 and divalent ions were rejected at reasonable efficiencies. According to these results, NF90 is more suitable as a pretreatment in desalination processes.

Wastewater treatment and reuse applications

Water scarcity in many parts of the world, along with improved living conditions and hygiene practices, has necessitated a search for new or alternative sources of water supply. Seawater and brackish water desalination, in addition to the reuse of domestic and industrial wastewaters, are important practices for optimal water usage, especially in water shortage regions and meeting the increasing global demand for potable water.

The United Nations identified in 2001, 'International Year of Freshwater', 18 countries that have a critical water shortage, mainly in the Middle East and North Africa and a few countries in Europe, Asia and the Caribbean. It is projected that in the year 2025 the above number of countries will increase to 29. Not only water quantity, but also water quality is important by means of human health in these countries. Membrane technology in the last years, is a promising contributor to solve water shortage and quality problems by wastewater reuse and seawater/brackish water desalination. Membrane technology can reduce the volume of wastewater generated and recover and recycle valuable components from the waste streams. Additionally, such a process occupies only a small portion of the floor space (Koyuncu *et al.*, 2001a,b). Municipal wastewater treatment applications for NF membranes are limited. However, it has a wide range of applications for industrial wastewaters. Some of these applications are summarized below.

Textile industry

The textile industry produces fabric materials by using synthetic and natural fibres and is originated by several production steps. Therefore, huge amounts of water and chemicals are used during production. Variations in production affect the characteristics of wastewater and at the end, highly coloured effluent streams, which are changing daily and seasonally and not acceptable by natural environment, are produced. Common water resources for textile industries are ground water, transportation by tankers from other places and municipal water. However, salinity of groundwater is increasing and water level is decreasing. This results in an increase in the unit water cost. Transportation by tankers is not safe and the cost is dependent on the oil prices which also changes very often. Municipal water price for the industries is around 2–3 euros/m^3 and this value is also very high. Based on this evaluation, industries should search for new water resources which are more cost effective than the options mentioned above. At this point water reuse can be a potential source for the textile industries.

Membrane processes can be used for the reuse of effluents from textile industries. Ultrafiltration (UF), nanofiltration (NF) and reverse osmosis (RO) membranes show great potential for direct reuse of dye bath wastewaters in the textile industry (Koyuncu, 2003a,b, Koyuncu *et al.*, 2004). Most of the divalent ions and dye molecules can be rejected by NF membranes, while water and sodium chloride pass. RO membranes can be used to remove salt. Additionally, UF membranes have potential as a pretreatment before RO and NF membranes (Koyuncu, 2002; Koyuncu and Topacik, 2002). The application of membrane filtration processes not only enables high removal efficiencies and water reuse, but also allows some of valuable waste constituents (Fersi, 2005; Koyuncu *et al.*, 2008b).

Dyehouse, rinsing and bleaching effluents are the three main most common pollution wastewater sources in the textile industry. Reactive dyes are soluble anionic dyes that contain one or more reactive groups capable of forming a covalent bond with the hydroxyl groups in the fibre that are not suitable for recycling (Erswell *et al.*, 1997). Thus, researchers have become intensive on the reuse of water and also recycling of other chemicals that is used in dyeing operations at recent times (Koyuncu *et al.*, 2001b; Gaeta and Fedele, 1991; Ciardelli *et al.*, 2000; Koyuncu, 2003b).

Nanofiltration membranes are becoming widely accepted in the textile industry wastewater treatment. Nanofiltration (NF) membranes allow the separation of low molecules weight organic compounds (200–1000 g/mol) (Akbari *et al.*, 2001) and divalent salts, with an appreciable softening effect (Marucci *et al.*, 2001; Koyuncu and Topacık, 2002; Koyuncu *et al.*, 2004). While nanofiltration membranes are able to reject dyes and other organic molecules, NaCl and other monovalent salts pass through the membrane, allowing the reuse of permeate water in dye bath preparation again (Fig. 5.11) (Koyuncu *et al.*, 2004). Dyehouse wastewaters remain at a high temperature after dyeing. At high temperatures, high flux values are obtained at the permeate stream due to the decreasing viscosity of permeating water (Wiesner and Aptel, 1996). Flux, salt and colour rejection graphs at pH 10 and 20 bar pressure over the temperature ranges of 22.5–37.5°C (Fig. 5.12). Flux value increased with increasing temperature, which was about 12 l/m^2h at 25°C and increased to 16 l/m^2h at 37°C (Koyuncu *et al.*, 2004). As shown in Fig. 5.12, flux value of NF effluent of dyehouse wastewater is very low. These effluents are difficult wastewaters to treat alone and a new tendency is to dilute these highly polluted wastewaters with diluted wastewaters such as rinsing water (Fig. 5.13) (Koyuncu *et al.*, 2008b). Removal efficiencies and flux values increased with the dilution of dyehouse effluents with rinsing water (Fig. 5.14). UF pretreatment had a slight effect on conductivity and TOC removals while colour removal was improved. UF pretreatment improved the colour rejection rates. Approximately more than 95% of removal efficiencies were obtained for NF and RO membranes.

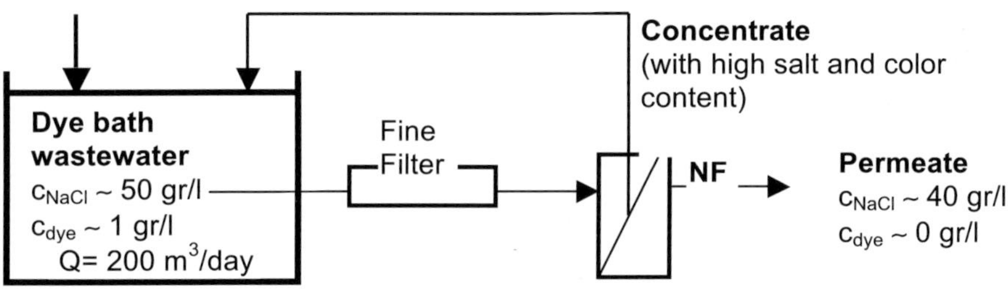

Figure 5.11 Application of NF membranes for the treatment of reactive dyehouse wastewater (Koyuncu *et al.*, 2004).

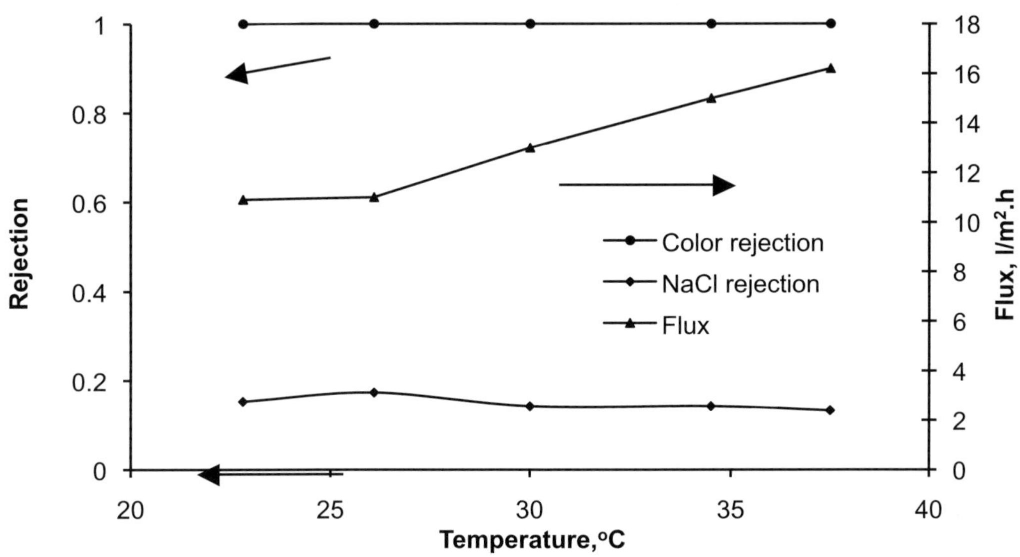

Figure 5.12 Flux, colour and salt rejections versus temperature with an NF membrane for dyehouse effluents (Koyuncu *et al.*, 2004).

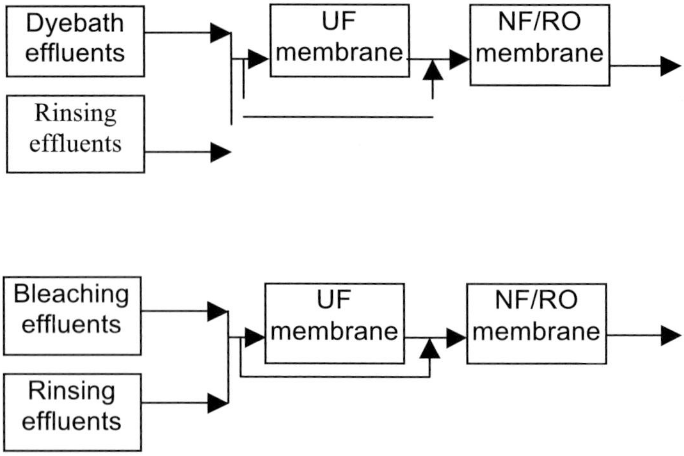

Figure 5.13 New wastewater reuse strategy in the textile industry (Koyuncu *et al.*, 2008b).

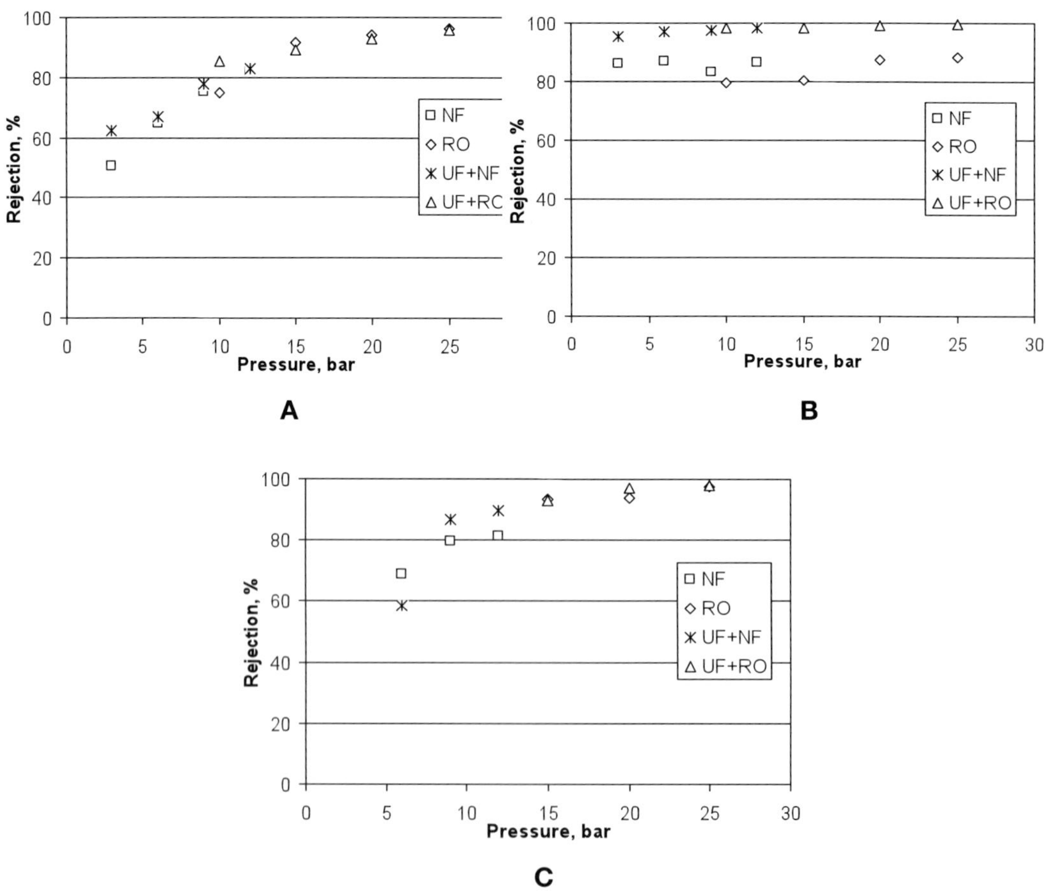

Figure 5.14 Removal efficiencies of (A) conductivity, (B) colour and (C) TOC for the mixture of dye bath (33.3%) and rinsing effluent (66.7%) (Koyuncu *et al.*, 2008b).

The reason of high TOC removal efficiency can be due to the removal of dye stuffs which are mainly in organic form. The flux attained with NF treatment was higher than RO in case of similar TOC, colour and conductivity removal rates. Therefore, it would be more rational to use NF treatment if the effluent will be reused in the dye bath process. However, if it will be used for process water containing no salt, RO treatment should be applied. It was also reported that mixing the bleaching and rinsing water promotes the treatment of bleaching effluent which is difficult to treat separately (Koyuncu *et al.*, 2008b).

Pulp and paper industry

Similar to the textile industry, the pulp and paper industry requires large amounts of water in bleaching and paper making processes. Membrane processes can be used either in the process or to reuse the effluent of the wastewater treatment plant (Koyuncu *et al.*, 1999; Yalcin *et al.*, 1999). Commonly, UF membranes were used to recycle the process water. However, permeate water quality limit the usage of this membrane. NF membrane has been tried as an alternative to UF membranes due to better water quality. COD reduction of 70–90%, AOX reduction of 90–97%, multivalent ion reduction of 90% and complete colour removal can be achieved with NF membranes (Nystrom *et al.*, 2005). Nystrom *et al.* (2005) summarized that the NF process can be applied as BAT (best available technology) to reduce water consumption which gives good quality of water in the mill in conjunction with pretreatment. Additionally, membrane fouling can be reduced using high shear modules. Thermophilic aerobic biological treatment, flocculation or UF pretreatment was suggested before using a NF membrane.

Food industry

Main application areas in the food industry are in sugar, dairy, oil and beverage industries (Bargeman *et al.*, 2005). UF and RO membranes have been used in the dairy industry since the 1970s (Koyuncu *et al.*, 2000). Recently, NF membranes have started to be utilized in the dairy industry mainly for the concentration and demineralization of whey and desalination and concentration of UF permeate for lactose production. Generally, low salt rejection NF membranes were used for this purpose (Helakorpi *et al.*, 2000). NF membranes can also be used for continuous cheese production (Fig. 5.15).

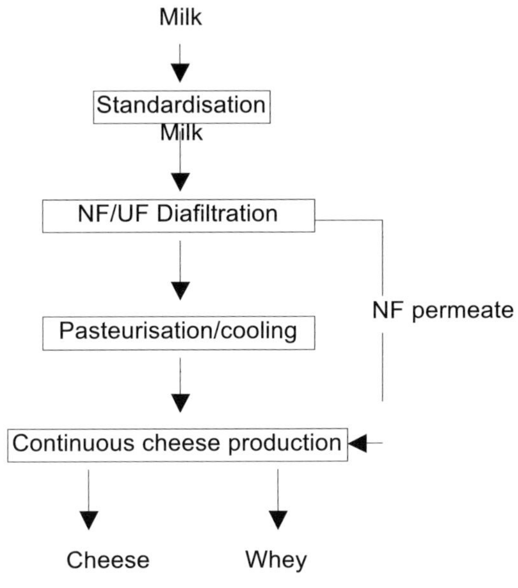

Figure 5.15 Continuous cheese production using NF membrane (Bargeman *et al.*, 2005).

NF membranes are also used for concentration and purification in the sugar industry and have an advantage in comparison to alternative processes. Concentration of dextrose syrup and demineralization of coloured brine from anion-exchange resin are the potential application areas of NF membranes in the sugar industry (Bargeman *et al.*, 2005).

Landfill leachate treatment

Treatment of leachate is very complicated and generally requires various process applications due to high concentrations of COD and nitrogen as well as colour. Several anaerobic and aerobic treatment systems have been studied for landfill leachate treatment. These systems have demonstrated high performance, although some problems have been detected depending on the leachate characteristics mainly due to the age of landfill (Keenan *et al.*, 1993; Ledakowicz and Kaczorek, 2001; Ozturk *et al.*, 2003). Therefore, the variations of flow rates and complex composition have to be considered during the application of a treatment process. Since biological processes alone cannot remove the residual and refractory organics, an additional treatment is often needed. For post-treatment of biologically pre-treated landfill leachates have physico-chemical processes widely been used (Ozturk *et al.*, 2003).

Several researchers have investigated NF and RO membranes for the treatment of landfill leachate. Performance of NF and RO membranes to remove the substances are given in Table 5.4 (Melin *et al.*, 2005). A NF membrane alone was also tried at pilot scale by Linde *et al.* (1995) to remove heavy metals. As given in Table 5.5, removal efficiencies of multivalent ions were high while removal efficiencies for monovalent ions were low. However, lower pressures were used compared to the RO membrane (Linde *et al.*, 1995). Nanofiltration membranes were also used in combination with RO membranes as shown in Fig. 5.16 (Peters, 1997) and removal efficiencies of this process are given in Table 5.6. Removal efficiencies increased in this configuration. MBR and NF membranes were also applied to landfill leachate treatment in several locations. A flow diagram of a landfill leachate treatment plant of Istanbul city is given in Fig. 5.17. Permeate quality of NF permeate reaches the discharge standards for landfill leachate as shown in Table 5.7.

Chemical industry

The chemical industry yields a wide variety of products. These products are converted from raw materials (oil, natural gas, air, water, metals and minerals) into more than 70,000 different products. Wide varieties of consumer goods, as well as thousands of products that are essential inputs to agriculture, manufacturing, construction and service industries are produced by chemicals. The modern

Table 5.4 Performance of NF and RO membrane for landfill leachate applications (Melin *et al.*, 2005)

Parameter	Unit	NF	RO
COD	%	90–95	92–97
AOX	%	85–93	90–96
NO_3-N	%	10–20	83–93
NH_4-N	%	10–20	86–94
Cl^-	%	5–10	98–99.6
SO_{4-}^{2}	%	94–98	99–99.9
Heavy metals	%	85–96	88–97
Permeate flux	$l/m^2.h$	20–40	10–20
Recovery	%	75–85	70–80
Pressure	Bar	10–30	40–65

Table 5.5 Pilot scale NF study to remove heavy metals from landfill leachate (Linde *et al.*, 1995)

Parameter	Feed	Permeate (AFC30 NF)	Permeate (AFC40 NF)
Ca (mg/l)	12,887	7918	9822
Cd (mg/l)	0.033	0.002	0.003
Cl^- (mg/l)	37,054	27,689	30.791
Cr (mg/l)	0.029	0.008	<0.007
Fe (mg/l)	85.8	19.74	25.85
K (mg/l)	3940	4004	3813
Mg (mg/l)	243.5	87.6	117.9
Na (mg/l)	6400	6.068	5874
Pb (mg/l)	0.609	<0.02	0.022
TOC (mg/l)	142	57	64

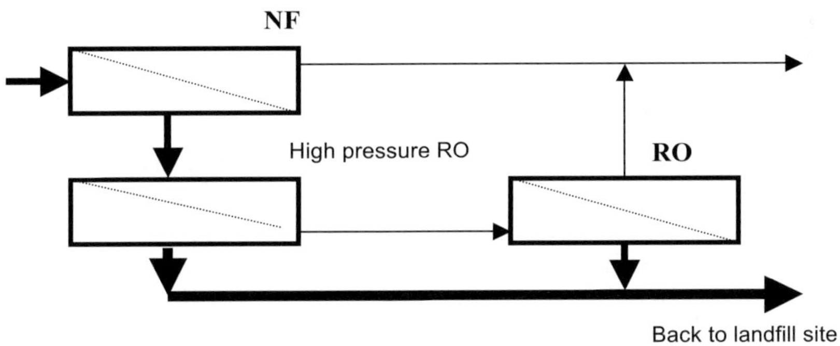

Figure 5.16 NF application combination with RO to treat landfill leachate (Peters, 1997).

Table 5.6 Permeate values and removal efficiencies of NF and RO membranes for Halle-Lochau landfill leachate (Peters, 1997)

Parameter	Feed	NF+ RO permeate	Removal efficiency (%)
BOD_5(mg/l)	591.5	239.6	59.5
COD (mg/l)	6400	<15	99.9
NH_4-N (mg/l)	2260	0.66	>99.2
Na (mg/l)	41,000	2.5	99.9
Ca (mg/l)	243	31	87.2
Mg(mg/l)	952	133	86
Cl^- (mg/l)	29,780	35,629	19.6
pH	6.23	6.31	–
Conductivity (mS/cm)	120	20	>98

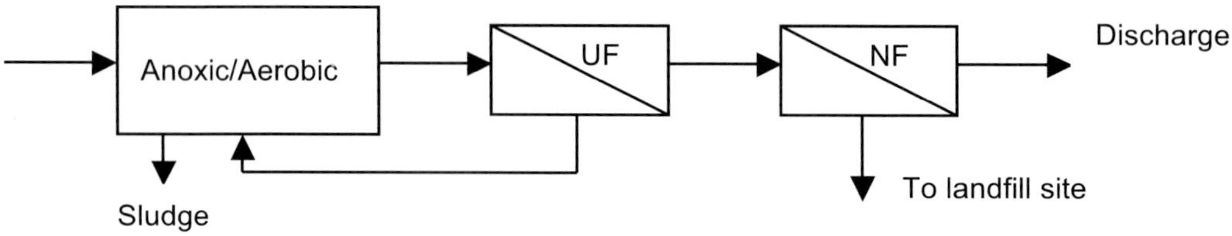

Figure 5.17 Flow diagram of landfill leachate treatment plant of Istanbul city.

Table 5.7 Permeate water quality parameters for the landfill leachate treatment plant of Istanbul city

Parameter	Feed	MBR (UF) permeate	NF Permeate
COD (mg/l)	10,000–20,000	1000–1500	200–300
NH_4-N (mg/l)	2750	<5	<1
PO_4-P (mg/l)	30–40	<5	<1
pH	8.2	8.2	-

chemical industry is divided into two sections: the first section is 'Production of Chemicals', including petrochemicals, acids, alkalis, pharmaceuticals, synthetic resins and plastic dusts, and the second section is 'Chemical Processing', including dyes, tires, drugs, foods and cosmetics. Wastewater can be generated from both sections. Conventional treatment technology is not always suitable for the treatment of this wastewater. Thus, it is required to use suitable treatment technology. Membrane technology can be used in both sections of the chemical industry. Recently, NF membranes has been preferred as a membrane technology due to the lower energy requirement (Park and Barnett, 2001; Zhang *et al.*, 2003; Mikulasek *et al.*, 2006; Banerjee *et al.*, 2007; He *et al.*, 2009).

NF membranes can be used in chemical production. Zhang *et al.* (2003) coated NF membrane with polyvinyl alcohol to reduce membrane fouling while concentrating antibiotics. Antibiotics solutions were concentrated from 12,000 to 110,000 U/ml at laboratory scale and from 80,000 to 340,000 U/ml at pilot scale. He *et al.* (2009) concentrated 5 dyes including CI Reactive Orange 12, CI Reactive Red 24, CI Reactive Blue 74, CI Reactive Black 5 and CI Reactive Blue 13 with a CA NF membrane. The schematic diagram of the dye concentration is shown in Fig. 5.18. Five dyes were rejected at levels higher than 99% by CA NF. Thus, dyes were more concentrated at high rejections. Banerjee *et al.* (2007) used a two-stage NF (NF^2) membrane to decline eosin dye concentration. After the two stage NF, eosin dye concentration declined from 70–200 mg/l to lower than 1 mg/l. Mikulasek *et al.* (2006) investigated salt-dye separation by a NF membrane. They observed a high concentration of dyes and salts in permeate streams. The phosphate and sulphate were almost completely retained while the nutrients and ammonia were well permeated. UF and NF membranes were compared for oil/water separation by Park and Barnett (2001).

Nanofiltration market

Nanofiltration membrane applications are increasing each year with the continued growth in regulations as shown in Fig. 5.19. The global market for nanofiltration membranes increased from $89.1 million in 2006 to an estimated $97.5 million by the end of 2007. It should reach $310.5 million by 2012 with a compound annual growth rate of 26.1%. The water treatment sector is projected to account for 72.7% of total revenues and estimated at $70.9 million in 2007 and expected to reach $238.2 million by 2012 (BCC Research, 2009).

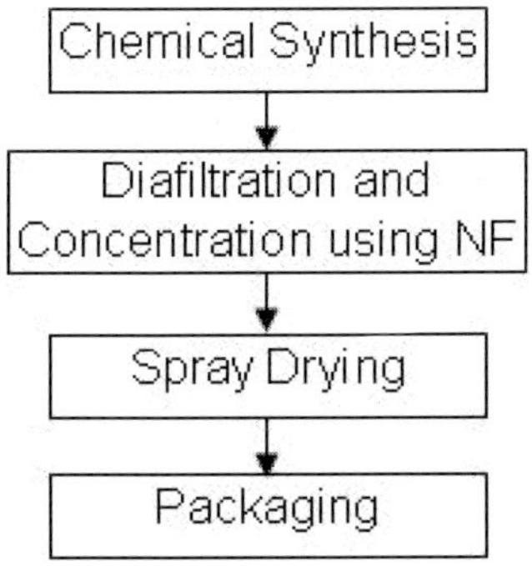

Figure 5.18 The membrane process for dye production (He *et al.*, 2009).

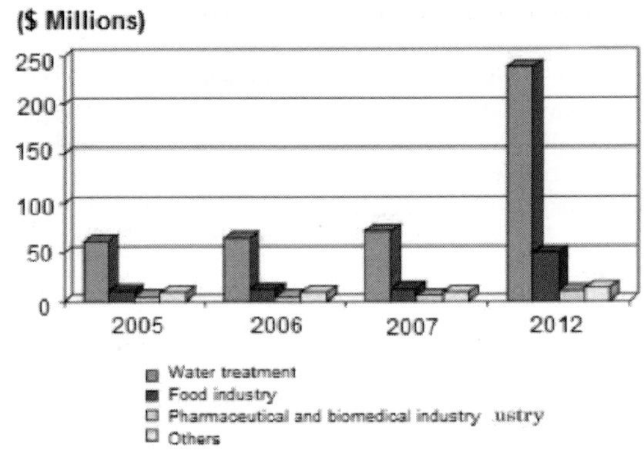

Figure 5.19 Global NF market (BCC Research, 2009).

Conclusions and future trends

The nanofiltration market is still growing and applications are increasing. NF membranes can be used either in the water treatment industry to supply potable water to the public or process water to the industry or in the wastewater treatment industry for water reuse or product recovery. It is going to play an important role in the water industry due to its low operating cost and selective removal capacity. A combination of NF with the product machines in the industry such as textile and pulp and paper industries, new process configurations, low fouling, low pressure and pH and temperature resistant NF membranes will find wide range of applications in the near future.

References

Ahmad, A.L., Tan, L.S., and Shukor, S.R. (2009). Modeling of the retention of atrazine and dimethoate with nanofiltration. Chem. Eng. J. *147*, 280–286.

Akbari, A., Desclaux, S., Remigy, J.C., and Aptel, P. (2001). Treatment of textile dye effluents using a new photografted nanofiltration membrane. Desalination *149*, 101–107.

Allgeier, S.C., and Summers, R.S. (1995). Evaluating NF for DBP control with the RBSMT. J. Am. Water Works Assoc. *87*, 87.

Al-Zoubi, H., and Omar, W. (2009). Rejection of salt mixtures from high saline by nanofiltration membranes. Korean J. Chem. Eng. *26*, 799–805.

Anne, C.O., Trebouet, D., Jaouen, P., and Quemeneur, F. (2001). Nanofiltration of seawater: fractionation of mono and multi-valent cations. Desalination *140*, 67–77.

A.W.W.A. (2007). Reverse Osmosis and Nanofiltration. In Manual of Water Supply Practices (Denver, USA).

Banerjee, P., DasGupta, S., and De, S. (2007). Removal of dye from aqueous solution using a combination of advanced oxidation process and nanofiltration. J. Hazard. Mater. *14*, 95–103.

Bargeman, G., Timmer, M., and Van der Horst, C. (2005). Nanofiltration in the food industry. In Nanofiltration-Principles and Applications, Schafer, A.I., Fane, A.G., and Waite, T.D., eds. (Elsevier, Oxford).

B.C.C Research (2009). http://www.bccresearch.com/report/NAN045A.html
Berg, P., Hagmeyer, G., and Gimbel, R. (1997). Removal of pesticides and other micropollutants by nanofiltration. Desalination *113*, 205–208.
Bergman, R.A. (1996). Cost of membrane softening in Florida, J. AWWA. *88*, 32–43.
Bhattacharya, A., and Ghosh, P. (2004). Nanofiltration and Reverse Osmosis Membranes: Theory and Applications in Separation of Electrolytes. Rev. Chem. Eng. *20*, 111–173.
Braeken, L., and Van der Bruggen, B. (2009). Feasibility of nanofiltration for the removal of endocrine disrupting compounds. Desalination *240*, 127–131.
Bodzek, M., and Dudziak, M. (2006). Removal of natural estrogens and synthetic compounds considered to be endocrine disrupting substances (EDs) by coagulation and nanofiltration, Pol. J. Environ. Stud. *15*, 35–40.
Boussahel, R., Bouland, S., Moussaoui, K.M., and Montiel, A. (2000). Removal of pesticide residues in water using the nanofiltration process. Desalination *132*, 205–209.
Bowen, W.R., and Welfort, J.S. (2005). Modelling the performance of nanofiltration membranes. In Nanofiltration: Principles and Applications, Schafer, A.I., Fane, A.G., and Waite, T.D., eds. (Elsevier, Oxford).
Bowen, W.R., and Mohammed, A.W. (1998). Diafiltration by nanofiltration: prediction and optimization. AICHE J. *44–48*, 1799–1812.
Cadotte, J.E., Cobian, K.E., Forester, R.H., and Petersen, R.J. (1976). NTIS report No: PB-253193.
Cadotte, J.E. (1977). Reverse Osmosis Membranes, US Patent 4039.
Cakmakci, M., Baspinar, A., Balaban, U., Uyak, V., Koyuncu, I., and Kinaci, C. (2008). Comparison of Nanofiltration and Adsorption Techniques to Remove Arsenic from Drinking Water, Membranes in Drinking Water Production and Wastewater Treatment (MDIW 08) Conference, 20–22 October, Toulouse, France.
Chellam, S., (2000). Effect of nanofiltration on trihalomethane and haloacetic acid precursor removal and speciation in waters containing low concentrations of bromide ion. Environ. Sci. Technol. *34*, 1813.
Ciardelli, G., Corsi, L., and Marcucci, M. (2000). Membrane separation for wastewater reuse in the textile industry. Resour. Conserv. Recy. *31*, 189–197.
Comerton, A.M., Andrews, R.C., and Bagley, D.M. (2009a). The influence of natural organic matter and cations on the rejection of endocrine disrupting and pharmaceutically active compounds by nanofiltration. Water Res. *43*, 613–622.
Comerton, A.M., Andrews, R.C., and Bagley, D.M. (2009b). The influence of natural organic matter and cations on the rejection of endocrine disrupting and pharmaceutically active compounds by nanofiltration. Water Res. *43*, 613–622.
Costa, A.R., and Maria Norberta de Pinho, M.N. (2006). Comparison of the performance of ultrafiltration and nanofiltration in surface water treatment. Desalination *199*, 73–75.
Diawara, C.K. (2008). Nanofiltration process efficiency in water desalination. Sep. Purif. Rev. *37*, 302–324.
Dresner, L. (1972). Some remarks on the integration of the extended Nernst–Planck equations in the hyperfiltration of multicomponent solutions. Desalination. *10*, 27–46.
E.E.C.D (European Economic Community Directive) (1998). Amended proposal for a Council Directive concerning the quality of water intended for human consumption-common position, In Proceedings of the Council of the European Union, Directive 80/778/EEC, Com (97) 228 final 95/0010 SYN, Brussels.
Eriksson, P. (1988). Nanofiltration extends the range of membrane filtration. Environ. Prog. *7*, 58–62.
Erswell, A., Brouckaert, C.J., and Buckley, C.A. (1988). The reuse of reactive dye liquors using charged ultrafiltration membrane technology. Desalination *70*, 157–167.
Fane, A.G. (2005). Module design and operation. In Nanofiltration: Principles and Applications, Schafer, A.I., Fane, A.G., and Waite, T.D., eds. (Elsevier, Oxford).
Fersi, C., Gzara, L., and Dhahbi, M. (2005). Treatment of textile effluents by membrane technologies. Desalination *185*, 399–409.
Gaeta, S.N., and Fedele, U. (1991). Recovery of water and auxiliary chemicals from effluents of textile dye houses. Desalination *83*, 183–194.
Geraldesa, V., Anila, A., Pinhoa, M.N., and Duarte, E. (2008). Dissolved air flotation of surface water for spiral-wound module nanofiltration pre-treatment. Desalination *228*, 191–199.
Gouellec, Y.A.L., Cornwell, D., Cheng, R.C., and Tseng, T. (2006). A Novel Approach to Seawater Desalination Using Dual-Staged Nanofiltration. (IWA Publishing, London, UK).
Harrison, C.J., Le Gouellec, Y.A., Cheng, R.C., and Childress, A.E. (2007). Bench-scale testing of nanofiltration for seawater desalination. J. Environ. Eng.-ASCE *133*, 1004–1014.
Hassan, A.M., Farooque, A.M., Jamaluddin, A.T.M., Al-Amoudi, A.S., Al-Sofi, M.A.K., Rubaian, A.F., Gurashi, M.M., Kither, N.M., Dalvi, A.G.I., and Al-Tisan, I.A.R. (1999). Optimization of NF pretreatment of feed to seawater desalination plants. IDA World Congress on Desalination Water Reuse, San Diego, USA.
He, Y., Li, G.M., Wang, H., Jiang, Z.W., Zhao, J.F., Su, H.X., and Huang, Q.Y. (2009). Experimental study on the rejection of salt and dye with cellulose acetate nanofiltration membrane. J. Taiwan Inst. Chem. Eng. *40*, 289–295.
Helakorpi, P., Mikkonen, H., Myllykoski, L., and Keiski, R. (2000). Nanofiltration in the Dairy Industry. Case Study: Effect of nanofiltration on lactose crystallization Proceedings of World Filtration Congress *8*, Brighton, pp. 914–917.

Hilal, N., Al-Zoubi, H., Darwish, N.A., Mohammad, A.W., and Abu Arabi, M. (2004). A comprehensive review of nanofiltration membranes: Treatment, pretreatment, modelling, and atomic force microscopy. Desalination *170*, 281–308.

Hilal, N., Al-Zoubia, H., Mohammad, A.W., and Darwishc, N.A. (2005). Nanofiltration of highly concentrated salt solutions up to seawater salinity. Desalination *184*, 315–326.

Hilal, N., Al-Zoubi, H., Darwish, N.A., and Mohammad, A.W. (2007). Performance of nanofiltration membranes in the treatment of synthetic and real seawater. Separ. Sci. Technol. *42*, 493–515.

Ikeda, K., Nakano, T., Kubota, T., and Yamamoto, S. (1986). New Composite Charged Membrane. Desalination *68*, 109–119.

Kang, M., Kawasaki, M., Tamada, S., Kamei, T., and Magara, Y. (2000). Effect of pH on the removal of arsenic and antimony using reverse osmosis membranes. Desalination *131*, 293–298.

Kazner, C., Lehnberg, K., Kovalova, L., Wintgens, T., Melin, T., Hollender, J., and Dott, W. (2008). Removal of endocrine disruptors and cytostatics from effluent by nanofiltration in combination with adsorption on powdered activated carbon. Water Sci. Technol. *58*, 1699–1706.

Keenan, P.J., Iza, J., and Switzenbaum, M.S. (1993). Inorganic solid development in a pilot scale anaerobic reactor treating municipal solid waste landfill leachate. Water Environ. Res. *65*, 181–188.

Khulbe, K.C., Feng, C.Y., and Matsuura, T. (2008). Synthetic Polymeric Membranes: Characterization by Atomic Force Microscopy (Springer, Germany).

Kim, S., and Kim, J. (2006). Rejection of endocrine disrupters contained in biologically treated sewage by nanofiltration. Korean J. Chem. Eng. *23*, 747–752.

Kim, H.-A., Jae-Hoon Choi, J.-H., and Takizawa, S. (2007). Comparison of initial filtration resistance by pretreatment processes in the nanofiltration for drinking water treatment. Separ. Sci. Technol. *56*, 354–362.

Kim, J.-H., Kim, S., Lee, C.H., Kwon, H.-H., and Lee, S. (2008). A novel nanofiltration hybrid system to control organic micro-pollutants: application of dual functional adsorbent/catalyst. Desalination *231*, 276–282.

Kim, J.H., Park, P.K., Lee, C.H., Kwon, H.H., and Lee, S. (2008). A novel hybrid system for the removal of endocrine disrupting chemicals: Nanofiltration and homogeneous catalytic oxidation. J. Membrane Sci. *312*, 66–75.

Kiso, Y., Nishimura, Y., Kitao, T., and Nishimura, K. (2000). Rejection properties of non-phenylic pesticides with nanofiltration membranes. J. Membrane Sci. *171*, 229–237.

Kiso, Y., Sugiura, Y., and Nishimura, K. (2001). Effects of hydrophobicity and molecular size on rejection of aromatic pesticides with nanofiltration membranes. J. Membrane Sci. *192*, 1–10.

Koffskey W.E., and Lykins, B.W. (1999). Disinfection/Disinfectant by-product optimization with ozone, biological filtration and chloramines. J. Water Supply Res. T. *48*, 92.

Korean Water (2010). http://www.korean-water.com/images/filtration-spectrum.gif

Kosutic, K., Furac, L., Sipos, L., and Kunst, B. (2005). Removal of arsenic and pesticides from drinking water by nanofiltration membranes. Separ. Sci. Technol. *42*, 137–144.

Koyuncu, I., Yalcin, F., and Ozturk, I. (1999). Color removal of high strength paper and fermentation industry effluents with membrane technology, Water Sci. Technol. *40*, 241–248.

Koyuncu, I., Turan, M., Topacik, D., and Ates, A. (2000). Application of low pressure nanofiltration membranes for the recovery and reuse of dairy industry effluents. Water Sci. Technol. *41*, 213–221.

Koyuncu I., Sevimli, M.F., Ozturk, I., and Aydin, A.F. (2001a). Application of membrane and ozonation technologies to remove color from agro-industry effluents. Water Sci. Technol. *43*, 233–241.

Koyuncu, I., Kural, E., and Topacik, D. (2001b). Pilot scale nanofiltration membrane separation for waste management in textile industry. Water Sci. Technol. *43*, 233–240.

Koyuncu, I. (2002). Reactive Dye Removal in Dye/Salt Mixtures by Nanofiltration Membranes Containing Vinylsulphone Dyes: Effects of Feed Concentration and Cross Flow Velocity. Desalination *143*, 243–253.

Koyuncu, I., and Topacik, D. (2002). Effect of organic ion on the seperation of salts by nanofiltration membranes. J. Membrane Sci. *195*, 247–263.

Koyuncu, I., and Topacik, D. (2003). Effects of operating conditions on the salt rejection of nanofiltration membranes in reactive dye/salt mixtures. Separ. Sci. Technol. *33*, 283–294.

Koyuncu, I. (2003a). Direct Filtration of Procion Dye Bath Wastewaters by Nanofiltration Membranes: Fouling and Rejection Characteristics. J. Chem. Technol. Biot. *78*, 1219–1224.

Koyuncu, I. (2003b). Influence of dyes, salts and auxiliary chemicals on nanofiltration of reactive dye baths: experimental observations and model verification. Desalination *154*, 79–88.

Koyuncu, I., Topacik, D., and Yuksel, E. (2004). Reuse of Reactive Dyehouse Wastewater by Nanofiltration: Process Water Quality and Economical Implications. Separ. Sci. Technol. *36*, 77–85.

Koyuncu, I., Brant, J.A., Lüttge, A., and Wiesner, M.R. (2006a). A Comparison of Vertical Scanning Interferometry (VSI) and Atomic Force Microscopy (AFM) for Characterizing Membrane Surface Topography. J. Membr. Sci. *278*, 410–417.

Koyuncu, I., Wiesner, M.R., Bele, C., Coriton, G., Djafer, M., and Cavard, J. (2006b). Bench-scale assessment of pretreatment to reduce fouling of salt rejecting membranes. Desalination *197*, 94–105.

Koyuncu, I., Arikan, O.A., Wiesner, M.R., and Rice, C. (2008a). Removal of hormones and antibiotics by nanofiltration membranes. J. Membrane Sci. *309*, 94–101.

Koyuncu, I., Demiral, N., Erdim, E., Soyer, E., and Arslan, A. (2008b). Reuse of Segregated Textile Effluents by Membrane Technologies, Proceedings of IWA Regional Conference on Membrane Technologies in Water and Waste Water Treatment, Moscow, Russia.

Lakshminarayanaiah, N. (1969). Transport Phenomena in Membranes (New York, Academic Press).

lbwater (2009). www.lbwater.org

Ledakowiwicz, S., and Kaczorek, K. (2001). Biodegradation of leachate from municipal landfill in Lodz enhanced by advanced oxidation processes. 8th International Waste Management and Landfill Symposium Proceedings, Cagliari, Sardinia, Italy.

Lee, S., and Lee, C.H. (2007). Effect of membrane properties and pretreatment on flux and NOM rejection in surface water nanofiltration. Separ. Sci. Technol. *56*, 1–8.

Levenstein, R., Hasson, D., and Semiat, R. (1996). Utilization of the Donnan effect for improving electrolyte separation with nanofiltration membranes. J. Membr. Sci. *116*, 77–92.

Linde, K., and Jönsson, A. (1995). Nanofiltration of salt solutions and landfill leachate. Desalination *103*, 223–232.

Linder, C., and Kedem, O. (2005). History of nanofiltration membranes 1960 to 1990. In Nanofiltration: Principles and Applications, Schafer, A.I., Fane, A.G., and Waite, T.D., eds. (Elsevier, Oxford).

Low, S.C., Cheng, L.P., and Hee, L.S. (2008). Water softening using a generic low cost nano-filtration membrane. Desalination *221*, 168–173.

Machenbach, I., Brouckaert, J.C., and Buckley, C.A. (1997). Nanofiltration of reactive dye effluent, 18th National Meeting at the South African Institution of Chemical Engineers, 16–18 April 1997.

Marconnet, C., Houari, A., Galas, L., Vaudry, H., Heim, V., and Di Martino, P. (2009). Biodegradable dissolved organic carbon concentration of feed water and NF membrane biofouling: a pilot train study. Desalination *242*, 228–235.

Marucci, M., Nosenzo, G., Capanelli, G., Ciabatti, I., Corrieri, D., and Ciardelli,G. (2001). Treatment and reuse of textile effluents based on new ultrafiltration and other membrane technologies. Desalination *138*, 75–82.

Mazzoni, C., and Bandini, S. (2006). On nanofiltration Desal-5 DK performances with calcium chloride-water solutions. Separ. Sci. Technol. *52*, 232–240.

Melin, T., Meier, J., and Wintgens, T. (2005). Nanofiltration in landfill leachate treatment. In Nanofiltration-Principles and Applications, Scha, A.I. ¨fer, Fane, A.G., and Waite, T.D., eds. (Elsevier, Oxford).

Mika, A.M., Ronald, F., Childs, R.F., and Dickson, J.M. (1999). Ultra-low pressure water softening: a new approach to membrane construction. Desalination *121*, 149–158.

Mikulasek, P., Kopecky, V., and Kusnierik, O. (2006). Salt removal from process solutions in liquid dye production by nanofiltration. Desalination *200*, 409–410.

Montovay, T., Assenmacher, M., and Frimmel, F.H. (1996). Elimination of pesticides from aqueous solution by nanofiltration. Magy. Kem. Foly. *102*, 241–247.

Mulder, M., Voorthuizen, E., and Peeters, J. (2005). Membrane characterization. In Nanofiltration: Principles and Applications, Schafer, A.I., Fane, A.G., and Waite, T.D., eds. (Elsevier, Oxford).

Nystrom, M., Nuortila-Jokinen, J.M. K., and Manttari, M.J. (2005). Nanofiltration in the pulp and paper industry. In Nanofiltration-Principles and Applications, Schafer, A.I., Fane, A.G., and Waite, T.D., eds. (Elsevier, Oxford).

Orecki, A., Tornaszewska, M., Karakulski, K., and Morawski, A.W. (2004). Surface water treatment by the nanofiltration method. Desalination *162*, 47–54.

Ouyang, L., Malaisamy, R., and Bruening, M.L. (2008). Multilayer polyelectrolyte films as nanofiltration membranes for separating monovalent and divalent cations. J. Membrane Sci. *310*, 76–84.

Ozturk, I., Altinbas, M., Koyuncu, I., Arikan, O., and Yangin, C. (2003). Advanced Physico-chemical Treatment Experiences on Young Landfill Leachates. Waste Manage *23*, 441- 446.

Park, E., and Barnett, S.M. (2001). Oil/water separation using nanofiltration membrane technology. Separ. Sci. Technol. *36*, 1527–1542.

Park, N., Kwon, B., Sun, M., Ahn, H., Kim, C., Kwoak, C., Lee, D., Chae, S., Hyung, H., and Cho, J. (2004). Application of various membranes to remove NOM typically occurring in Korea with respect to DBP, AOC and transport parameters. Desalination *178*, 161–169.

Perry, M., and Linder, C. (1989). Intermediate reverse osmosis ultrafiltration (RO UF) membranes for concentration and desalting of low molecular weight organic solutes. Desalination *71*, 233–245.

Peters, T. (1997). Treatment of landfill leachate by reverse osmosis. Proceedings Sardinia, *97*, Sixth International Landfill Synposium, Cagliari, Italy.

Pronk, W., Palmquist, H., Biebow, M., and Bollera, M. (2006). Nanofiltration for the separation of pharmaceuticals from nutrients in source-separated urine. Water Res. *40*, 1405–1412.

Radjenovic, J., Petrovic, M., Ventura, F., and Barcelo, D. (2008). Rejection of pharmaceuticals in nanofiltration and reverse osmosis membrane drinking water treatment. Water Res. *42*, 3601–3610.

Ratasuk, C., Ratanatamskul, C., and Ratasuk, N. (2009). Removal of haloacetic acids by nanofiltration. J. Environ. Sci. *21*, 96–100.

Rautenbach, R., and Gröschl, A. (1990). Separation potential of nanofiltration membranes. Desalination *77*, 73–84.

Reiss, C.R., Taylor, J.S., and Robert, C. (1999). Surface water treatment using nanofiltration pilot testing results and design considerations. Desalination *125*, 97–112.

Royalsoc (2000). Endocrine disrupting chemicals (EDCs). The Science Advice Section at the Royal Society (www.royalsoc.ac.uk), Registered Charity No 207043.

Saitua, H., Campderros, M., Cerutti, S., and Perez, P.A. (2005). Effect of operating conditions in removal of arsenic from water by nanofiltration membrane. Desalination *172*, 173–180.

Sarkar, B., Venkateswralu, N., Rao, R.N., Bhattacharjee, C., and Kale, V. (2007). Treatment of pesticide contaminated surface water for production of potable water by a coagulation–adsorption–nanofiltration approach. Desalination *212*, 129–140.

Sato, Y., Kang, M., Kamei, T., and Magara, Y. (2002). Performance of nanofiltration for arsenic removal. Water Res. *36*, 3371–3377.

Schaep, J., Van der Bruggen, B., Uytterhoeven, S., Croux, R., Vandecasteele, C., Wilms, D., Van Houtte, E., and Vanlerberghe, F. (1998). Removal of hardness from groundwater by nanofiltration. Desalination *119*, 295–302.

Schafer, A., Fane, A.G., and Waite, D. (2005). Nanofiltration: Principles and Applications, (Elsevier, Oxford, UK.).

Schirg, P., and Widmer, F. (1992). Characterization nanofiltration membranes for the separation of aqueous dye-salt solutions. Desalination *89*, 89–107.

Seidel, A., Waypa, J.J., and Elimelech, M. (2001). Role of charge (Donnan) exclusion in removal of arsenic from water by a negatively charged porous nanofiltration membrane. Environ. Eng. Sci. *18*, 105–113.

Shih, M.-C. (2005). An overview of arsenic removal by pressure-driven membrane processes. Desalination *172*, 85–97.

Singer, P.C. (1999). Formation and Control of Disinfection By-products in Drinking Water, AWWA, Denver, CO.

Spiegler, K.S., and Kedem, O. (1966). Thermodynamics of hyperfiltration (RO): Criteria for efficient membranes. Desalination *1*, 311.

Strantz Jr., J., and Brehm, W.J. (1990). Treatment for reverse osmosis membranes, US Patent 4938872.

Thorsen, T., and Fløgstad, H. (2006). Nanofiltration in drinking water treatment: Literature Review. Techneau, D5.3.4B report (www.techneau.org/fileadmin/files/Publications/.../D5.3.4b.pdf).

Tsuru, T., Urairi, M., Nakao, S., and Kimura, S. (1991a). Negative Rejection of Anions in the Loose Reverse Osmosis Separation of Mono- and Divalent Ion Mixtures. Desalination *81*, 219–227.

Tsuru, T., Nakao, S., and Kimura, S. (1991b). Calculation of Ion Rejection by Extended Nernst–Planck Equation with Charged Reverse Osmosis Membranes for Single and Mixed Electrolyte Solutions. J. Chem. Eng. JPN. *24*, 511–517.

Tsuru, T., Urairi, M., Nakao, S. and Kimura, S. (1991c). Reverse Osmosis of Single and Mixed Electrolytes with Charged Membranes: Experiment and Analysis. J. Chem. Eng. JPN. *24*, 518–524.

Uddin, M.T., Mozumder, M.S.I., Islam, M.A.S., Deowan, A., and Hoinkis, J. (2007). Nanofiltration membrane process for the removal of arsenic from drinking water. Chem. Eng. Technol. *30*, 1248–1254.

Urase, T., Oh, J., and Yamamoto, K. (1998). Effect of pH on rejection of different species of arsenic by nanofiltration. Desalination *117*, 11–18.

USEPA (1998). Disinfectants and Disinfection By-products Rule

USEPA (2000). Technologies and Costs for Removal of Arsenic from Drinking Water, EPA/815/R-00/*028*, Washington

USEPA (2003).National primary drinking water regulations: Stage 2 disinfectants and disinfection by-products (D/DBP), Final rule, *68*, 2003.

Uyak, V., Koyuncu, I., Oktem, I., Cakmakci, M., and Toroz, İ. (2008). Removal of trihalomethanes from drinking water by nanofiltration membranes. J. Hazard. Mater. *152*, 789–794.

Xia, S.J., Dong, B.Z., Zhang, Q.L., Xu, B., Gao, N., and Causseranda, C. (2007). Study of arsenic removal by nanofiltration and its application in China. Desalination *204*, 374–379.

Xu, X., and Spencer, H.G. (1997). Dye-salt separation by nanofiltration using weak acid polyelectrolyte membranes. Desalination *114*, 129–137.

Van der Bruggen, B., Everaert, K., Wilms, D., and Vandecasteele, C. (2001). Application of nanofiltration for the removal of pesticides, nitrate and hardness from ground water: Retention properties and economic evaluation. J. Membrane Sci. *193*, 239–248.

Van der Bruggen, B., and Vandecasteele, C. (2003). Removal of pollutants from surface water and groundwater by nanofiltration: overview of possible applications in the drinking water industry. Environ. Pollut. *122*, 435–445.

Van der Bruggen, B., Koninckx, A., and Vandecasteele, C. (2004a). Separation of monovalent and divalent ions from aqueous solution by electrodialysis and nanofiltration. Water Res. *38*, 1347–1353.

Van der Bruggen, B., Jeong Hwan Kim, J.H., DiGiano, F.A., Geens, J., and Vandecasteele, C. (2004b). Influence of MF pretreatment on NF performance for aqueous solutions containing particles and an organic foulant. Separ. Sci. Technol. *36*, 203–213.

Van der Bruggen, B., Manttari, M., and Nystrom, M. (2008). Drawbacks of applying nanofiltration and how to avoid them: A review. Separ. Sci. Technol. *63*, 251–263.

Van der Bruggen, B., and Geens, J. (2008). Nanofiltration, In Advanced Membrane Technology and Applications, Norman, N.L., Anthony, G.F., Fane, W.S., Ho, W., and Matsuura, T., eds. (John Wiley and Sons Inc., USA).

Vankelecom, I., Smet, K., Gevers, L.E.M., and Jacobs, P.A. (2005). Nanofiltration membrane materials and preparation. In Nanofiltration: Principles and Applications, Schafer, A.I., Fane, A.G., and Waite, T.D., eds. (Elsevier, Oxford).

Walha, K., Ben Amar, R., Firdaous, L., Quemeneur, F., and Jaouen, P. (2007). Brackish groundwater treatment by nanofiltration, reverse osmosis and electrodialysis in Tunisia: performance and cost comparison. Desalination *207*, 95–106.

Wang, Z., Zhao, Y., Wang, J., and Wang, S. (2005). Studies on nanofiltration membrane fouling in the treatment of water solutions containing humic acids. Desalination *178*, 171–178.

Wiesner, M.R., and Aptel, P. (1996). Mass transport and permeate flux and fouling in pressure-driven processes. In Water Treatment Membrane Handbook, Malleviella, J., ed. (McGraw-Hill Com, New York).

Wiesner, M.R., Barron, A.R., and Rose, J. (2008). Membrane Processes. In Environmental Nanotechnology, M.R Wiesner and J-Bottero, Y., ed. (McGraw Hill, New York).

Yalcın, F., Koyuncu, I., Ozturk, I., and Topacık, D. (1999). Pilot scale UF and RO studies on water reuse in corrugated board industry. Water Sci. Technol. *40*, 303–310.

Yangali-Quintanilla, V., Sadmani, A., McConville, M., Kennedy, M., and Amy, G. (2009). Rejection of pharmaceutically active compounds and endocrine disrupting compounds by clean and fouled nanofiltration membranes. Water Res. *43*, 2349–2362.

Yoon, Y., and Lueptow, R. (2005). Removal of organic contaminants by RO and NF membranes, J. Membrane Sci. *261*, 76–86.

Yoon, Y., Westerhoff, P., Snyder, S.A., and Wert, E.C. (2006). Nanofiltration and ultrafiltration of endocrine disrupting compounds, pharmaceuticals and personal care products. J. Membrane Sci. *270*, 88–100.

Yoon, Y., Westerhoff, P., Snyder, S.A., Wert, E.C., and Yoon, J. (2007). Removal of endocrine disrupting compounds and pharmaceuticals by nanofiltration and ultrafiltration membranes. Desalination *202*, 16–23.

Zhang, W., He, G.H., Gao, P., and Chen, G.H. (2003). Development and characterization of composite nanofiltration membranes and their application in concentration of antibiotics. Separ. Sci. Technol. *30*, 27–35.

Zhang, Y., Van der Bruggen, B., Chen, G.X., Braeken, L., and Vandecasteele, C. (2004). Removal of pesticides by nanofiltration: effect of the water matrix. Separ. Sci. Technol. *38*, 163–172.

Reverse Osmosis: Membranes, Materials, Applications and Nanotechnology

6

Jesús Hierrezuelo-León, Elena Garrido-Pérez and J. Manuel López-Romero

Abstract

This chapter provides a review about the membrane separation technologies focusing on reverse osmosis (hyperfiltration) and nanofiltration. The first one is based on the basic principle of osmotic pressure, while the latter makes use of molecule size for separation. Recent advances on nanotechnology are opening a range of possibilities in membrane technologies. This chapter also reviews some of these aspects: new membrane preparation and cleaning methods, new surface and interior modification possibilities, the use of new nanostructured materials, and new characterization techniques.

Introduction

Membrane separation processes have nowadays an important place in separation technologies with a broad range of applications. The basic property of a membrane for a separation process is its ability to control the permeation rate of one or more chemical species or components of a mixture, giving a total separation in the case of ideal systems. Among the different membrane processes, reverse osmosis (RO) is used for water desalting, in that case, pressurized water containing dissolved salts contacts the feed side of the membrane while water depleted of salt is withdrawn as a low-pressure permeate (Baker, 2004).

Historically, the study of membrane processes is dated in the XVIII, but membranes found their first significant application in the testing of drinking water at the end of World War II in Germany and other European countries, where filtering water was a need for safety reasons. However, the major advance in the field of water desalting was in the early 1960s, due to the development of anisotropic or asymmetric defect-free cellulose acetate RO membranes (Loeb and Sourirajan, 1963), which presented high flux and salt rejection.

Asymmetric membranes can be considered as two homogeneous layers with different structures and transport properties. The a dense and thin (less than 0.1 μm) active layer with the adequate separation characteristics, and a thick and porous substructure or support for mechanical stability; the dense surface layer and its porous support may be obtained from the same material (asymmetric membranes) or from different materials (composite membranes). Fig. 6.1 shows a scheme of concentration profiles across an ideal bilayer membrane at steady state under pressure (left) and concentration (right) gradients without rejection in the porous layer (where $\sigma_p = 0$) and concentration-polarization effect.

Since the transport rate through a membrane is inversely proportional to membrane/dense-layer thickness, the small thickness of the dense layer in the anisotropic membranes provides higher fluxes than isotropic ones. In fact, the development of the anisotropic RO membranes has allowed the transformation of membrane separation processes from laboratory to industrial scale. These membranes were the industry standard through the 1960s to the mid 1970s, when the interfacial polymerization (IP) method of producing composite membranes was developed (Cadotte, 1985). Later, in 1978, a fully aromatic

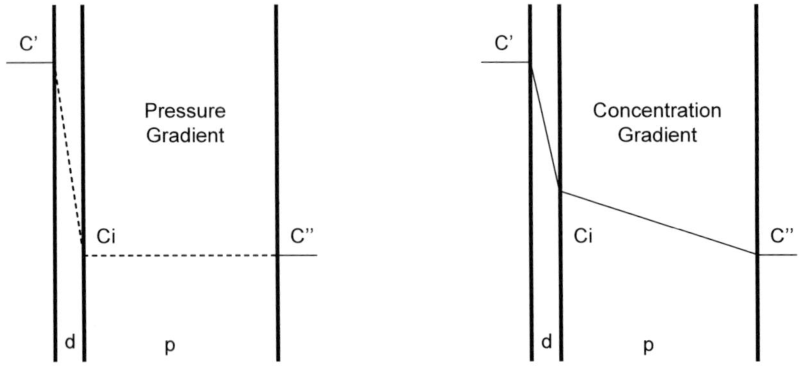

Figure 6.1 Scheme of concentration profiles across an ideal bi-layer RO membrane at steady state under pressure (left) and concentration (right) gradients without rejection in the porous layer and no concentration-polarization effect.

interfacial composite membrane based on the reaction of phenylene diamine and trimesoyl chloride (TMC) was developed (Cadotte, 1981; Larson *et al.*, 1981) and became the new industry standard. The most recent development, beginning in the 1980s, was the introduction of low-pressure nanofiltration membranes used to separate trace amounts of salts and other dissolved solutes from already good-quality water to produce ultrapure water (Kamiyama *et al.*, 1984; Eriksson, 1988). In recent years, the interfacial composite membranes supplied in spiral wound module form have displaced the anisotropic cellulose acetate membrane in most applications (Bray, 1968; Westmoreland, 1968).

RO, also called hyperfiltration, is a class of membrane process closely related to nanofiltration (NF). However, RO is based upon the fundamental pursuit for balance, a fluid containing a lower concentration will move through the membrane into the fluids containing a higher concentration of dissolved solids. Nanofiltration is a pressure related process, during which separation takes place, based on the molecule size. Membranes and nanofibre materials bring about the separation. The technique is mainly applied for the removal of organic substances, such as micro pollutants, pesticides and multivalent heavy ions from water, and generally, nanofiltration membranes have a moderate retention for univalent salts (Sutherland, 2008).

Nanotechnology has produced entirely new classes of functional materials whose application to desalination and water purification need further exploration. The filtration and desalination membranes modified or fabricated from carbon nanotubes (Holt *et al.*, 2006), zeolite nanoparticles and films, or lipid layers offer exciting new possibilities. RO/NF membranes modified by nanotechnology methods may exhibit improved mechanical, chemical, and thermal stability as well as improved separation, reaction, and sorption capacity. These membranes are examples of nanoscale functional materials. This chapter gives an overview on principles, materials, applications, preparation techniques and modifications made by nanotechnology procedures of RO/NF membranes.

Theory principles

Salt and water permeate RO membranes follows a solution-diffusion transport mechanism. The water flux (Jv) depends on the pressure and concentration gradients across the membrane. At a low applied pressure, when the pressure difference across the membrane (dp) is less than the osmotic pressure differential across the membrane ($d\pi$), water flows from the dilute to the concentrated salt solution side of the membrane by normal osmosis. Otherwise, water flows from the concentrated salt-solution side.

Volume and solute fluxes across RO membranes are usually related to local membrane parameters by using the Spiegler–Kedem equations for dense (d) and porous (p) layers (Fig. 6.1). The salt rejection of a single electrolyte has been described by the three coefficients, water permeability (Pw), salt permeability (Ls) and reflection coefficient (σ):

$$Jv = -P_w^j \left[\left(\frac{dP}{dx} \right) - \sigma \left(\frac{d\pi}{dx} \right) \right] \tag{6.1}$$

$$Js = -L_s^j \left(\frac{dc_s}{dx} \right) + (1-\sigma)\, J_v c_s \tag{6.2}$$

where P^j_w and L^j_s are the specific hydraulic and the local solute permeabilities, respectively, and j = d, p denotes the dense and porous membrane layers.

Another important parameter for the characterization of transport across RO membranes is the salt rejection which is related to both the feed solution concentration (Cf) and the resulting permeate solution (Cp):

$$S = 1 - (Cp/Cf) \tag{6.3}$$

It follows from these equations that the water flux is proportional to the applied pressure but the salt flux is independent. This means that the membrane becomes more selective as the pressure increases.

Applied pressure and salt concentration in the feed solution are important operating parameters affecting membrane water flux and salt rejection (Cadotte *et al.*, 1980). As can be observed in Fig. 6.2, water flux and salt rejection increase as the pressure is increased, but the first one does it linearly while the last one tends to a limit value. Increasing the salt concentration in the feed solution effectively increases the osmotic pressure; consequently, at a constant feed pressure, the water flux falls with increasing salt concentration as can be seen in Fig. 6.3 (Jonsson and Benavente, 1992). These measurements were carried out with a polyamide/polysulphone RO membrane from DDS (HR-95).

Temperature also affect water flux and salt rejection but in a more complex mechanism. Transport of both salt and water is an activated process, and both increase exponentially with increasing temperatures. The effect on the water flux is quite dramatic, however the effect on the salt flux is even more marked. This means that salt rejection coefficient declines slightly as temperature increases (Baker, 2004).

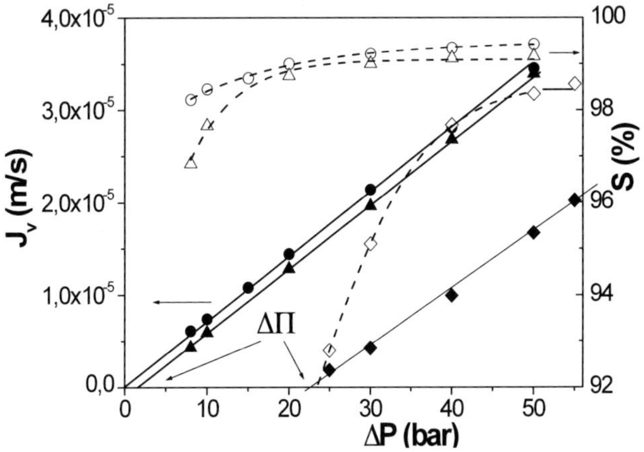

Figure 6.2 Effect of pressure on volume flow, J_v (solid symbols), and rejection, S (open symbols) for a polyamide/polysulphone RO membrane from DDS (HR-95) for different NaCl feed concentrations. (●) and (○) cf = 0.005 M; (▲) and (△) cf = 0.05 M; (◆) and (◇) cf = 0.5 M.

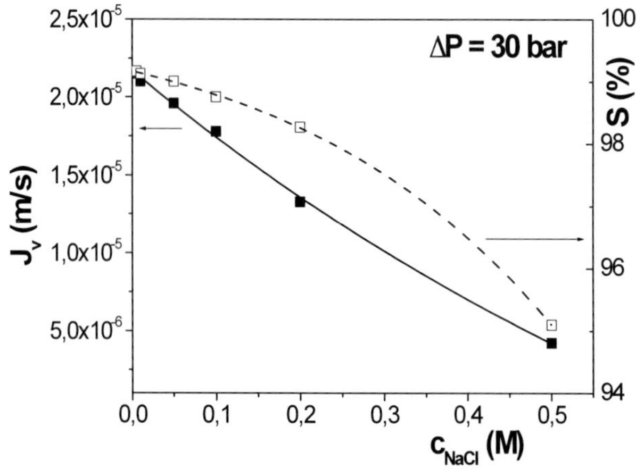

Figure 6.3 Effect of feed concentration on volume flow, J_v (dense symbol), and rejection, S (open symbol), for the HR-95 RO membrane from DDS for measurements carried out at 30 bar.

In reference to the analysis of experimental properties of membranes for RO and NF, efforts have been made for predicting rejection of neutral organic compounds. The application of the artificial neural network (ANN) models based on quantitative structure–activity relationship (QSAR) results in good correlations and standard deviation of error of 5% (Yangali-Quintanilla *et al.*, 2009).

Membrane types

The driving force on the development of RO membrane materials and membrane preparation methods has been the seawater desalination, which requires membranes with salt rejections of greater than 99.3% to produce an acceptable permeate containing less than 500 ppm salt. The most important commercial membranes are cellulose membranes and interfacial composite membranes.

Cellulose membranes

Even when interfacial composite membranes have surpassed cellulose derivative membranes, some of them, like cellulose acetate membranes, present positive properties. They are easy to make, mechanically tough and resistant to degradation by chlorine and other oxidants. As cellulose can tolerate up to 1 ppm of chlorine, chlorination can be used to sterilize feed water. As result of this, cellulose triacetate membranes are still used on the desalination of seawater with a salt rejection of around 99.5%.

The water and salt permeability of cellulose acetate membranes are extremely sensitive to the degree of acetylation of the membrane polymer (Reid and Breton, 1959; Lonsdale, 1966; Rosenbaum *et al.*, 1967). Most commercial cellulose acetate membranes use a polymer containing about 40 wt% acetate with a 2.7 degree of acetylation, achieving 98–99% sodium chloride rejection. The freshly formed cellulose membranes have almost no rejection of sodium chloride because they appear to have a fine microporous structure. The rejection can be greatly improved by nanofunctionalization and annealing procedures that eliminate the micropores and produce a dense structure. Nanofunctionalization can be carried out by using nanoparticles or a lipid layer, while the annealing temperature of the step determines the final properties (Fig. 6.4). Nanoparticles are usually prepared by the ultrasound-high pressure homogenization method. Nuclear magnetic resonance (NMR), transmission electron microscopy, Raman, atomic force microscopy (AFM) or impedance spectroscopy (IS) techniques gives significant information of the membrane composition and structure (Cahill *et al.*, 2008).

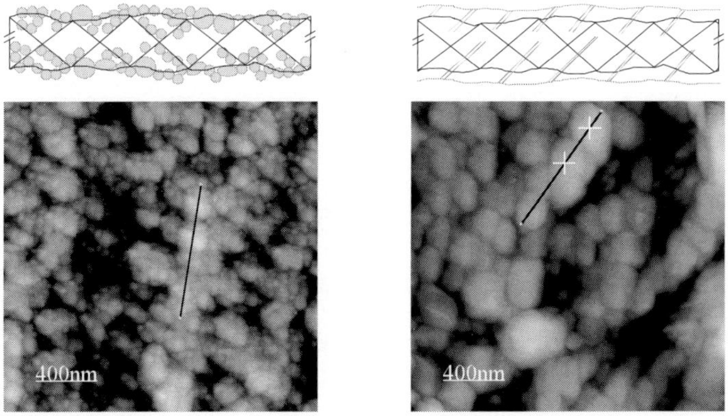

Figure 6.4 High hydrophilic cellulose membranes nanofunctionalized with lipid nanoparticles (NPs, image left) and a lipid layer (right) resulting in a stable hybrid material.

Interfacial composite membranes

High-flux and high-rejection RO membranes can be made by IP (Cadotte, 1981, 1985; Larson *et al.*, 1981). RO membranes manufactured by IP are called thin film composite (TFC) membranes. This procedure can be used to prepare both RO and polymeric nanofiltration membranes. In the preparation process, two monomers dissolved in immiscible liquids (e.g. water and an organic solvent) are used. A porous ultrafiltration (UF) membrane is first impregnated with the aqueous solution of the first monomer and then put in contact with the organic solution of the second monomer. The rapid polymerization reaction at the interface between the two solutions (corresponding to the surface of the porous membrane) creates a polymer nanofilm. Typical membranes, tested with 3.5% sodium chloride solutions, have salt rejection of 99.5% at 800 psi; this is less than half the salt passage of the cellulose acetate membranes and twice the water flux. The rejection of low molecular-weight dissolved organic solutes by interfacial membranes is also far better than cellulose acetate. These membranes experiment a rapid and permanent loss in selectivity that results from exposure to even ppb levels of chlorine or hypochlorite disinfectants (Petersen and Cadotte, 1990). Although the chlorine resistance of interfacial composite membranes has been improved, these membranes still cannot be used with feed water containing more than a few ppb of chlorine.

The polyethylenimine cross-linked with toluene 2,4-diisocyanate membrane (NS100) was the first interfacial composite membrane. It achieved seawater desalination characteristics of more than 99% rejection at 1500 psi (Cadotte, 1977). The first all-monomeric interfacial membrane was the piperazine cross-linked with trimesoyl chloride (NFT40 and NTR7250) (Cadotte, 1981; Kamiyama *et al.*, 1984). It has only modest seawater desalination properties but is a good brackish water membrane and has more chlorine resistance due to the absence of secondary amine bonds.

FT-30/SW-30 membranes were of commercial success for RO (Cadotte, 1981). In the FT-30, the thin film active layer is aromatic polyamide that is prepared via IP of *m*-phenylenediamine in the aqueous phase and TMC in the organic phase (Zhou *et al.*, 2006). FT-30 membranes commercially produced in spiral-wound give 99.3% salt rejection in seawater desalination at 800 psi. By tailoring the preparation techniques, brackish water or seawater membranes can be manufactured. Many attempts have been made to improve the TFC membrane performance through membrane modification or addition of different additives. TFC-RO flux can be increased up to 30–70% by soaking the freshly prepared TFC membranes with solutions containing various organic species, including glycerol, sodium lauryl sulphate, and the salt of triethylamine with camphorsulphonic acid (Kuehne *et al.*, 2001). More recently, the water flux of RO-TFC membranes was doubled without affecting the salt rejection using zeolite nanoparticles in preparing mixed thin film nanocomposite (TFN, Fig. 6.5) membranes (Jeong *et al.*, 2007). Additionally, an aqueous solution composed of poly(vinyl

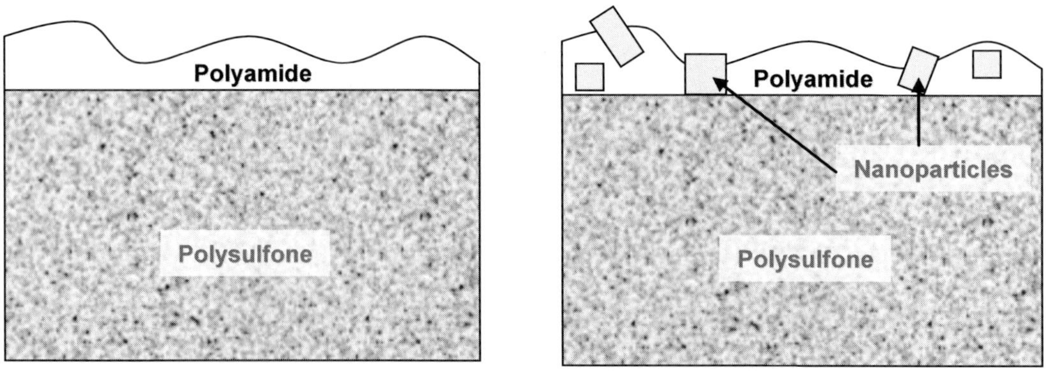

Figure 6.5 Conceptual illustration of TFC membrane (A) and TFC modified with nanoparticles (TFN).

alcohol) (PVA), and a buffer solution was applied as a post-treatment step during the preparation of the TFC membranes to improve the physical properties (improved abrasion resistance) and enhance flux stability. On the other hand, high flux TFC membrane at low pressure can be produced by using PVA-based amine compound having a side chain amino group as the aqueous phase monomer (Hirose, 2005). The modification of TFN membranes with zeolite nanocrystals alter the polyamide thin film structure, increasing the membrane permeability when small zeolite crystals are present and improving the membrane surface properties with the large ones (Lind *et al.*, 2009).

Other membranes

Other materials used in RO are polyamide derivatives including aliphatic and aromatic substituents (Endoh *et al.*, 1977), or those produced by condensation of furfuryl alcohol with sulphuric acid (Cadotte, 1975). These membranes are not made by the interfacial composite process. They have seawater salt rejections up to 99.6% at 800 psi. Unfortunately, they are even more sensitive to oxidants such as chlorine or dissolved oxygen than the interfacial composite, losing their excellent properties after a few hundred hours of operation unless the feed water is completely free of those oxidant agents.

It is worth to mention that electrospun nanofibrous membranes (ENMs) possess several attractive attributes making them very attractive in separation technology, such as high porosity, pore sizes ranging from tens of nanometres to several micrometres, interconnected open pore structure, high permeability for gases and a large surface area per unit volume (Fig. 6.6). In particular, they have been highly successful in developing high-performance air filters. However, the application of ENMs has yet to make similar breakthrough in other avenues of separation, especially in pressure-driven liquid separations as micro- (MF), ultra- (UF) or nano-(NF) filters/membranes. The problem with electrospun nanofibres is that it is difficult to handle as it usually accumulates electrostatic charges during the electrospinning process (Gopal *et al.*, 2006).

Polymeric membranes having advanced functions in the membrane separation processes for liquid mixtures in RO, NF, ultrafiltration or microfiltration are in continuous development. Important approaches towards this aim include the synthesis of novel polymers with well-defined structure as 'designed' membrane materials and advanced surface nanofunctionalization of the membranes, the use of templates for creating 'tailored' barrier or surface structures for membranes and the preparation of composite membranes for the synergistic combination of different functions by different polymeric materials (Ulbricht, 2006).

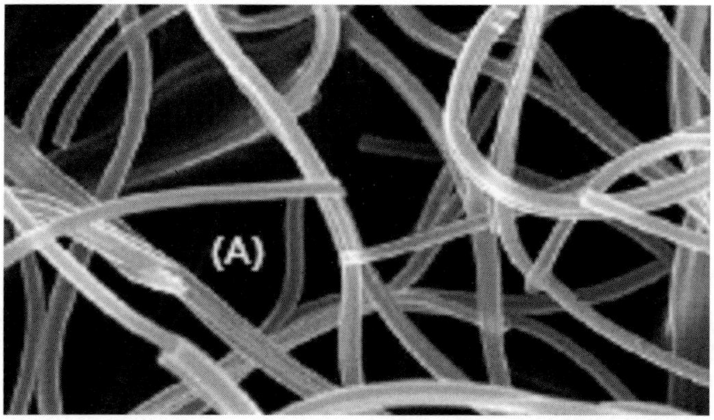

Figure 6.6 Electrospun membranes.

Membranes for reverse osmosis: nanofiltration

Nanofiltration membranes exhibit intermediate properties between RO membranes, with a salt rejection of more than 90%, and molecular weight less than 50, and ultrafiltration membranes with a salt rejection less than 5%.

They are also called low-pressure membranes, usually operating at 50–200 psi, and they have low sodium chloride rejections (between 20 and 80%) but high water permeabilities and molecular weight cut-offs for dissolved organic solutes of 200–1000 Da. Because of that and other advantages such as high-permeate flux, and high retention of multivalent ion salts, relatively low investment or low operation and maintenance cost, they are gaining interest worldwide.

Most NF membranes developed to date have a TFC structure, due to key advantages compared to asymmetric membranes. These membranes have a thin and dense active layer that control membrane performance (permeability and selectivity), and a much thicker porous substrate that provides mechanical support to the active layer with low resistance to permeate flow. Each layer can be optimized for the desired combination of permeate flux and solute rejection. In order to achieve high values of permeability and selectivity, the active layer should be ultra-thin and hydrophilic.

A variety of polymers and formation procedures has been employed for this purpose. A composite membrane is obtained by forming an ultra-thin dense layer on a porous support. There are different techniques for preparation of thin film composite membranes, e.g. plasma initiated polymerization, photo-initiated polymerization, photo-grafting, electron beam irradiations, dip-coating or IP (Wei *et al.*, 2008). Among all these techniques, IP is the most commonly used for preparing such composite membrane.

Transport property (flux and rejection) of a nanofiltration membrane is mostly determined by the membrane's intrinsic properties like surface charge, morphology, hydrophilicity/hydrophobicity, pore size and their geometry and thickness of the thin film. All of these properties are influenced by the membrane preparation conditions, like polymerization reaction time, curing temperature, curing time, monomer type and concentration.

The preparation procedure used in IP can result in acid groups attached to the polymeric backbone. Neutral solutes such as lactose are not affected by the presence of the charged groups and the membrane rejection increases in proportion to solute size (van der Bruggen, 1999). The salt rejection depends of both molecular size and Donnan exclusion effects caused by the acid groups attached to the polymer backbone. Charged groups tend to exclude ions of the same charge, particularly multivalent ions while being freely permeable to ions of the opposite charge, particularly multivalent ions. In that way, there are anionic nanofiltration membranes that have positive groups attached and cationic nanofiltration membranes (Peters *et al.*, 1982). This type of membranes have very high rejections and permeances of salt al low salt concentrations, but lose their selectivity in salt concentrations above 1000–2000 ppm in feed water.

Many RO and NF thin film composite membranes are based on interfacially polymerized polyamides, but they are generally susceptible to oxidative/chlorine attack. Several poly(esteramide)-based TFC membranes showed improved oxidative (chlorine) resistance compared to polyamide membranes, but low passage for monovalent salts (Jayarani *et al.*, 2000). Polysulphone supported TFC membranes consisting of aromatic polyester or a copolymer of an aromatic polyester and aromatic polyamide were synthesized interfacial reaction of 4,4′-dihydroxybiphenyl and *m*-phenylenediamine (MPDA) in the aqueous phase and TMC in the organic phase, resulting in a TFC membrane with NaCl rejection of 96.5% (Kim *et al.*, 1997). Similarly, it was found that polyesteramide membranes synthesized from MPDA and TMC incorporating either *m*-aminophenol, hydroquinone or bisphenol-A had RO-type rejection profiles, with NaCl rejections in the range of 95–98% (Jayarani *et al.*, 2000).

Membranes of polyesteramide-based copolymers with the two bulky diols phenolphthalein and tetrabromobisphenol-A have been synthesized (Razdan and Kulkarni, 2004). The membranes were prepared interfacially on nanoporous polysulphone supports by reacting one of these diols along with MPDA in aqueous phase and TMC in petroleum ether. It was expected that the high excluded volume of the bulky substituents on these bisphenols would hinder polymer packing in the separating layer resulting in a more 'open' membrane. The change in diol chemistry was also expected to affect the hydrophilicity of the membrane as well as its surface charge and p*K*, with corresponding effects on the membrane's ability to separate monovalent salts from divalent salts containing co-ions.

Another procedure to synthesize RO membranes with 5-chloroformloxy-iso-phthaloyl chloride and *m*-phenylenediamine through an IP technique on the polysulphone supporting film has been developed (Zhou *et al.*, 2005).

A homologous series of thin film composite membranes made by IP of TMC and *m*-phenylenediamine have also been developed. Membrane flux and rejection as well as the chemical and mechanical properties of the unsupported IP film were evaluated as a function of monomer concentration (Roh *et al.*, 2006).

Results demonstrate that variation in the acid chloride concentration has a more pronounced effect on the thin film material properties than corresponding variations in the amine concentration. Increasing the TMC concentration from 0.1 to 1.0 wt/v% led to a marked increase in the IP film thickness as well as surface hydrophilicity. In contrast, the thickness and surface hydrophilicity changes were relatively small when the MPDA concentration was varied from 0.1 to 1.0 wt/v%. However, because the water flux mainly depends upon the IP film thickness and surface hydrophilicity and these two parameters have counterbalancing effects, the decrease in water flux with increasing MPDA concentration was pronounced but a corresponding increase in the TMC concentration had a relatively smaller effect. Overall, these experiments provided an improved basis for understanding the relationship between the specific polymer properties of the active thin layer and IP-TFC membrane performance.

A novel TFC membrane for nanofiltration was developed by the IP of triethanolamine and TMC on the polysulphone supporting membrane (Tang *et al.*, 2008). Furthermore, the nanofiltration properties of the TFC membrane were tested by examining the separating performance of various salts at 0.6 MPa operating pressure. The rejection to different salt solutions decreased as per the order of Na_2SO_4 (82.2%), $MgSO_4$ (76.5%), NaCl (42.2%) and $MgCl_2$ (23%). The investigation of the flux for NaCl solution at different pH showed that the polyester NF composite membrane is also particularly suitable for treating acidic feeds: the flux increased from 8.4 to 11.5 L/m^2 h when pH of the feed decreased from 9 to 3. Additionally, the TFC membrane exhibits good long-term stability.

A large number of TFC membranes have been successfully developed from different polymers such as polyurea, polyamides, polyurea-amides, polyimide, polyester, polysulphonamide, etc. In recent years, studies have been focused on choosing or synthesizing new monomers with special functional groups to prepare the TFC membrane. For instance, a novel nanofiltration membrane was prepared with poly (amidoamine) (PAMAM) dendrimer and TMC by *in situ* IP on ultrafiltration membrane (Li *et al.*, 2006). The membrane surface charge can be adjusted according to the amino groups and tertiary amino groups on the membrane. Poly(amide–imide) composite membrane containing poly(amide-amic acid) (PAA) ultrathin barrier layer was prepared by the IP of diamine amic acids and isophthaloyl chloride (Buch and Reddy, 2005). The membrane displayed high water permeation due to the fact that the membrane contained a less compact and highly hydrophilic PAA barrier layer.

There have been reports on the use of mixed acid chlorides like 5-isocyanato-phthaloyl chloride and phthaloyl chloride (Zhou *et al.*, 2006b) or mixed amines like piperazine and 3,5-diaminobenzoic acid (Ahmad *et al.*, 2004; Ahmad and Ooi, 2005), *m*-phenylenediamine and *m*-phenylenediamine-5-sulphonic acid (Yong *et al.*, 2006) for thin film formation. The objective of using such a mixed cross-linking monomer was to increase the flux of the membrane by increasing surface hydrophilicity which was enhanced by the presence of a higher number of acid groups without compromising the rejection.

Today, the main NF module geometries used are spiral-wound or tubular modules. The advantages of spiral modules are the large surface area of the membrane per m^3 and the low cost of the module. On the other hand, they are extremely susceptible to floating matter, have high energy losses caused by friction in the spacers and it is not possible to reverse the flush (Futselaar *et al.*, 2002). Tubular modules do not require strong pre-treatment and back flushing is possible, but they have low packing density and are expensive (Frank *et al.*, 2001). Efforts are recently being done to develop capillary modules to bridge the gap between the two module types (Futselaar *et al.*, 2003). A hollow fibre module would give a better packing density than the tubular module and allows a less demanding pretreatment and maintenance than the spiral-wound module. This module would be a solution for the treatment of low quality water sources by direct capillary nanofiltration. It is also reported in the literature (van der Meer and van Dijk, 1997) that an optimized hollow fibre nanofiltration module would give a 100% increase in performance compared to an optimized spiral-wound module.

A manufacturing method of a hollow fibre nanofiltration membrane that contains a polyolefin hollow fibre microporous base film as a support layer, a polyvinyl alcohol transition layer, and a polyamide nanofiltration separation layer has been developed (Shen, 2009). The polyvinyl layer is arranged between the hollow fibre microporous base film and the polyamide layer (obtained by performing interface polycondensation between multi-amine and multi-acyl halide). The membrane can be periodically washed with gas and back-washed to maintain the membrane performance and has a high ability for removing pollutants in wastewater.

Alternatively to the IP process, Wand *et al.* (2009) reported the synthesis of a novel aromatic polyamide nanofiltration membrane by phase inversion method using poly(*m*-phenyleneisophthalamide) as the membrane material. This is the main technique used to prepare polymeric membranes. In the thermal phase inversion, a polymer is dissolved in a hot solvent and cast to form a film. Through cooling, a de-mixing process takes place and a porous structure is obtained. This production process is used to produce polypropylene membranes both as a film and as a hollow fibre. In the wet phase inversion method, a polymer is dissolved in a solvent and cast to form a film which is then dipped in a coagulation bath where the membrane precipitates.

Membranes for reverse osmosis: hyperfiltration

Hyperfiltration membranes are used to separate ionic solutes from organic solvent solutions. These types of separations are difficult due to the high osmotic pressure that must be overcome and because they require membranes that are sufficiently solvent-resistant to be mechanically stable, but are sufficiently permeable for good fluxes to be obtained. In addition, different membranes must be developed for each category of solvent. The first dense, solution-diffusion, hyperfiltration membranes did not appear until the late 1990s when a cross-linked silicone composite membrane was developed with uses in hyperfiltration of non-polar solvents (Linder *et al.*, 1993; Machado *et al.*, 1999). The solvent-permeable hyperfiltration membranes Starmem® (White *et al.*, 1993), asymmetric polyimide phase-inversion membranes, found their first large-scale commercial use in refineries separations (Gould *et al.*, 1996; Bhore *et al.*, 1999; White and Nitsch, 2000), followed by removal of bromine and bromate ions from drinking water, reaching the hyperfiltration process 89% and 97% rejection, respectively (Marhaba and Medlar, 1993).

Hyperfiltration is mostly used to eliminate water and salts during natural actives purification processes (Zhao, 2009). However, it has been also used for the L-citrulline extraction from fruits with a nanometre membrane which retains mol wt. of 1000–500 mu, regulating the pH from 3 to 5 (Zhao, 2009). In petrochemistry or refinery operations, feed streams containing aromatics such as benzene, toluene, or xylenes can be selectively enriched or depleted in these compounds by hyperfiltration. The use of polyimide membranes with various permeate fluxes results in different selectivities. Spiral-wound elements were designed and built containing these membranes. Tests of these modules at 800 psi and 50°C for up to 2000 h with a toluene recycle stream from a refinery showed stable operation in hot hydrocarbons. The high efficiencies of these modules can allow for high stage cuts (>50%) from aromatic process streams. With these membranes acting as organic solvent nanofiltration devices, the permeation properties can be interpreted through a solution of the diffusion equation (White and Wildemuth, 2006).

The RO hyperfiltration has been applied to the recovery of waste sulphuric acid from industrial production of lead acid storage battery polar plates. The recovery process comprises filtering the waste sulphuric acid by a hyperfiltration membrane, followed by nanofiltration to obtain a filter liquor with iron content of 0–5 ppm. The process reduces sulphuric acid treatment costs and the environment pollution (Li *et al.*, 2006).

Several reviews regarding the fundamentals and materials for membrane technologies for purification of water including hyperfiltration have been recently published (Paul, 1997; 2000).

Membrane fouling control and cleaning

Membrane technology has emerged, in the last years, as the main contributor to solve water shortage problems. Recently, nanofiltration and RO membrane processing have been considered as a promising technology for drinking water production, wastewater treatment, and water reuse. NF/RO membrane processing removes ions and organic chemicals, and its treatment efficiency and performance are stable and predictable. NF and RO systems have shown to be adequate for producing clear water from various water sources (Ray *et al.*, 1991; Lee *et al.*, 2001a).

Compared to conventional water treatment processes, membrane technology offers the advantage to specifically remove contaminants from the water depending on the pore size and the surface physical properties of the membrane; thus offering a targeted product quality according to needs. Despite a fast growing market, membrane technology still has to reach the level of application of many competing water treatment processes and it has found limitations due to technical and economical issues, for instance the price of membrane systems, which has strongly decreased over the last decade, is still high. Furthermore, cost related to operation and maintenance could be significant. Above all, fouling represents the major constraint to more cost effective and therefore expanded application of membrane technology in water treatment.

Membrane fouling has a great impact on process performance. Plant operators have to define the best operating conditions to minimize membrane fouling due to extending membrane life and thereby minimizing replacement costs. Fouling as a result from foulant accumulation on the membrane surface is the major cause of RO system failure. RO membrane fouling is a complex phenomenon involving the deposition of several different but related types of foulants on the membrane surface. RO system fouling problems are becoming more prevalent as the use of low quality feed water increases. In addition, water treated with various organic chemicals pose very serious and challenging fouling problems.

The types of foulants most commonly encountered in RO systems include (i) inorganic fouling (scaling), (iii) biological fouling and (iv) organic fouling. Fouling control involves pretreatment of the feed water as well as a regular cleaning.

Scaling of RO membrane surfaces is caused by the precipitation of sparingly soluble salts from concentrated brine. Scales are hard mineral deposits that precipitate from the feed stream onto the membrane surface. As salt-free water is removed from the permeate, the concentration of ions in the feed increases until at some point the solubility limit is exceeded. The salts that cause scale formation are ubiquitous in most water environments. They include calcium sulphate ($CaSO_4$), calcium carbonate ($CaCO_3$), and silica (SiO_2) (Lee and Lee, 2005).

Several studies have been attempted to elucidate the fouling phenomena and scale formation mechanism. Supersaturation, due to the concentration polarization of the salts near the high-pressure side of the membrane surface, has shown to result in the crystallization of soluble salts. Various parameters are involved in this process. For example, temperature (Troup and Richardson, 1978), pH (Klepetsanis and Koutsoukos, 1989), the types of pretreatment, other metal ions (Hamdona *et al.*, 1993), and the selection of membrane module have been reported to influence the scale formation kinetics (Kasper, 1993).

Scale formation is a complex process in which both crystallization and hydrodynamic transport mechanisms are involved. Two pathways for crystallization have been identified (Pervov, 1991; Lee *et al.*, 1999), first the surface (heterogeneous) crystallization and, second, bulk (homogeneous) crystallization. In surface crystallization, flux decline results from the blockage of the membrane surface by lateral growth of the scale deposit on the membrane. In bulk crystallization, crystals formed in the bulk solution sediment on the membrane surface leading to a flux decline.

Since scale formation not only lowers the flux and rejection of NF/RO permeates but also shortens the membrane life, it has always been recognized as a serious constraint in designing and operating NF/RO systems. Attempts to reduce scale formation in NF/RO membrane system include increasing the fluid velocity (Lee and Lee, 2000), adding antiscalants (Belfort, 1984; Kronmiller, 1984), performing chemical pretreatments (Watson, 1990), using ion exchange prior to NF/RO (Watson, 1990), combining a crystallizer with NF/RO (Rautenbach and Habbe, 1991; Pulles *et al.*, 1992; Bremere *et al.*, 1998), and applying an online microfilter (Lee *et al.*, 1999).

The most popular way to control scale formation is to inject antiscalants (that interferes with the precipitation of the insoluble salt and maintain the salt in solution even when the solubility limit is exceeded) as sulphuric acid and SHMP (sodium hexametaphosphate) into NF/RO feed stream. Calcium carbonate scale, by far the most common problem, is easily controlled by acidifying the feed water or by using an ion exchange water softener to exchange calcium for sodium. Silica can be particularly troublesome scalant because no effective antiscalant or dispersant is available.

However, conventional antiscaling chemicals cause some serious problems. Handling of an acid such as H_2SO_4 poses a safety hazard. When overdosed, it also causes corrosion of piping material, leading to loss of salt rejection and water output by plugging the membrane. SHMP, being a phosphate, acts as food for bacteria that can thrive in the phosphate environment and plug the membrane. SHMP also undergoes hydrolysis or reversion in an aqueous solution to orthophosphate form that combines with calcium ions to form an insoluble sludge of calcium orthophosphate, thereby itself acting as a scalant.

Therefore, several studies have been performed to develop advanced antiscalants that provide superior performance as well as chemical safety and stability. Moreover, the performance of antiscalants in membranes systems is unpredictable. Processing conditions influencing the permissible water recovery level are numerous and include the supersaturation level of the concentrate, the concentration polarization level, the nature of antiscalant used to suppress the scaling propensity and the antiscalant dosage level. Recently, it was reported that natural organic matters (NOM) interfere with antiscalant to alter the scaling kinetics, making the prediction of antiscalant effect even more difficult. At the present state of knowledge, it is impossible to predict theoretically the limiting recovery level under given process conditions. Alternatively to antiscalants, increasing fluid motion near the membrane surface is another way to slow down the fouling scale formation.

Since the control of scale formation by adding antiscalants or increasing cross-flow velocity has many limitations, a more challenging way of controlling scale control in NF/RO processing would be to develop a membrane system that can be operated regardless of scale formation.

Of particular interest is a microfiltration–NF/RO hybrid system consisting of an on-line microfilter and a NF or RO membrane system (Lee *et al.*, 1995, 1999; Fig. 6.7). The basic idea underlying the MF–NF hybrid system is that flux decline by scale formation may be prevented by suppressing membrane fouling by bulk and surface crystallization. Filtering the crystal particles continuously in the retentate flow using a microfilter can eliminate the fouling due to bulk crystallization. On the other hand, reducing the concentration polarization, the major driving force for surface crystal growth, may minimize fouling due to surface crystallization. Assuming that fouling due to bulk crystallization is successfully prevented by the on-line MF, it is likely that the overall performance of the MF–NF/RO hybrid system depends on how efficiently surface crystallization is controlled.

Colloidal fouling is due to the presence of suspended solids in water, such as mud and silt, and tends to cause gross plugging of the device rather than fouling of the membrane surface. Typical sources are organic colloids, iron corrosion products, precipitated iron hydroxide, algae and fine particulate matter. With the parameter SDI (silt density index) it is possible to achieve a good prediction of the likelihood of a particular feed water to cause fouling by silt. An SDI higher than 5 is unacceptable and indicates that additional pretreatment is required. It has been studied by measuring salt transport across the membrane (Wang *et al.*, 2007) the coupled effects of scaling and colloidal deposition on the performance of RO membranes.

Bio-fouling is a special case of particulate fouling that involves living organisms and can be a serious problem. Biological material growing on membrane surfaces not only causes loss of flux but may physically degrade certain types of membranes. Cellulose membranes are an ideal nutrient for bacteria and can be completely destroyed by a few weeks of uncontrolled bacterial attack. Therefore, the feed water must always be sterilized. Composite membranes are generally more resistant and only needs periodic treatments with a bactericide or biocide as a method of control. Biocide application is a well established technology to inhibit bio-fouling and nuisance organism growth in the desalination process. Its selection and application should be optimized according to raw water characteristics and the membranes used.

Disinfection pretreatment process can be divided into physical and chemical methods. The most common chemical biocide is chlorine (or hypochlorite) as it has a high inactivation efficiency and organic matter removal besides relative low cost. However, it can produce chemical corrosion of a membrane, disinfection by-product formation, and microorganism outgrowth because of the following quenching process using sodium bisulphite. Chloramines have also proved to be effective for

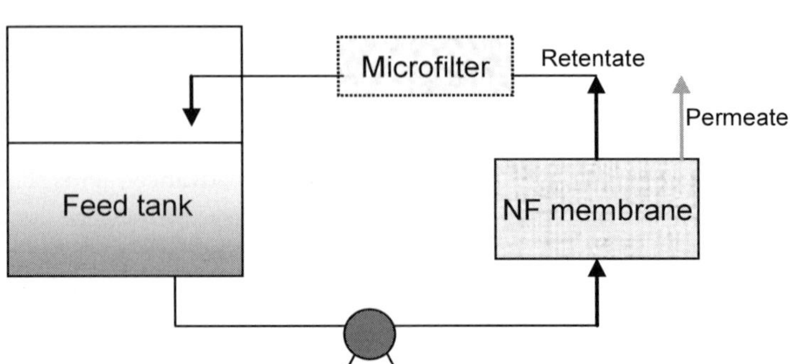

Figure 6.7 MF–NF/RO hybrid system.

inhibiting biofouling and are less harmful to the membrane than chlorine. Chlorine dioxide (ClO_2) is gaining attention for its lower corrosion and disinfection by-product formation in permeate.

UV treatment is not regarded as a chemical method but rather a physical process. UV irradiation at 254 nm breaks the bacterial DNA genetic code and inhibits their reproduction. UV treatment does not produce disinfection by-products, is independent of pH, and eliminates toxic and hazardous chemical usage. UV treatment is operator friendly, has a short contact time (e.g. less than 20–30 s with a low-pressure lamp), and it has a smaller footprint than the other disinfection system. Regardless of its advantages, ozone and hydroxyl radicals (product of ozone degradation) produced during UV irradiation are such strong oxidants and are known to break down the membrane surface.

Ozonation can also be included in the physical process. However, the disadvantages are the hypobromous acid or hypobromite anion generation during ozonation of seawater (Perrins *et al.*, 2006) which deteriorate RO membrane surface, and the production of carcinogenic bromate since seawater is rich in bromide. Furthermore, the breakdown of larger molecular weight organic matter is favoured for microbial growth and could accelerate biofouling (Oh *et al.*, 2007). Combination of ozone and biological activated carbon (BAC) reduces fouling potential after biological filter is matured (Koyuncu *et al.*, 2006). Ozonation and biological filters also enhance organics removal.

Sequential disinfection is a combination of two or more biocides. This new disinfection technique is noteworthy because of the synergetic effect of using several disinfectants (Rand *et al.*, 2007). This reduces the amount of used disinfectant so significantly that it decreases the disinfection by-product formation and membrane chemical corrosion. UV followed by chlorine injection could quench remaining chlorine to some extent, which could reduce the sodium bisulphite dosing (Worley *et al.*, 2003).

Organic fouling is the attachment of materials such as oil or grease onto the membrane surface. It is more common in industrial applications. The removal of the organic material from the feed water is achieved by filtration or carbon adsorption.

Fouling is a major difficulty in industrial wastewater treatment. The fouling of NF membranes used for the textile dye wastewater treatment in textile industry has been studied (Lau and Ismail, 2009). Fouling is often a weakness of NF for complex textile manufacturing. Dyes can produce a colloidal fouling layer, which further introduces an undesirable flux decline in the operation. Heavy membrane fouling is expected since dyes can be accumulated on the active layer of polyamide NF membranes by chemical bonds in either the ionic or covalent bond, depending on pH or the class of dye. Generally, the phenomenon of membrane fouling is inevitable, but it is reversible by using feed pre-treatment, modifying the membrane surface or by controlling membrane cleaning procedures.

Organic fouling cleaning can be made by physical methods as intermittent back-washing, application of critical flux, critical TMP, intermittent suction operation, low TMP, high cross flow velocity and hydrodynamic shear stress scouring. On the other hand, chemical cleaning agents can be acids (strong or weak), alkalis (NaOH), detergents, enzymes, complexing agents (EDTA) and disinfectants. Furthermore, membrane manufacturers recommend chemical cleaning agents since they have the ability to recover completely the initial membrane permeability and require less energy consumption compared to physical methods. Nevertheless, chemical treatments are relatively expensive and may cause severe membrane damage and produce toxic by-product waste (Soung *et al.*, 2002).

Some researchers (Mo *et al.*, 2008) reported that each class of dyes could cause membrane fouling. The difference between these dyes was the thickness or hardness of dye cake layer accumulated on the membrane surface. They suggested that a pretreatment with a chemical coagulant (alumina) was required to decrease the extent of membrane fouling. In comparison, the degree of flux decline of pretreated wastewater over the operation time became smaller than when original wastewater was used. The results showed that approximately 20% improvement of flux was observed using the pretreatment system prior to NF.

It was also concluded that membrane fouling caused by dye absorption is reversible, but it was highly dependent on membrane cleaning (Shu *et al.*, 2000). The use of NaCl in the dyeing process could enhance the penetration of reactive dye into membranes, which results in NF membranes heavily coloured after the experiments (Sungpet *et al.*, 2004). This, however, subsequently led to a flux decline in the membrane process. Interestingly, the use of chemical cleaning with 0.2 wt% HNO_3 followed by 0.5 wt% NaOH could recover 80–100% of the flux where the chemical cleaning procedures were periodically carried out.

On the other hand, membrane fouling caused by the adsorption or pore blocking of organic compounds on the membranes had a large influence on the permeate flux due to the high concentration of organic compounds used in textile dyeing (van der Bruggen *et al.*, 2005). In the application of NF membranes for the treatment of exhausted dye baths, 26–46% of the irreversible flux declines were reported as the relative difference between the initial pure water flux and pure water flux measured after the experiment. Although absorbed organic compounds might be desorbed by rinsing water, it was still difficult for those dyes which were supposed to attach strongly to the membranes.

The attention of both membrane producers and researchers is focused on the study of both traditional and novel materials, and on the preparation of antifouling membranes. For the preparation of a membrane with antifouling properties the recent research trend is moving in two directions: (i) starting from materials with antifouling properties and (ii) providing antifouling properties through chemical treatments of the membrane after synthesis or after module construction.

In terms of membrane nanomodification, negatively charged membranes have the potential of reducing fouling, especially in the presence of negatively charged colloids in the feed. On the contrary, NF membrane surfaces modified by coating with a neutral polyvinyl alcohol (PVA), reduce surface charge and membrane surface roughness, so that the ionic bridge of the cations between the membrane and dye could be reduced (Kim *et al.*, 2006). The fouling could easily have resulted from the divalent cations in the effluent using a typically higher surface charge of NF. The results confirmed that the PVA-coated NF membranes were successful in increasing fouling resistance and subsequently reducing membrane fouling.

The antifouling NF membrane was introduced by coating the fluoride polymer support membrane with an amphiphilic graft copolymer to obtain, poly(vinylidene fluoride)-*graft*-poly(oxyethylene) methacrylate, PVDF-*g*-POEM (Asatekin *et al.*, 2006). The higher fouling resistance and higher water production can be attributed to both the nanoscale dimensions of the hydrophilic channels through the coating and to the neutral charge of POEM which creates the barrier to the adsorption. Recently, the amphiphilic combination of copolymers in preparation of the polyacrylonitrile-*graft*-polyethylene oxide membrane (PAN-*g*-PEO) was also found to have excellent antifouling characteristics by coating it on a PAN ultrafiltration membrane. The authors attributed this to the surface segregation and local orientation of PAN-*g*-PEO molecules at the membrane surface and pore walls, forming a dense brush layer as the barrier to the adsorption.

A new generation of low fouling nanofiltration membranes has been described (Bartels *et al.*, 2007). These membranes are composite polyamide type, similar to existing RO membranes, but are chemically treated to adjust the hardness rejection. This treatment also imparts fouling resistance. This low fouling nature can be attributed to the smoothness of the surface and the near neutral surface charge.

Applications

Reverse osmosis applications

Fifty per cent of the RO systems currently installed are desalinating brackish or seawater; another 40% are producing ultrapure water for the electronics, pharmaceutical and power generation industries and the remainder is used in small niche applications such as pollution control and food processing (Williams *et al.*, 1992).

Since the seawater salt rejection for cellulose acetate membranes is about 97–99% the application was limited. Polyamide hollow fine fibres and interfacial composites are membranes suitable for desalination. In general, membranes are not competitive for large seawater desalination plants, and multistage flash evaporation is usually used for plants larger than 40 million l/day capacity. In the 1 to 40 million l/day range, membranes are more competitive, and flexibility of membrane systems as well as their easy start-up/shut-down and turndown capability are advantageous (Baker, 2004). However, production of ultrapure water for the electronics industry is an established and growing application of RO. The usual feed is municipal drinking water, which often contains less than 200 ppm dissolved solids.

In principle, pollution control should be a major application for RO. In practice, membrane fouling, causing low plant reliability has inhibited its widespread use in this area. The most common applications are special situations in which the chemicals separated from the water are valuable.

Nanofiltration applications

NF offers several advantages such as low operation pressure, high flux, high retention of multivalent anion salts and an organic molecular above 300 mu, relatively low investment and low operation and maintenance costs. Because of these advantages, the applications of NF have increased worldwide. NF membranes applications in treatment of water have been reviewed (Hilal *et al.*, 2004). NF can either be used to treat all kinds of water including (i) ground, (ii) surface and)iii) wastewater or (iv) used as a pretreatment for desalination.

Ground water

Softening of ground water by using NF membranes has been studied by many investigators. NF membranes have been used for hardness removal (Schaep *et al.*, 1998). Retentions higher than 90% were found for multivalent ions, whereas monovalent ions were retained for about 60–70%. Comparison of NF membranes with pellet softening and granular activated carbon for softening water has been made (Sombekke *et al.*, 1997). Both methods gave good results; however, NF membranes have several advantages from the point of view of health aspects and lower investment costs. Removal of natural organic matter (NOM) and disinfection by-product (DBP) precursors were also studied (Gorenflo *et al.*, 2002). Both RO and NF are used in the US for removal of NOM (Jacangelo *et al.*, 1997). To date, NF has been employed primarily for groundwater containing relatively low total dissolved solids but with high total hardness, colour, and DBP precursors.

The removal of pesticides and some micropollutants from ground water has been studied by many researchers. Is has been shown that the NF-70 membrane can reject pesticides such as atrazine, simazine, diuron and isoproturon over 90% (van der Bruggen *et al.*, 1998). Also the removal of fluoride and aluminium from groundwater sources is effective (Kettunen and Keskitalo, 2000). In addition to the removal of hardness, nitrates iron and strontium, membranes for fluoride removal are also available (Pervov *et al.*, 2000).

Surface water

Surface waters often have a changing chemistry or composition due to seasonal changes or after dilution with rain. NF is a reliable option for surface water treatment, although the focus is here rather on removal of organics than on softening. NF membranes were also found to be able to remove viruses and bacteria from surfaces quite successfully. Using a novel NF membrane (NF 200), removal of organics up to 96% in a large NF plant was achieved (De Witte, 1997).

Wastewater

Removal of chromate, from water using a TFC-NF membrane is an alternative to the conventional methods of Cr(VI) removal solution (Hafiarle *et al.*, 2000). The results showed that the rejection depended on the ionic strength and pH. Better retention was obtained at basic pH (up to 80% at a pH of 8). Results also showed that NF is a very promising method of treatment for wastewater charged with hexavalent chromium. The same membrane was used to treat opium alkaloid processing industry effluents (Koyuncu, 2003). NF can also remove lead up to 90% from a nitric acid solution (Jakobs and Baumgarten, 2002). Also, purification of phosphoric acid by NF (DS5DL NF) was achieved by using different industrial commercial membranes (González *et al.*, 2002).

NF as pretreatment for desalination

Feed pretreatment is one of the major factors determining the success or failure of a desalination process. NF was used as pretreatment for seawater reverse osmosis (SWRO), minimizing hardness, microorganisms and turbidity (Hassan *et al.*, 1998). NF pretreatment of seawater in desalination plants prevents SWRO membrane fouling by the removal of turbidity and bacteria and prevents scaling by removal of scale forming hardness ions. The introduction of NF for water pretreatment leads to an improvement in the performance by maintaining almost invariant the energy requirement (Criscuoali and Drioli, 1999).

Within the framework of limited water resources and increasing costs, industries need to lower water consumption and to look at process water and effluent treatment for water recycling or reuse. The food industry was one of the first industries to introduce membrane filtration into their commercial processes since membrane processes are potentially non-destructive, relatively energy efficient (no phase change), or even cheaper than conventional treatment.

Several works dedicated to the treatment of food industry streams show that using membrane operations for producing purified water from effluents or process waters is convenient. In the meat industry, NF was used for treating cooling water from sausage products to produce drinking quality water for water reuse. NF of wastewater from the fishmeal industry permits recycling of the treated water. In the beverage industry, water from bottle washing water processed by NF and low-pressure RO is similar to water of drinking quality; and a hybrid process using a membrane bioreactor and two-stage NF was developed to treat spent process water from a fruit juice company for purified water reuse. In the vegetable industry, soybean soaking water was recovered using a combined NF-RO for water treatment.

NF and RO treatment of dairy model process waters for the production of reusable purified water has been studied (Vourch *et al.*, 2005). The dairy industry has high water consumption and generates a large amount of effluent from 0.2 to 10 l per litre of processed milk. The process water generated during the starting, equilibrating, interrupting and rinsing steps of the different plant units contribute to effluent production. The treatment of the process waters to produce purified water for reuse could lower the effluent volume and the total water consumption of the dairy factory. Many membrane operations have been described for the treatment of dairy effluents: one-step operations based on UF, NF or RO or two-stage operations such as UF+RO, NF+NF or RO+RO. NF and RO are convenient operations for the production of water for reuse in dairy plants from low contaminated process water (Chmiel *et al.*, 2000).

NF technology can be also applied to separate amphoteric materials as amino acids and proteins, since most commercial NF membranes showed diverse rejection performance for solutes at the different pH values.

Application of polymeric NF membranes in the specific waste stream in the textile industry has been reviewed (Woei-Jye and Ismail, 2009). The textile industry typically generates large volumes of wastewater containing complex contaminants from its daily operation. Textile wastewater typically consists of different types of dyes, detergents, grease and oil, heavy metal, inorganic salts and fibres in amounts depending on the processing regime. It is necessary that as much of this waste as possible is recycled instead of being disposed of in landfill sites. This operation also allows for the recovery of the valuable chemical components and water from a number of different textile process streams. Most of the conventional technologies seem unable to provide sufficient treatment for the effluents. Therefore, it is generally accepted that NF membranes can offer solutions for the problem. Of these NF membranes, the TFC-NF is the most widely used. Some of the commercial membranes achieve maximum separation of dye and salts while some achieve higher flux. This is because of the large variability of the parameters of textile wastewater and the NF membranes chosen. The results of NF performance evaluation have proven that NF membranes are the suitable separation process to be employed for the treatment of textile wastewater and, as a general trend, showed an acceptable rejection (Akbari *et al.*, 2002).

There are a number of studies that reported on the effects of different operating conditions of textile effluents on NF performance. However, in most cases, it was just carried out at laboratory or pilot-plant scale.

In a study at pilot-plant scale, commercial NF membranes such as NF-45 and DK-1073 exhibited good performance in terms of dye retention (Lopes *et al.*, 2005). The maximum dye rejection was up to 99.2% and 99.8% respectively, with an initial dye concentration of 450–500 mg/l. Meanwhile, the performance of MPS-31 was also investigated and gave results of dye retention which varied from 90.1% to 97.3%. It is attributed to the dye rejection to a secondary layer formed by retained dye on the membrane surface (Sungpet *et al.*, 2004). Secondary layers formed by dye and the Donnan effect may be responsible for dye removal instead of membrane molecular weight cut off. Fouling layer occurred as a result of the absorption of dye onto the membrane, resulting in an increase of dye rejection.

Based on the numerous studies conducted so far, NF membranes have proved applicable in dealing with textile wastewater which is highly coloured as well as highly loaded with monovalent and/or divalent salts. However, it is difficult to draw a general conclusion on the feasibility and the efficiency of NF for dyeing effluent treatment in the textile industry due to the large variability of textile wastewater parameters and the quality of the permeate required. Generally, it can be concluded that NF offers many more advantages compared to conventional treatment methods and the other categories of membrane technologies. However, improvements on these membranes are still needed in order to further enhance its performance before NF becomes a dominant commercialized wastewater treatment system in a large scale industrial plant.

Solvent-resistant NF membranes have a strong potential for a variety of applications ranging from pharmaceutical to chemical to food industries in non-aqueous systems (Bhanushali and Bhattacharyya, 2003). With increasing energy costs, industries are looking into alternate processes for recovery of valuable chemicals. Pharmaceutical industries, for example, spend most of their research efforts in downstream processing to recover valuable drugs. Membranes and membrane processes offer an excellent avenue in downstream processing with their versatility and low energy costs.

Solvent lube oil dewaxing (Gould *et al.*, 2001) processes are commonly used in refinery operations. Current recovery processes employ dissolution in solvent blends followed by precipitation of the wax by cooling. ExxonMobil in conjunction with W.R. Grace Company developed a membrane-based process in which the membrane reduces the residence time for the solvent mixtures which allows faster processing of the solvent. The membrane was typically operated at −10°C (14°F) and a pressure of 600 psi (41 bar) with 20 wt% lube oil in the feed. The cold solvent was directly recovered from the permeate and recycled back into the process. The membrane used in this process was a developmental asymmetric polyimide-based membrane that can withstand the corrosive solvent blend (methyl-ethyl-ketone and toluene) used in the application. The authors reported better than expected flux and rejection behaviour. The initial data gives better than 95% rejection of lube oil and better than 99% pure solvent under the operating conditions stated above. The process has been scaled-up from a laboratory scale to a commercial scale by ExxonMobil in Beaumont, Texas, and is called the MAX-DEWAX process. The authors also reported an increase in the base oil production by 25 vol% and that energy consumption per unit volume of product was reduced by about 20%.

Homogeneous catalyst recovery is another important area in chemical/pharmaceutical industries. Use of organometallic catalysts is extremely common in such industries and their recovery is essential because of their high costs. Membrane processes like NF can be used in such applications to recover the catalyst, which can then be reused without sacrificing activity and product quality. This would also enable wide use of homogeneous catalysis (such as Ru or Pd containing organometallic catalysts) to eliminate mass-transfer limitations for the reaction rate. Several researchers have ventured into this area to recover these valuable organometallic catalysts (Scarpello *et al.*, 2002).

For example, the reaction of styrene and iodobenzene to form *trans*-stilbene with a palladium-based Heck catalyst (Bhanushali and Bhattacharyya, 2003) by using polyimide membranes and studying the reaction in three different organic solvents (ethyl acetate, MTBE, and THF) to match catalyst and membrane properties. It has also been studied the phase transfer catalyst system in the reaction of bromoheptane and potassium iodide to form iodoheptane (Luthra *et al.*, 2001). In this reaction, tetraoctylammonium bromide was used as the phase transfer catalyst and toluene the organic solvent. After the reaction, the iodoheptane and the bromoheptane partition into the organic phase and this organic phase is the feed to the NF membrane (142A, polyimide-based 220 Da cut-off), which concentrates the catalyst in the retentate.

Conclusions

The demand on seawater desalination units and ultrapure water for the industry is continuously increasing. Nowadays, in RO the targets are the improvement of chlorine resistance of current composite membranes, control of the membrane fouling, improvement on the membrane cleaning, and on the recovering of organic solutes from water. These achievements will come through the nanotechnology innovation and modification on the membrane surface and interior. In fact, the increasing demands on the development of innovative new technologies and materials for production of potable water in both developed and developing countries have aimed current investigation at nanotechnology as an important solution for problems involving water purification (Savage and Diallo, 2005; Bottero *et al.*, 2006).

Nanomaterials and nanostructures have nanoscale dimensions that range from 1 to 100 nm, and often exhibit novel and significantly changed physical, chemical and biological properties as a result of their structure, larger surface area per unit of volume and quantum effects that occur at the nanoscale. Thus, novel nanoreactive RO/NF membranes have been developed for use in water filtration process (Stanton *et al.*, 2003). The RO and nanotechnology connection is now well established.

References

Ahmad, A.L., and Ooi, B.S. (2005). Properties and performance of thin film composites membrane: study on trimesoyal chloride content and polymerization time. J. Membr. Sci. *255*, 267–277.

Ahmad, A.L., Ooi, B.S., Mohammad, A.W., and Choudhary, J.P. (2004). Composite nanofiltration polyamide membrane: a study on diamine ratio and its performance evaluation. Ind. Eng. Chem. Res. *43*, 8074–8082.

Ahmad, A.L., Ooi, B.S., Mohammad, A.W., and Choudhary, J.P. (2004). Effect of constricted polymerization time on nanofiltration membrane characteristic and performance: a study using Donnan steric pore flow model. J. Appl. Polym. Sci. *94*, 394–399.

Akbari, Remigy, J.C., A., and Aptel, P. (2002). Treatment of textile dye effluent using a polyamide-based nanofiltration membrane. Chem. Eng. Proc. *41*, 601–609.

Asatekin, A., Menniti, A., Kang, S., Elimelech, M., Morgenroth, E., and Mayes, A.M. (2006). Antifouling nanofiltration membranes for membrane bioreactors from self-assembling graft copolymers. J. Membr. Sci. *285*, 81–89.

Baker, R.W. (2004). Membrane technology and applications (Wiley)

Bartels, C., Wilf, M., Casey, W., and Campbell, J. (2008). New generation of low fouling nanofiltration membranes, Desalination *221*, 158–167.

Belfort, G. (1984). Synthetic Membrane Processes – Fundamentals and Water Applications, (Academic Press, Orlando, USA).

Bhanushali, D., and Bhattacharyya, D. (2003). Advances in solvent-resistant nanofiltration membranes. Experimental observations and applications. Ann. N.Y. Acad. Sci. *984*, 159–177.

Bhore, N., Gould, R.M., Jacob, S.M., Staffield, P.O., McNally, D., Smiley, P.H., and Wildemuth, C.R. (1999). New membrane process debottlenecks solvent dewaxing unit. Oil Gas J. *97*, 67–75.

Bottero, J.Y., Rpse, J., and Wiesner, M.R. (2006). Nanotechnologies. Tools for sustainability in a new wave of water treatment processes. Integr. Environ. Assess. Manag. *2*, 391–395.

Bray, D.T. (1968). Reverse Osmosis purification apparatus, US Patent *3,417*,870.

Bremere, I., Kennedy, M.D., Johnson, A., Emmerik, R.V., Witkamp, G.-J., and Schippers, J.C. (1998). Increasing conversion in membrane filtration systems using a desupersaturation unit to prevent scaling. Desalination *119*, 199–204.

Buch, P.R., and Reddy, A.V.R. (2005). Synthesis, characterization and thermal properties of the soluble aromatic poly(amide imide)s. Polymer *46*, 5524–5534.

Cadotte, J.E. (1975). Reverse Osmosis Membrane, US Patent 3.926.798.

Cadotte, J.E. (1977). Reverse Osmosis Membrane, US Patent 4.039.44.

Cadotte, J.E., Petersen, R.J., Larson, R.E., and Erikson, E.E. (1980). A new thin film seawater Reverse Osmosis membrane, Presented at 5th Seminar on Membrane Separation Technology, Clemson University, Clemson, SC.

Cadotte, J.E. (1981). Interfacially Synthetized Reverse Osmosis Membrane, US Patent 4.277.344.

Cadotte, J.E. (1985). Evaluation of composite reverse osmosis membrane. In Materials Science of Synthetic Membranes, D.R. Lloyd, ed. ACS Symposium Series Number *269*, American Chemical Society, Washington, DC.

Cahill, D., Freger, V., and Kwkak, S.Y. (2008). Microscopy and microanalysis of reverse-osmosis and nanofiltration membranes. MRS Bulletin *33*, 27–32.

Chmiel, H., Mavrov, V., and Belieres, E. (2000). Fil. Sep. *37*, 24–27.

Criscuoali, A., and Drioli, E. (1999). Energetic and exergetic analysis of an integrated membrane desalination system. Desalination *124*, 243–249.

De Witte, J. (1997). Surface water potabilisation by means of a novel nanofiltration element. Desalination *108*, 153–157.

Endoh, R., Tanaka, T., Kurihara, M., and Ikeda, K. (1977). New polymeric materials for Reverse Osmosis membranes. Desalination *21*, 35–41.

Eriksson, P. (1988). Water and salt transport through two types of polyamide composite membranes. J. Membr. Sci. *36*, 297–303.

Frank, M., Bargeman, G., Zwijnenburg, A., and Wessling, M. (2001). Capillary hollow fiber nanofiltration membranes. Sep. Purif. Technol. *22–23*, 499–506.

Futselaar, H., Schonewille, H., and van der Meer, W. (2002). Direct capillary nanofiltration–A new high-grade purification concept. Desalination *145*, 75–80.

Futselaar, H., Schonewille, H., and van der Meer, W. (2003). Direct capillary nanofiltration for surface water. Desalination *157*, 135–136.

González, M., Navarro, R., Saucedo, I., Avila, M., Revilla, J., and Bouchar, C. (2002). Purification of phosphoric acid solutions by reverse osmosis and nanofiltration. Desalination, *147*, 315–320.

Gopal, R., Kaur, S., Ma, Z., Chan, C., Ramakrishna, S., and Matsuura, T. (2006). Electrospun nanofibrous membrane. J. Membr. Sci. *218*, 581–586.

Gorenflo, A., Velázquez-Padrón, D.T., and Frimmel, F.H. (2002). Nanofiltration of a german groundwater of high hardness and NOM content: performance and costs. Desalination *151*, 253–265.

Gould, R.M., and Nitsch, A.R. (1996). Lubricating oil dewaxing with membrane separation of cold solvent, US Patent *5,494,566*.

Gould, R.M., White, S.L., and Wildemuth, C.R. (2001). Membrane separation in solvent lube dewaxing. Environ. Prog. *20*, 12–16.

Hafiarle, H., Lemordant, D., and Dhahbi, M. (2000). Removal of hexavalent chromium by nanofiltration. Desalination *13*, 305–312.

Hamdona, S.K., Nessim, R.B., and Hamza, S.M. (1993). Spontaneous precipitation of calcium sulphate dihydrate in the presence of some metal ions. Desalination *94*, 69–80.

Hilal, N., Al-Zoub H., Darwish, N.A., Mohammad, A.W., and Arabi, M.A. (2004). A comprehensive review of nanofiltration membranes: treatment, pretreatment, modelling, and atomic force microscopy. Desalination *170*, 281–308.

Hirose, M. (2005). Highly permeable composite reverse osmosis membrane and method of producing the same, US Patent 6.837.381.

Holt, J.K., Park, H.G., Wang, Y.M., Stadermann, M., Artyukhin, A.B., Grigoropoulos, C.P., Noy, A., and Bakajin, O. (2006). Fast mass transport through sub-2-nanometer carbon nanotubes. Science *312*, 1034–1040.

Jacangelo, J.G., Trussell, R.R., and Watson, M. (1997). Role of membrane technology in drinking water treatment in the United States. Desalination *113*, 119–127.

Jakobs, D., and Baumgarten, G. (2002). Nanofiltration of nitric acidic solutions from picture tube production, Desalination *145*, 65–68.

Jayarani, M.M., Rajamohanan, R.R., Kulkarniand, S.S., and Kharul, U.K. (2000). Synthesis of model diamide, diester and esteramide adducts and studies on their chlorine tolerance. Desalination *130*, 1–16.

Jeong, B.H., Hoek, E.M.V., Yan, Y., Subramani, A., Huang, X., Hurwitz, G., Ghosh, A.K., and Jawor, A. (2007). Interfacial polymerization of thin film nanocomposites: a new concept for reverse osmosis membranes. J. Membr. Sci. *294*, 1–12.

Jonsson, G., and Benavente, J. (1992). Determination of some transport coefficients for the skin and porous layer of a composite membrane. J. Membr. Sci. *69*, 29–42.

Kamiyama, Y., Yoshioki, N., Matsui, K., and Nakagome, K. (1984). New thin film composite Reverse Osmosis membranes and spiral wound molecules. Desalination *51*, 79–85.

Kasper, D.R. (1993). Pre- and posttreatment processes for membrane water treatment systems. AWWA Membrane Technology Conference Proceedings, Baltimore, Md, USA.

Kim J.J., Chang, K., and Kwak, S.Y. (1997). US Patent 5.593.588.
Kim, I.-C., and Lee, K.-H. (2006). Dyeing process wastewater treatment using fouling resistant nanofiltration and reverse osmosis membranes. Desalination *192*, 246–251.
Klepetsanis, P.G., and Koutsoukos, P.G. (1989). Precipitation of calcium sulfate dihydrate at constant calcium activity. J. Crystal Growth *98*, 480–486.
Koyuncu, A. (2003). An advanced treatment of high-strength opium alkaloid processing industry wastewaters with membrane technology: pretreatment, fouling and retention characteristics of membranes. Desalination *155*, 265–275.
Koyuncu, A., Wiesner, M.R., Bele, C., Coriton, G., Djafer, M., and Cavard, J. (2006). Bench-scale assessment of pretreatment to reduce fouling of salt-rejecting membranes. Desalination *197*, 94–105.
Kronmiller, D.L. (1984). What every reverse osmosis water system manager should know. Desalination *98*, 401–411.
Kuehne, M.A., Song, R.Q., Li, N.N., and Petersen, R.J. (2001). Flux enhancement in TFC RO membranes. Environ. Prog. *20*, 23–30.
Larson, R.E., Cadotte, J.E., and Petersen, R.J. (1981). The FT-30 seawater Reverse Osmosis membrane-element Test Results. Desalination *38*, 473–483.
Lau, W.-J., and Ismail, A.F. (2009). Polymeric nanofiltration membranes for textile dye wastewater treatment: Preparation, performance evaluation, transport modeling and fouling control- a review. Desalination *245*, 321–348.
Lee, S., Kim, J.-Y., Lee, C.-H., Lee, M.-C., and Chung, K.-J. (1995). Flux improvement in nanofiltration through a hybrid system with microfiltration. Euromembrane 95, University of Bath, U.K.
Lee, S., Kim, J., and Lee, C.H. (1999). Analysis of $CaSO_4$ scale formation mechanism in various nanofiltration modules. J. Membr. Sci. *163*, 63–74.
Lee, S., and Lee, C.H. (2000). Effect of operating conditions on $CaSO_4$ scale formation mechanism in nanofiltration for water softening. Wat. Res. *34*, 3854–3866.
Lee, S., and Lee, C.H. (2005). Water Science & Technology, Q IWA Publishing, vol *51*, N° *6–7*, pp 267–275.
Lee, S., and Lueptow, R.M. (2001a). Toward a reverse osmosis membrane system for recycling space mission wastewater. Life Support Bio. Sci. *7*, 151–161.
Lee, S., and Lueptow, R.M. (2001b). Reverse osmosis filtration for space mission wastewater: membrane properties and operating conditions. J. Membr. Sci. *182*, 77–90.
Li, H., Xu, X., Yang, G., and Xie, X. (2006). Process and device for recovery of waste sulfuric acid from production of lead acid storage battery polar plate. Faming Zhuanli Shenqing Gongkai Shuomingshu. Patent application CN 2004–10060943.
Li, L.C., Wang, B.G., Tan, H.M., Chen, T.L., and Xu, J.P. (2006). A novel nanofiltration membrane prepared with PAMAM and TMC by in situ interfacial polymerization on PEK-C ultrafiltration membrane. J. Membr. Sci. *269*, 84–90.
Lind, M.L., Ghosh, A.K., Jawor, A., Huang, X., Hou, W., Yang, Y., and Hoek, E.M.V. (2009). Influence of Zeolite Crystal Size on Zeolite-Polyamide Thin Film Nanocomposite Membranes. Langmuir *25*, 10139–10145.
Linder, C., Nemas, M., Perry, M., and Katraro, P. (1993). Silicone-derived solvent stable membranes, US Patent *5,265,734*.
Loeb, S., and Sourirajan, S. (1963). Seawater demineralization by means of an osmotic membrane. In Saline Water Conversion II, Advances in Chemistry Series N° *38*, American Chemical Society, Gould, R.F., ed. (Washington DC), pp. 117–132.
Lonsdale, H.K. (1966). Properties of cellulose acetate membranes. In Desalination By Reverse Osmosis, Merten, E., ed. (MIT Press, Cambridge, MA), pp. 93–160.
Lopes, C.N., Petrus, J.C.C., and Riella, H.G. (2005). Color and COD retention by nanofiltration membranes. Desalination *172*, 77–83.
Luthra, S.S., Yanga, X., and Freitas Dos Santos, L.M. (2001). Phase-transfer catalyst separation and reuse by solvent resistant nanofiltration membranes. Chem. Commun. *16*, 1468–1469.
Machado, D.R., Hasson, D., and Semiat, R. (1999). Effect of solvent properties on permeate flow through nanofiltration membranes. Part I: Investigation of parameters affecting solvent flux. J. Membr. Sci. *163*, 93–101.
Marhaba, T.F., and Medlar, S.J. (1993). Treatment/removal of bromate and bromide ions from drinking water. Proceedings-Water Quality Technology Conference *PT1*, 297–320.
Mo, J.H., Lee, Y.H., Kim, J., Jeong, J.Y., and Jegal, J. (2008). Treatment of dye aqueous solutions using nanofiltration polyamide composite membranes for the dye wastewater reuse. Dyes Pigments *76*, 429–434.
Oh, B.S., Jang, H.Y., Jung, Y.J., and Kang, J.W. (2007). Microfiltration of MS2 bacteriophage: Effect of ozone on membrane fouling. J. Membr. Sci. *306*, 244–252.
Paul, D.H. (1997). A review of membrane water treatment. Ultrapure Water *14*, 39–42.
Paul, D.H. (2000). Back to fundamentals of membrane technology. Ultrapure Water *17*, 71–43.
Perrins, J.C., Cooper, W.J., van Leeuwen, J.H., and Herwig, R.P. (2006). Ozonation of seawater from different locations: formation and decay of total residual oxidant–implications for ballast water treatment. Mar. Pollution Bull. *52*, 1023–1033.

Pervov, A.G. (1991). Scale formation prognosis and cleaning procedure schedules in reverse-osmosis systems operation. Desalination *83*, 77–118.

Pervov, A.G., Dudkin, E.V., Sidorenko, O.A., Antipov, V.V., Khakhanov, S.A., and Makarov, R.I. (2000). RO and NF membrane systems for drinking water production and their maintenance techniques. Desalination *132*, 315–321.

Peters, J.M.M., Boom, J.P., Mulder, M.H.V., and Strathmann, H. (1982). Retention measurements of nanofiltration membranes with electrolyte solutions. J. Membr. Sci. *145*, 199–205.

Petersen, R.J., and Cadotte, J.E. (1990). Thin film composite Reverse Osmosis membranes. In Handbook of Industrial Membrane Technology, Porter, M.C., ed. (Noyes Publications, Park Ridge, NJ), pp. 307–348.

Pulles, W., Juby, G., and Busby, R. (1992). Development of the slurry precipitation and recycle reverse osmosis (SPARRO) technology for desalinating scaling mine waters. Wat. Sci. Tech. *25*, 177–192.

Rand, J.L., Hofmann, R., Alam, M.Z.B., Chauret, C., Cantwell, R., Andrews, R.C., and Gaynon, G.A. (2007). A field study evaluation for mitigating biofouling with chlorine dioxide or chlorine integrated with UV disinfection. Water Res. *41*, 1939–1948.

Rautenbach, R., and Habbe, R. (1991). Seeding technique for zero-discharge processes, adaption to electodialysis. Desalination *84*, 153–161.

Ray, R.J., McCray, S.B., and Newbold, D.D. (1991). Small-scale membrane systems for the recovery and purification of water. Sep. Sci. Tech. *26*, 1155–1176.

Razdan, U., and Kulkarni, S.S. (2004). Nanofiltration thin-film-composite polyesteramide membranes based on bulky diols. Desalination *161*, 25–32.

Reid, C.E., and Breton, E.J. (1959). Water and ion flow across cellulosic membranes. J. Appl. Polym. Sci. *1*, 133–137.

Roh, I.L., Greenberg, A.R., and Khare, V.P. (2006). Synthesis and characterization of interfacially polymerized polyamide thin films. Desalination *191*, 279–290.

Rosenbaum, S., Mahon, H.I., and Cotton, O. (1967). Permeation of water and sodium chloride through cellulose acetate. J. Appl. Polymer Sci. *11*, 2041–2049.

Savage, N., and Diallo, M.S. (2005). Nanomaterials and water purification. Opportunities and challenges. J. Nanopart. Res. *7*, 331–342.

Scarpello, J.T., Nair, D., and Freitas Dos Santos, L. (2002). The separation of homogeneous organometalic catalysts using solvent-resistant nanofiltration membranes. J. Membr. Sci. *203*, 71–85.

Schaep, J., van der Bruggen, B., Uytterhoeven, S., Croux, R., van der Casteele, D., Wilms, D., van Houtte, E., and Vanlerberghe, L. (1998). Removal of hardness from groundwater by nanofiltration. Desalination *119*, 295–302.

Shen, L. (2009). Faming Zhuanli Shenqing Gongkai Shuomingshu, Patent Application: CN 2008–10059988, 20080307.

Shu, L., Fane, A.G., Waite, T.D., Pailthorpe, M.T., and Bliss, P.J. (2000). I: Proc. World Filtration Congress, London, pp. 647–650.

Sombekke, H.D.M., Voorhoeve, D.K., and Hiemstra, P. (1997). Environmental impact assessment of groundwater treatment with nanofiltration. Desalination *113*, 293–296.

Soung, I.C., Clech, L.P., Bruce, J., and Judd, S. (2002). Membrane fouling in membrane bioreactors for wastewater. ASCE. *121*, 1018–1102.

Stanton, B.W., Harris, J.J., Miller, M.D., and Bruening, M.L. (2003). Ultrathin, multilayered polyelectrolyte films as nanofiltration membranes. Langmuir *19*, 7038–7042.

Sutherland, K. (2008). Development in filtration: what is nanofiltration? Filtration and Separation *45*, 32–35.

Sungpet, R., Jiraratananon, R., and Luangsowan, P. (2004). Treatment of effluents from textile-rinsing operations by thermally stable nanofiltration membranes. Desalination *160*, 75–81.

Tang, B., Huo, Z., and Wu, P. (2008). Study on a novel polyester composite nanofiltration membrane by interfacial polymerization of triethanolamine (TEOA) and trimesoyl chloride (TMC). J. Membr. Sci. *320*, 198–205.

Troup, D.H., and Richardson, J.A. (1978). Scale nucleation on a heat transfer surface and its prevention. Chem. Eng. Commun. *2*, 167–180.

Ulbricht, M. (2006). Advanced functional polymer membranes. Polymer *47*, 2217–2262.

Van der Meer, W.G.J., and van Dijk, J.C. (1997). Theoretical optimization of spiral-wound and capillary nanofiltration modules. Desalination *113*, 129–146.

Van der Bruggen, A., Schaerp, A., Wilms, D., and Vandecasteele, C. (1999). Influence of molecular size, polarity and charge on the retention of organic molecules by nanofiltration. J. Membr. Sci. *156*, 29–33.

Van der Bruggen, A., Cornelis, G., Vandecasteele, C., and Devreese, I. (2005). Fouling of nanofiltration and ultrafiltration membranes applied for wastewater regeneration in textile industry. Desalination *175*, 111–119.

Vourch, M., Balannec, B., Chaufer, B., and Dorange, G. (2005). Nanofiltration and Reverse Osmosis of model process waters from the dairy industry to produce water for reuse. Desalination *172*, 245–256.

Wang, F., Tarabara, A., and Volodymyr, V. (2007). Abstracts of Papers, 233rd ACS National Meeting, Chicago, IL, United States, March 25–29.

Watson, B.W. (1990). High recovery reverse osmosis. Desalination *78*, 91–97.

Wei, X.-Z., Zhu, L.-P., Deng, H.-Y., Xu, Y.-Y., Zhu, B.-K., and Huang, Z.-M. (2008). New type of nanofiltration membrane based on crosslinked hyperbranched polymers. J. Membr. Sci. *323*, 278–287.

Westmoreland, J.C. (1968). Spirally wrapped Reverse Osmosis membrane cell, US Patent 3.367.504.

White, L.S., Wang, I.F., and Minhas, B.S. (1993). Polyimide membranes for separation of solvents from lube oil, US Patent 5.264.166.

White, L.S., and Nitsch, A.R. (2000). Solvent recovery from lube oil filtrates with polyimide membranes. J. Membr. Sci. *179*, 267–300.

White, L.S., and Wildemuth, C.R. (2006). Aromatics enrichment in refinery streams using hyperfiltration. Ind. Engineering Chem. Res. *45*, 9136–9143.

Williams, M.E., Bhattacharyya, D., Ray, R.J., and McRay, S.B. (1992). Selected applications of Reverse Osmosis. In Membrane Handbook, Ho, W.S.W. and Sirkar, K.K., eds. (Van Nostrand Reinhold, New York), pp. 312–354.

Woei-Jye, L., and Ismail, A.F. (2009). Polymeric nanofiltration membranes for textile dye wastewater treatment: Preparation, performance evaluation, transport modelling, and fouling control – a review. Desalination *245*, 321–348.

Worley, J.L., Dietrich, A.M., and Hoehn, R.C. (2003). Dechlorination techniques to improve sensory odor testing of geosmin and 2-MIB. J. AWWA *95*, 109–117.

Yangali-Quintanilla, V., Verliefde, A., Kim, T.-U., Sadmani, A., Kennedy, M., and Amy, G. (2009). Artificial neural network models based on QSAR for predicting rejection of neutral organic compounds by polyamide nanofiltration and reverse osmosis membranes. J. Membr. Sci. *342*, 251–262.

Yong, Z., Yu, S., Liu, M., and Gao, C. (2006). Polyamide thin film composite membrane prepared from m- phenylenediamine and m- phenylenediamine –5-sulfonic acid. J. Membr. Sci. *270*, 162–168.

Zhao, J. (2008). Application of integrated membrane separations in sewage treatment by fine chemicals production. Jingxi Yu Zhuanyong Huaxuepin *16*, 18–19.

Zhao, L. (2009). Industrial method for extracting L-citrulline from watermelon or seeding watermelon and their system and leaf. Zhuanli Shenqing Gongkai Shuomingshu. Patent application CN 2007–10120714.

Zhou, Y., Yu, S., Liu, M., and Gao, C. (2005). Preparation and characterization of polyamide-urethane thin-film composite membranes. Desalination *180*, 189–196.

Zhou, Y., Yu, S., Liu, M., and Gao, C. (2006a). Polyamide thin film composite membrane prepared from *m*-phenylenediamine and *m*-phenylenediamine-5-sulfonic acid. J. Membr. Sci. *270*, 162–168.

Zhou, Y., Yu, S., Liu, M., Chen, H., and Gao, C. (2006b). Effect of mixed cross-linking agents on performance of thin-film-composite membranes. Desalination *192*, 182–189.

Electrospinning Nanofibres for Water Treatment Applications

7

Eugene Smit

Abstract
Electrospinning is a highly versatile technique that can be used to create ultrafine fibres of various polymers and other materials, with diameters ranging from a few micrometers down to tens of nanometres. The non-woven webs of fibres formed through this process typically have high specific surface areas, nano-scale pore sizes, high and controllable porosity and extreme flexibility with regard to the materials used and modification of the surface chemistry of the fibres. This chapter describes the combination of these features in the application of electrospun nanofibres in a variety of water treatment applications, including filtration, solid phase extraction and reactive membranes.

Introduction

Electrospinning and nanofibres are two terms that most readers have probably come across by now. Briefly, electrospinning is a technique in which high-voltage is used to form extremely fine fibres from various natural and synthetic polymers. Although not strictly **nano**materials, workers employing electrospinning frequently refer to the fibres formed through this process as nanofibres, even though their diameters typically range between 50 nm and 2 μm. For a more general overview of the electrospinning technique the reader is referred to several excellent reviews and books on the subject (Huang *et al.*, 2003; Ramakrishna *et al.*, 2005, 2006; Greiner and Wendorff, 2007; Andrady, 2008).

Since the revival of electrospinning in the mid-1990s (Doshi and Reneker, 1995) the number of publications and patents in this field have grown exponentially (Greiner and Wendorff, 2007). This rapidly growing interest in the electrospinning technique and electrospun nanofibres stems from the list of unique and ideal features of electrospun fibres and their non-woven webs, and includes the following:

- Owing to their typically small diameters, electrospun fibres have high surface-to-volume ratios and high specific surface areas (theoretical specific surface area for smooth 50-nm fibres $\approx 80\,m^2/g$).
- The generally nano-scale pore sizes and the pore size distribution of electrospun nanofibre webs can be easily controlled through manipulation of electrospinning process parameters (Barhate *et al.*, 2006).
- Electrospun nanofibre webs have very high porosity, often ~80% (that means 20% fibres and 80% open space), and both the fibres and the voids between the fibres form interconnected structures. These, fully accessible, interconnected pore structures can facilitate the transfer of various sizes of molecules with much better efficiency than conventional porous materials (Yoon *et al.*, 2008).
- Various additives, including salts, proteins, pharmaceuticals and nanoparticles, can be incorporated into electrospun fibres and their webs, simply by pre-mixing them with the fibre spinning solution (Chronakis, 2005; Sawicka and Gouma, 2006; Tan and Lim, 2008).

- Various strategies exist for manipulating the surface chemistry of polymeric nanofibres in order to obtain desired interactions between the nanofibres and their environment (Gopal *et al.*, 2007b).
- Electrospinning can be used to make nanofibres from a growing list of materials, including various natural and synthetic polymers (Greiner and Wendorff, 2007), ceramics (Ramaseshan *et al.*, 2007), and even metals (Bognitzki *et al.*, 2006; Wu *et al.*, 2007).

This list of attractive features, combined with the practical simplicity and versatility of the electrospinning process, effectively gives researchers a handy 'toolbox', which can be used for developing a host of nanofibre-based materials science solutions to various problems. The diverse fields in which electrospun nanofibres are currently being evaluated for application include filtration (liquids and gases), biomedical applications (wound dressings, tissue engineering scaffolds, and drug release materials), fibre-reinforced composites, nanosensors, energy (batteries, fuel cells, and solar cells), absorbent textiles, and sound absorption materials.

When investigating the possible uses of electrospun nanofibres in water treatment applications, a good starting point is to evaluate how each of the features (or tools) of the electrospinning 'toolbox' described above can be used to solve common problems in water treatment. For instance, the small pore sizes obtained in electrospun nanofibre webs could potentially be utilized in micro- or nano-filtration applications, while the surface chemistry of the fibres could be manipulated in order to obtain antimicrobial properties to prevent biofilm formation in commercial water treatment membranes. In the next section an overview is given of various recent studies on water treatment in which the different tools of the electrospinning 'toolbox' have been utilized, among others, in filtration, affinity membranes, and photolysis of organic pollutants.

Applying the electrospinning 'toolbox' in water treatment

When we talk about water treatment, we are usually referring to the removal of impurities. These impurities can be in the form of dispersed solids, dissolved salts and heavy metal compounds, miscible and/or immiscible natural and/or synthetic organic compounds, and biological contaminants like bacteria and viruses. The size of these contaminants can range from a few micrometres (e.g. bacteria) down to tenths of a nanometre (ions).

In devising nanofibre-based strategies for removal of these impurities, one of the first obvious ideas would be to utilize the small pore sizes, high specific surface areas and high porosity of electrospun nanofibre webs to separate solid impurities from the liquid phase through **filtration**. A second strategy could be to utilize the high specific surface areas, high nanofibre web porosity, the ability to manipulate the fibre surface-chemistry, and the broad choice of fibre forming materials, in order to make affinity membranes for **selective adsorption and/or absorption** of dissolved or dispersed contaminants. One could also combine the choice of a suitable nanofibre-forming material, the option of fibre surface-chemistry modification, and the ability to include various additives into electrospun fibres, with the high specific surface area of nanofibre webs and create **reactive membranes** that can neutralize or decompose impurities. In the following sub-sections, these and other water treatment applications of electrospun nanofibres are discussed with reference to studies that recently appeared in literature.

Filtration

Traditional drinking water treatments, consisting of chemical disinfection, flocculation, sedimentation and filtration, have been shown to be ineffective in removing some contaminants, and in some cases can even lead to production of new harmful compounds (like trichloromethane from the reaction between chlorine and natural organic matter) (Bottino *et al.*, 2009). Various membrane-based filtration methods, with different separation ranges, have been proposed as possible alternatives for treatment of water for drinking purposes. For easy reference, a comparison of the impurities removed, their typical particle sizes, and the operating pressures of the traditional water treatment process and various membrane processes are given in Table 7.1.

The four main relevant types of filtration mechanisms that can apply to nanofibre filter media are: surface straining, depth straining, depth filtration and cake filtration (Barhate *et al.*, 2008). In *surface straining*, the filter traps contaminants on the outside of the filter medium. Larger particles cannot pass through the filter media, but smaller particles do pass through. In *depth straining*, the filter media is relatively thick, compared to surface filter media, and the pores are variable along the flow path. Particles enter the filter through initial pores but then get trapped inside the filter when the pore diameters become smaller than the particle. In *depth filtration*, various physical mechanisms are used to remove particles from a fluid, even though the particle may be smaller than the pore diameter at any point in the pore structure. These physical mechanisms include inertial impaction, direct interception, diffusional (Brownian) deposition, electrostatic attraction and gravity settling. In *cake filtration*, particles initially deposit on the filter medium and then over time successive layers of particles deposit on top of preceding layers to form a filter cake.

The unique tools of the electrospinning toolbox have been combined in various ways to create membranes that utilize these mechanisms for micro-, ultra-, and nanofiltration.

Microfiltration

The ability to control pore size distribution from sub-micron levels to a few micrometres whilst maintaining a high level of porosity could make electrospun nanofibre membranes good candidates for liquid microfiltration (Yoon *et al.*, 2008).

In one of the early published studies evaluating electrospun nanofibre membranes for microfiltration (Gopal *et al.*, 2006), complete self-supporting nanofibre web membranes were electrospun from poly(vinylidene difluoride) (PVDF), a model membrane material commonly used in commercial micro- and ultrafiltration membranes. The nanofibres in the membrane web had an average diameter of 380 nm, the membrane thickness was 300 μm, and the pore size in the web was 10.6–4.0 μm. The membrane's ability to function as a microfiltration membrane was evaluated using a simple polystyrene (PS) microparticle model. Separation of 500 ppm water dispersions of 10 μm particles was carried out first, followed by 5 μm particles and finally 1 μm particles. The main objective of the experiment was to evaluate the effect of particle size on flux through the filter, as well as the degree of membrane fouling.

Table 7.1 Comparison of various process parameters of traditional sand filtration and various membrane filtration processes

	Traditional process	Membrane process			
	Sand filtration	Microfiltration	Ultrafiltration	Nanofiltration	Reverse osmosis
Smallest impurities removed	Suspended matter	Colloids Bacteria	Macromolecules Viruses	Small organic molecules Divalent ions	All dissolved species
Particle sizes removed	>1 mm	0.1–10 μm	0.001–0.5 μm	0.005–0.001 μm	0.0001–0.001 μm
Operating pressure (bar)	0.1 to 2	0.2 to 2	2 to 10	8 to 20	10 to 80
Typical flux (l/m^2/h)	2000 to 10,000	100 to 1000	50 to 200	20 to 50	10 to 50

Separation factors of 96% and 91% were observed for the 10 µm and 5 µm particles respectively and stirring of the fluid in the membrane module was sufficient to lift the PS particles from the membrane and completely recover the flux. This indicated a *surface straining* mechanism for 10 and 5 µm particles through this membrane. Very different filtration behaviour was observed when 1 µm particles were filtered. A high separation factor of 98% and an instantaneous flux reduction was observed at the onset of the experiment and even at reduced particulate concentrations (100 ppm) the flux was still greatly reduced and could not be recovered. This change in filtration behaviour for the 1 µm particles could be explained by a *cake filtration* mechanism. The small size of the microparticles allowed them to pack together closely in a dense 'cake' which, while reducing the effective pore size and increasing the separation factor, also led to irreversible fouling of the membrane as can be seen in the FESEM micrograph of the membrane in Fig. 7.1.

In a later similar study in the same group (Gopal *et al.*, 2007a), the potential use of polysulphone (PSU) nanofibre membranes as pre-filters was evaluated. Nanofibre membranes were identified as ideal pre-filters (or depth filters) based on their extremely high internal specific surface area, which would lead to high dirt-loading capacity. Application of such microfiltration pre-filters upstream from ultrafiltration, nanofiltration and reverse osmosis processes can reduce or prevent fouling of the membranes in the downstream process and improve process performance and membrane life time.

In this study, a PSU nanofibre membrane was formed through electrospinning of a 20 wt% PSU solution in DMF. The 470 nm diameter fibres were spun to form a free-standing 135-µm-thick web, which was heat treated to increase its structural integrity. The mean flow pore of the web was 2.1 µm and the bubble point was 4.6 µm.

Aqueous suspensions (100 ppm) of PS microparticles, 10, 8, 7, 3, 2, 1, 0.5 and 0.1 µm in size, were consecutively filtered through the same membrane, and the effect of particle size on separation factor and flux recovery was evaluated. Similar to their previous study, the membrane showed >99% rejection rate for the particles that were significantly larger than the pore size in the membrane (i.e. 10, 8 and 7 µm) and flux could easily be recovered through stirring of the solution in the membrane module. This indicated that the nanofibre membrane acted as a screen filter utilizing a *surface straining* mechanism.

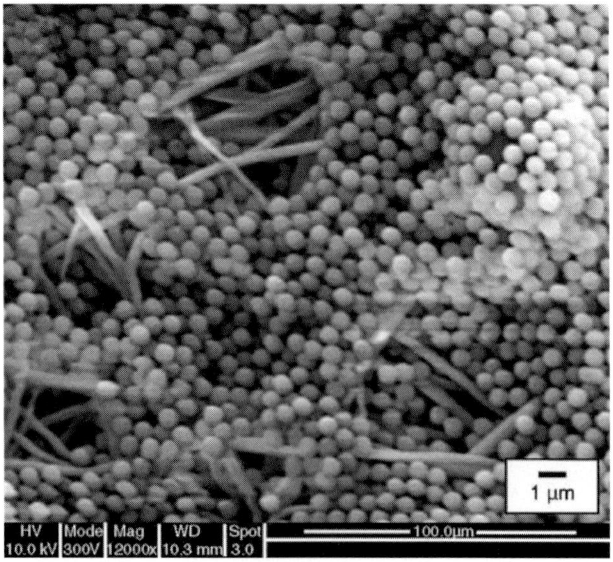

Figure 7.1 Electrospun PVDF fiber membrane irreversibly fouled with 1 µm PS microparticles. Reprinted from (Gopal *et al.*, 2006), Copyright (2006), with permission from Elsevier.

When 3 µm particles were filtered, the separation factor dropped to 92% and a drop in flux recovery was observed. The bubble point of 4.6 µm implied that some of the 3 µm particles could pass through the membrane and be detected in the permeate. It is likely that some entrapment of the particles in the membrane occurred, which would explain the slight reduction in flux.

When the 2 µm particles were filtered, a severe and immediate drop in flux was observed. Another significant drop in flux was also observed after the 1 µm separation run. Similar to what was observed in their previous study, filtration of microparticles with sizes in the same range as the average pore diameter in the membrane caused immediate and irreversible fouling of the membrane through build-up of a dense microparticle 'cake'. The dense packing of the microparticles in the 'cake' layer reduced the effective pore size of the nanofibre membrane significantly at its surface. This led to high separation factors for 0.5- and 0.1-µm particles (88% and 89% respectively), even though the particles were much smaller than the actual pores of the nanofibre membrane.

A separate study was performed to evaluate the performance and selectivity of the nanofibre membrane for 1, 0.5 and 0.1 µm particles in which the effect of the cake layer described above was eliminated. Instead of consecutive filtrations through the same membrane, a fresh membrane was used for each separation experiment. While the 1 µm particles once again caused an immediate sharp drop in flux, as a result of cake-layer formation, filtration of the 0.5 µm particles led to a gradual decrease in flux and eventual permanent fouling of the membrane. So while the 1 µm particles were separated through a *surface straining* mechanism, it could be observed that the smaller particles (0.5 and 0.1 µm) entered the membrane readily and were attracted onto the surfaces of the fibres through *depth filtration* by mechanisms such as direct interception, inertial impaction and diffusion.

In more recent study at the University of Tennessee (Desai *et al.*, 2009), electrospun chitosan fibre membranes were evaluated for their physical filtration, heavy metal absorption, and antimicrobial properties. Unfortunately, the pressure across the membrane could not be carefully controlled by the system used and this led to tearing of the fibre mat, resulting in reduced filtration efficiency. The experiments were repeated without applying vacuum to the filter media for filtration, while varying the planar density of the electrospun chitosan nanofibre web on the spunbonded polypropylene substrate fabric between 1 gsm and 3 gsm. Fifty and seventy per cent filtration efficiency, respectively, were obtained for a nanofibre web with 92 nm average fibre diameter and 1.562 µm maximum pore size.

It can be concluded that electrospun nanofibre membranes can be employed as microfiltration filters or pre-filters for the removal of microparticles from fluids. When the particles are larger than the pores within the membrane, it acts as a screen filter (through a *surface straining* mechanism) and no fouling is observed, or the fouling can be easily removed by stirring of the filtered solution above the membrane. However, when the particle size is similar to the average pore size within the filter, rapid irreversible fouling and particle cake formation occurs. Although this can lead to high separation factors through reduction of the effective pore size, this simultaneously leads to rapidly reduced flux which cannot be recovered. When particles fall in the sub-micron range, some entrapment of the particles within the nanofibre membrane still occurs through *depth filtration* mechanisms, but lower rejection rates and associated higher flux is typically observed.

Ultrafiltration and nanofiltration

Ultrafiltration (UF) is used to filter out particulates with sizes ranging from 1 to 100 nm. This implies that, in order to use electrospun nanofibre webs directly for ultrafiltration, the surface pore size of the web needs to be less than 0.1 µm (Yoon *et al.*, 2008). Although it is possible to create electrospun nanofibre webs with such small pore sizes, this is not a practical solution for ultrafiltration membranes due to the rapid fouling of these webs. The rapid fouling rate in neat nanofibre membranes leads to high replacement frequency, which in turn leads to increased cost.

The group of Chu and Hsiao at Stony Brook University have proposed the use of a new type of thin-film composite (TFC) membrane to resolve the rapid fouling problem of neat nanofibre ultrafiltration membranes (Yoon *et al.*, 2008). Traditionally, TFC membranes consist of three layers: a substrate made from a non-woven fabric (e.g. spunbond or meltblown), an asymmetric porous mid-layer (usually created by phase inversion) and a barrier layer (made by coating or interfacial polymerization). In such a three-layered structure, the thin barrier layer is the primary component for determining selectivity of the membrane and prevention of surface fouling, while the asymmetric porous mid-layer and the underlying non-woven fabric layer act as porous supports for the thin barrier layer.

In a new type of TFC membrane an electrospun nanofibrous web is used as the porous mid-layer instead of the typical asymmetric porous membrane. Since nanofibre webs can have surface porosities of up to 80%, while the typical asymmetric porous membranes made through phase inversion have surface porosities of only about 20%, this leads to a significant increase in flux through the membrane. Fig. 7.2 shows the three-tier conceptual design of these membranes (Wang *et al.*, 2005). The micron-size-range pores in the nanofibrous layer permit the top-coating layer to be much thinner than when using conventional mid-layer substrates. The water flux through the composite membrane is mainly dependent on properties of the top coating layer, specifically its hydrophilicity and its thickness (as thin as possible). The fouling of the membrane is minimized by the thin non-porous hydrophilic coating layer, which prevents particles from adsorbing on the surface or moving into the pores of the mid- and support layers.

The applicability and efficacy of this three-tier approach have been evaluated in several studies for different combinations of mid- and top-layer materials as briefly summarized below.

In one of the first published studies utilizing this three-tier composite structure (Wang *et al.*, 2005), TFC membranes were prepared with a nanofibrous mid-layer prepared by electrospinning poly(vinyl alcohol) (PVOH) followed by chemical cross-linking with glutaraldehyde (GA). The top coating consisted of a composite of cross-linked PVOH hydrogel with a surface-oxidized multi-walled carbon nanotube (MWNT) filler (Fig. 7.3). This was compared to membranes with a similar PVOH nanofibrous mid-layer with a top coating consisting of polyether-*b*-polyamide copolymer (Pebax) with MWNT filler. The average porosity of the PVOH nanofibrous mid-layer was 82% after cross-linking. Evaluation of the filtration properties of these TFC membranes were performed using oil/water emulsion [soya bean oil, 1350 ppm; non-ionic surfactant (Dow Corning, 193 fluid), 150 ppm in water]. The cross-linked PVOH TFC membrane exhibited excellent organic solids rejection

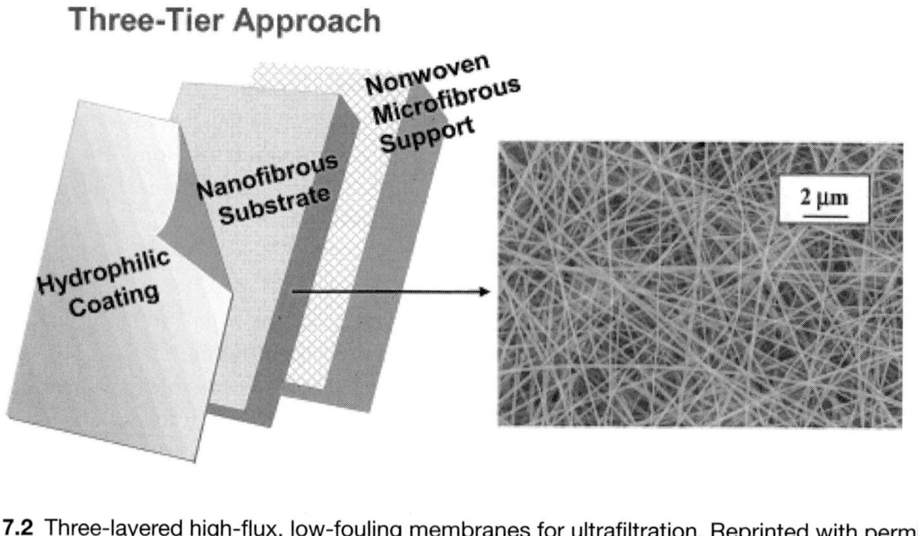

Figure 7.2 Three-layered high-flux, low-fouling membranes for ultrafiltration. Reprinted with permission from (Wang *et al.*, 2005). Copyright, 2005 American Chemical Society.

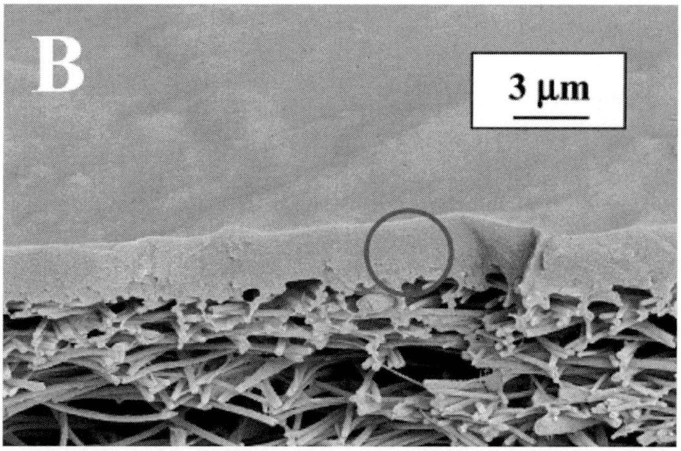

Figure 7.3 SEM cross-sectional image showing electrospun PVOH coated with 10wt% MWNT/PVOH hydrogel nanocomposite. Reprinted with permission from (Wang *et al.*, 2005). Copyright, 2005 American Chemical Society.

rates (99.8%) without appreciable fouling at a flux rate up to 330 l/m^2 h. Increasing the concentration of the surface-oxidized MWNT in the coating layer generally improved the flux rate and this was attributed to the generation of more effective hydrophilic nano-channels for water passage through the composite membrane.

Later studies by this group differed primarily with regard to the materials used in making the mid-and top-layers of the TFC. In (Wang *et al.*, 2006) the TFC consisted of a nanofibrous mid-layer of cross-linked PVOH, while PVOH hydrogel coatings with various degrees of cross-linking were evaluated for the top-layer. In (Yoon *et al.*, 2006) the mid-layer was made from electrospun PAN, and the top layer was made of a chitosan coating. In (Tang *et al.*, 2009) UV-reactive PVOH was photo-cross-linked to form the top-layer on a mid-layer consisting of electrospun PVOH that had been chemically cross-linked using glutaraldehyde. In (Yoon *et al.*, 2009a) the top-layer consisted of cross-linked PVOH supported on a PAN nanofibre mid-layer. In all instances the filtration properties of these TFC membranes were evaluated using oil/water emulsions.

Although there were small differences between the TFC membranes in these studies, relating to the different materials used and the degrees of porosity of the mid-layer nanofibre webs, they generally had three things in common:

- very high flux (usually up to several times higher than conventional UF membranes),
- excellent rejection ratios (>99%), and
- very good fouling resistance.

The same group recently went a step further and applied this three-tiered approach to produce a high flux **nano**filtration (NF) membrane (Yoon *et al.*, 2009b) consisting of a mid-layer support of electrospun PAN nanofibres with a top-layer that was produced by interfacial polymerization of polyamides containing different ratios of piperazine and bipiridine. The nanofiltration test of the membrane, using a divalent salt solution ($MgSO_4$), showed over 2.4 times more permeate flux than conventional TFC membranes with the same chemical composition, while maintaining the same rejection rate (ca. 98%).

Selective adsorption and/or absorption

As stated earlier, the high specific surface area of electrospun nanofibre webs, the high nanofibre web porosity, the ability to manipulate the fibre surface chemistry, the broad choice of fibre-forming materials, and the ability to blend various additives into electrospun nanofibres simply by premixing them with the spin solution can potentially be used to make affinity membranes for **selective adsorption or absorption** of dissolved or dispersed contaminants. This selective adsorption or absorption can potentially be used for removing and recovering precious and/or toxic contaminants in water. Alternatively these selective materials may be used as solid-phase extraction sorbents for pre-concentration of trace pollutants for analysis in environmental samples.

Although the specific surface area of electrospun nanofibres (typically 10–80 m^2/g) is not as high as the specific surface areas of well-known particulate absorbents like activated charcoal (upwards of 500 m^2/g), the fibrous form of the electrospun nanofibres significantly simplifies handling of the material and in many applications the additional filtration step required for particulate absorbents can be eliminated. This reduction in process complexity, combined with the various other materials and process features of electrospinning, as discussed above, can potentially offset the lower specific surface area compared to particulate absorbents.

Although the high specific surface area of electrospun nanofibres has been one of the main drivers behind different applications for quite some time, and while the combination of this feature with the broad materials choices and possibilities of blended additives naturally leads to the concept of affinity membranes, research activity on this specific topic has only recently started to pick up. To date, two different approaches have been taken to obtain affinity membranes using electrospun fibres. In the first approach, the material from which the electrospun fibres were formed was chosen such that it would have a natural affinity for the species to be absorbed. In the second approach nanoparticles, with an affinity for the species to be absorbed, was blended into the fibre spinning solution prior to electrospinning. In this case the nanofibre web served as a highly porous, high specific surface area carrier for the absorbing component. The different studies employing these approaches are briefly discussed below.

In one of the first studies combining high specific surface area of electrospun fibres with an affinity of the fibre forming polymer for the species to be absorbed, electrospun webs of wool keratose (WK), silk fibroin (SF) and their blend (WK/SF) were evaluated for adsorption of Cu^{2+} as model heavy metal ion contaminant in solution (Ki *et al.*, 2007). Wool keratose, being the oxidized form of wool protein, was expected to contain cysteic acid residues ($-SO_3H$), yielding a capability to bind to positively charged metal ions. The WK was blended with the SF in order to improve the electrospinnability of the WK and to improve the mechanical properties of the resulting nanofibre webs. After the initial adsorption tests, recycling tests (desorption and to re-adsorption) were also performed to evaluate the practical applicability of the membrane. The average diameters for the SF and WK/SF blend fibres were 270 nm and 280 nm respectively, which would yield approximate specific surface areas on the order of 11 m^2/g (assuming a WK/SF density of approximately 1.3 g/cm^3). The adsorption tests indicated a Cu^{2+} adsorption capacity (at pH 7) of 1.65 µg/mg and 2.88 µg/mg for the SF and WK/SF blend fibres respectively and the recycling tests showed that the adsorption capacity of the WK/SF blend fibres were maintained above 90% over six recycling cycles.

In another study (Qi *et al.*, 2008), electrospun nanofibres of poly(styrene-*co*-methacrylic acid), poly(styrene-*co*-*p*-styrene sulphonate), and polystyrene were evaluated as solid-phase extraction (SPE) sorbents to extract nitrobenzene, 2-naphthol, benzene, *n*-butyl *p*-hydroxybenzoate, naphthalene, and *p*-dichlorobenzene from environmental water samples by HPLC. For comparison, commonly used octadecylsilica (ODS) SPE cartridges were also evaluated in terms of absolute recovery, sensitivity, and reproducibility for the compounds investigated. The very promising results indicated that, in contrast to the ODS, the nanofibre sorbents showed far better adsorption properties leading to higher absolute recovery, lower detection limits, and significantly smaller sample volumes (200 µl vs. 100 ml). The nanofibre-based method also required a shorter analysis time.

In two separate recent studies chitosan, an abundant natural polymeric material available from renewable resources, which has the ability to chelate metal ions, was evaluated for adsorption of heavy metals from aqueous solution. In the first study (Haider and Park, 2009), freestanding chitosan electrospun nanofibre webs gave equilibrium adsorption capacities of 485.44 mg/g for Cu(II) and 263.15 mg/g for Pb(II) respectively. In the second study (Desai *et al.*, 2009), Cr(VI) binding capacities of up to 35 mg Cr/g chitosan were obtained with electrospun nanofibre filter media consisting of a 90:10 chitosan/PEO blend on a spunbonded polypropylene substrate. In both studies it was concluded that the nanofibrous form of the chitosan, and the associated high specific surface area, led to a significant increase in heavy metal ion adsorption capacity over existing commercially available particulate adsorbent materials.

The second approach to making affinity membranes from electrospun fibres, by incorporating nanoparticles with an affinity for the species to be absorbed into the electrospun fibres was evaluated in two different recent studies with differing results. In the first study (Hota *et al.*, 2008), nanoparticles of boehmite, hydrated alumina (AlOOH) with known metal-ion binding properties, were encapsulated in electrospun fibres of polycaprolactone (PCL) and Nylon-6 and the resulting fibres were evaluated for its Cd(II) adsorption capabilities. It was observed that encapsulation of the boehmite nanoparticles in the electrospun fibres led to a 30–40% decrease in sorption capacity (0.001–0.003 mg/g) as compared to the pure boehmite nanoparticles (~0.21 mg/g). However, the authors of this study argued that the reduced sorption capacity of the fibre-encapsulated nanoparticles could potentially be offset by the fact that the said encapsulation would prevent release of such particles into the environment and that it could avoid or reduce the costs associated with separation of nanomaterials from treated water.

In the other far more successful study employing this approach, molecularly imprinted nanoparticles were encapsulated into the electrospun nanofibres (Yoshimatsu *et al.*, 2008). The composite nanofibres could be used as an affinity membrane to selectively extract trace amounts of propranolol (1 ng/ml) in tap water. The impressive selectivity of this material was emphasized by the fact that, without this highly selective solid-phase extraction method this level of detection could not be achieved even with sophisticated analytical equipment such as HPLC-MS/MS.

Reactive membranes

In the preceding two sections on filtration and affinity membranes, contaminants in water were essentially removed by keeping the contaminant trapped in one place whilst allowing the water to pass on. In these approaches the contaminants therefore remain chemically and physically unchanged. An alternative strategy is to combine the wide choice of a nanofibre-forming materials, the possibility of fibre surface-chemistry modification, and the ability to include various additives into electrospun fibres, with the high specific surface area of nanofibre webs in order to create reactive membranes that can chemically or physically react with potentially harmful contaminants and decompose them or convert them into more benign forms.

In a fashion similar to that discussed for affinity membranes, two different approaches can be taken to creating reactive membranes. The first strategy involves utilizing nanofibres that are made from a material that is inherently reactive to the contaminants in the water, whilst the second strategy consists of incorporating reactive components into fibres that are electrospun from polymers that may be otherwise unreactive to the contaminant species.

Note – Although antimicrobial membranes also fall within the category of reactive membranes, the use of antimicrobial electrospun nanofibres in water treatment was discussed in detail in the earlier chapter by M. Botes and T.E. Cloete and so will not be discussed here. However, it can be briefly mentioned that both of the strategies described above have been utilized for making antimicrobial nanofibres. For example, electrospun chitosan nanofibres have been evaluated for their inherent antimicrobial properties (Desai *et al.*, 2009), while nanofibre webs of nylon 6, which in itself does not specifically have antimicrobial activity, were prepared and shown to be highly antimicrobial after inclusion of silver nanoparticles in the fibres (Park *et al.*, 2009).

One very promising field in which reactive nanofibre membranes have been evaluated is in the photocatalytic degradation of various organic pollutants in water. In one recent study (Doh *et al.*,

2008), nanofibres of TiO_2 with anatase crystalline phase were prepared by the, now well-known, two-step procedure consisting of electrospinning suitable precursor materials and followed by calcining at high temperatures (550°C). When these nanofibres were placed in an aqueous solution containing Basic Blue 26 as a model dye pollutant, irradiation with UV light led to photolytic degradation of the dye with 25.3% of the dye being degraded after three hours of UV illumination. In an attempt to increase the photolytic activity of the membrane, the TiO_2 nanofibres were coated with additional 25 nm TiO_2 nanoparticles through a sol–gel process. The photolytic activity of this composite membrane was significantly enhanced, showing degradation of 78.7% of the dye after three hours of UV illumination. This increased photolytic activity was attributed to the increased effective surface area in the composite membrane.

In another very recent study (Shang *et al.*, 2009), composite nanofibres of TiO_2 and Bi_2WO_6 with a three-dimensional hierarchical heterostructure were created in order to obtain nanofibrous membranes with visible-light (as opposed to UV light) photolytic activity. After formation of the TiO_2 nanofibres through the well-known two-step procedure (electrospinning and calcining), precursors to the Bi_2WO_6 were mixed together and autoclaved in the presence of the TiO_2 nanofibres. The resulting heterostructure consisted of Bi_2WO_6 nanoplates growing aslant on the surface of the primary TiO_2 nanofibres as shown in Fig. 7.4. The visible-light photocatalytic degradation of rhodamine B (taken as a model pollutant in water) reached nearly 100% after 30 minutes of visible-light irradiation and comparison with the reaction of Bi_2WO_6/TiO_2 powder in bulk showed that the reaction of the nanofibrous heterostructure was 12 times faster than that of the bulk material.

Sensors

The ability to measure the efficacy of a specific water treatment strategy and the resulting quality of the water is almost as important as the treatment itself. With this in mind, highly sensitive nanosensors could potentially be made by combining the high specific surface areas, fibre surface-chemistry modification, and the right choice of fibre-forming material available from the electrospinning toolbox. Although there have been numerous studies evaluating the potential use of electrospun nanofibrous materials in gas and biological sensing applications (Ramaseshan *et al.*, 2007), it appears that the field is still wide open for development of sensor applications in water treatment.

Conclusion

The versatility of the electrospinning process and the attractive features of electrospun nanofibre webs have proven to be ideally suited to a diverse range of applications in water treatment. It was shown that the right combinations of high specific surface area, nano-scale pore sizes, controllable porosity, and flexibility with regard to the materials used and surface chemistry of the fibres can lead

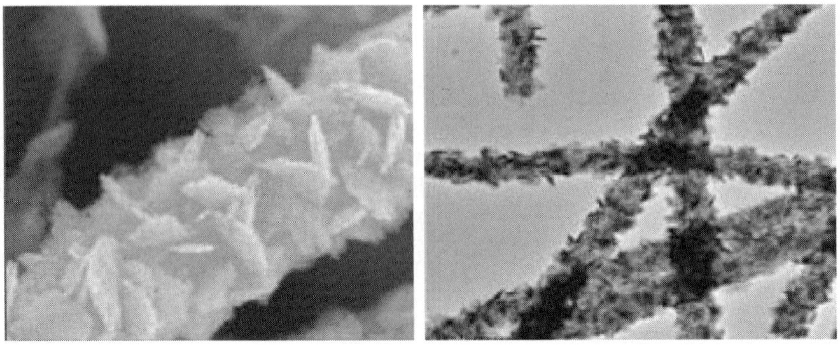

Figure 7.4 SEM and TEM micrographs showing heterostructure consisting of Bi_2WO_6 nanoplates growing aslant on the surface of the primary TiO_2 nanofibres. Reprinted with permission from Shang *et al.*, 2009. Copyright, 2009 American Chemical Society.

to significant improvements in filtration, solid phase extraction and reactive membranes. In future, the seemingly infinite number of possible combinations of the various 'tools' of the electrospinning process is very likely to yield further improvements to existing water treatment strategies, as well as completely new processes that do not yet exist.

References

Andrady, A.L. (2008). Science and technology of polymer nanofibers (Hoboken, N.J., Wiley).

Barhate, R.S., Loong, C.K., and Ramakrishna, S. (2006). Preparation and characterization of nanofibrous filtering media. J. Membrane Sci. *283*, 209–218.

Barhate, R.S., Sundarrajan, S., Pliszka, D., and Ramakrishna, S. (2008). Finechemical processing – The potential of nanofibers in filtration. Filtr. Separat. *May*, 32–35.

Bognitzki, M., Becker, M., Graeser, M., Massa, W., Wendorff, J.H., Schaper, A., Weber, D., Beyer, A., Gölzhäuser, A., and Greiner, A. (2006). Preparation of sub-micrometer copper fibers via electrospinning. Adv. Mater. *18*, 2384–2386.

Bottino, A., Capannelli, G., Comite, A., Ferrari, F., Firpo, R., and Venzano, S. (2009). Membrane technologies for water treatment and agroindustrial sectors. CR Chim. *12*, 882–888.

Chronakis, I.S. (2005). Novel nanocomposites and nanoceramics based on polymer nanofibers using electrospinning process. A review. J. Mater. Process. Tech. *167*, 283–293.

Desai, K., Kit, K., Li, J., Davidson, P.M., Zivanovic, S., and Meyer, H. (2009). Nanofibrous chitosan non-wovens for filtration applications. Polymer *50*, 3661–3669.

Doh, S.J., Kim, C., Lee, S.G., Lee, S.J., and Kim, H. (2008). Development of photocatalytic TiO_2 nanofibers by electrospinning and its application to degradation of dye pollutants. J. Hazard. Mater. *154*, 118–127.

Doshi, J., and Reneker, D.H. (1995). Electrospinning process and applications of electrospun fibers. J. Electrostat. *35*, 151–160.

Gopal, R., Kaur, S., Feng, C.Y., Chan, C., Ramakrishna, S., Tabe, S., and Matsuura, T. (2007a). Electrospun nanofibrous polysulfone membranes as pre-filters: Particulate removal. J. Membrane Sci. *289*, 210–219.

Gopal, R., Kaur, S., Ma, Z., Chan, C., Ramakrishna, S., and Matsuura, T. (2006). Electrospun nanofibrous filtration membrane. J. Membrane Sci. *281*, 581–586.

Gopal, R., Zuwei, M., Kaur, S., and Ramakrishna, S. (2007b). Surface Modification and Application of Functionalized Polymer Nanofibers. In Molecular Building Blocks for Nanotechnology (Berlin/Heidelberg, Springer), pp. 72–91.

Greiner, A., and Wendorff, J.H. (2007). Electrospinning: A fascinating method for the preparation of ultrathin fibers. Angew. Chem. Int. Ed. *46*, 5670–5703.

Haider, S., and Park, S.-Y. (2009). Preparation of the electrospun chitosan nanofibers and their applications to the adsorption of Cu(II) and Pb(II) ions from an aqueous solution. J. Membrane Sci. *328*, 90–96.

Hota, G., Kumar, B.R., Ng, W.J., and Ramakrishna, S. (2008). Fabrication and characterization of a boehmite nanoparticle impregnated electrospun fiber membrane for removal of metal ions. J. Mater. Sci. *43*, 212–217.

Huang, Z.-M., Zhang, Y.-Z., Kotaki, M., and Ramakrishna, S. (2003). A review on polymer nanofibers by electrospinning and their applications in nanocomposites. Compos. Sci. Tech. *63*, 2223–2253.

Ki, C.S., Gang, E.H., Um, I.C., and Park, Y.H. (2007). Nanofibrous membrane of wool keratose/silk fibroin blend for heavy metal ion adsorption. J. Membrane Sci. *302*, 20–26.

Park, S.-W., Bae, H.-S., Xing, Z.-C., Kwon, O.H., Huh, M.-W., and Kang, I.-K. (2009). Preparation and Properties of Silver-Containing Nylon 6 Nanofibers Formed by Electrospinning. J. Appl. Polym. Sci. *112*, 2320–2326.

Qi, D., Kang, X., Chen, L., Zhang, Y., Wei, H., and Gu, Z. (2008). Electrospun polymer nanofibers as a solid-phase extraction sorbent for the determination of trace pollutants in environmental water. Anal. Bioanal. Chem. *390*, 929–938.

Ramakrishna, S., Fujihara, K., Teo, W.-e., Lim, T.-c., and Zuwei Ma (2005). An introduction to electrospinning and nanofibers (Singapore, World Scientific Publishing).

Ramakrishna, S., Fujihara, K., Teo, W.-E., Yong, T., Ma, Z., and Ramaseshan, R. (2006). Electrospun nanofibers: solving global issues. Mater. Today *9*, 40–50.

Ramaseshan, R., Sundarrajan, S., Jose, R., and Ramakrishna, S. (2007). Nanostructured ceramics by electrospinning. J. Applied Phys. *102*, 1111011–1111017.

Sawicka, K.M., and Gouma, P. (2006). Electrospun composite nanofibers for functional applications. J. Nanopart. Res. *8*, 769–781.

Shang, M., Wang, W., Zhang, L., Sun, S., Wang, L., and Zhou, L. (2009). 3D Bi_2WO_6/TiO_2 Hierarchical Heterostructure: Controllable Synthesis and Enhanced Visible Photocatalytic Degradation Performances. J. Phys. Chem. *113*, 14727–14731.

Tan, E.P.S., and Lim, C.T. (2008). Structure and mechanical properties of electrospun nanofibers and nanocomposites. Nanotechnol. Tissue Eng. 221–242.

Tang, Z., Wei, J., Yung, L., Ji, B., Ma, H., Qiu, C., Yoon, K., Wan, F., Fang, D., Hsiao, B.S., and Chu, B. (2009). UV-cured poly(vinyl alcohol) ultrafiltration nanofibrous membrane based on electrospun nanofiber scaffolds. J. Membrane Sci. *328*, 1–5.

Wang, X., Chen, X., Yoon, K., Fang, D., Hsiao, B.S., and Chu, B. (2005). High Flux Filtration Medium Based on Nanofibrous Substrate with Hydrophilic Nanocomposite Coating. Environ. Sci. Technol. *39*, 7684–7691.

Wang, X., Fang, D., Yoon, K., Hsiao, B.S., and Chu, B. (2006). High performance ultrafiltration composite membranes based on poly(vinyl alcohol) hydrogel coating on crosslinked nanofibrous poly(vinyl alcohol) scaffold. J. Membrane Sci. *278*, 261–268.

Wu, H., Zhang, R., Liu, X., Lin, D., and Pan, W. (2007). Electrospinning of Fe, Co, and Ni Nanofibers: Synthesis, Assembly, and Magnetic Properties. Chem. Mater. *19*, 3506–3511.

Yoon, K., Hsiao, B.S., and Chu, B. (2008). Functional nanofibers for environmental applications. J. Mater. Chem. *18*, 5326–5334.

Yoon, K., Hsiao, B.S., and Chu, B. (2009a). High flux ultrafiltration nanofibrous membranes based on polyacrylonitrile electrospun scaffolds and crosslinked polyvinyl alcohol coating. J. Membrane Sci. *338*, 145–152.

Yoon, K., Hsiao, B.S., and Chu, B. (2009b). High flux nanofiltration membranes based on interfacially polymerized polyamide barrier layer on polyacrylonitrile nanofibrous scaffolds. J. Membrane Sci. *326*, 484–492.

Yoon, K., Kim, K., Wang, X., Fang, D., Hsiao, B.S., and Chu, B. (2006). High flux ultrafiltration membranes based on electrospun nanofibrous PAN scaffolds and chitosan coating. Polymer *47*, 2434–2441.

Yoshimatsu, K., Ye, L., Lindberg, J., and Chronakis, I.S. (2008). Selective molecular adsorption using electrospun nanofiber affinity membranes. Biosens. Bioelectron. *23*, 1208–1215.

Potential Risks of Using Nanotechnology in Water Treatment on Human Health

8

Michele de Kwaadsteniet and T. Eugene Cloete

Abstract

The risk assessment of nanoparticles and nanomaterials is of key importance for the continuous development in the already striving new field of nanotechnology. Humans are increasingly being exposed to nanoparticles and nanomaterials, placing stress on the development and validation of reproducible toxicity tests. Tests currently used include genotoxicity and cytotoxicity tests, and *in vivo* toxicity models. The unique characteristics of nanoparticles and nanomaterials are responsible for their toxicity and interaction with biological macromolecules within the human body. This may lead to the development of diseases and clinical disorders. A loss in cell viability and structure can also occur in exposed tissues as well as inflammation and granuloma formation. The future of nanotechnology depends on the responsible assessment of nanoparticles and nanomaterials.

Introduction

Nanotechnology is expanding an astounding rate with a 600 percent increase in peer-reviewed articles being published in the period from 2000 to 2007 (Ostrowski *et al.*, 2009). Nanotechnology is a multidisciplinary field with biologists, engineers, chemists, physicists and toxicologists involved employed by the industry, government and academics. Nanotoxicology is however falling behind with only a small part of research funding and time being spent on this area (Fischer and Chan, 2007; Hutchison, 2008). Assessing the potential risks of nanoparticles and nanomaterials is of cardinal importance for human and environmental health and public acceptance.

The public's confidence in nanotechnology can easily be broken by a safety incident. In 2006, the international media raised alarms about nanotoxicology after a German bathroom cleaner, Magic Nano, was pulled off the market within 3 days. This product was responsible for severe respiratory problems in more than 100 people, of whom six were hospitalized with pulmonary oedema (Weiss, 2006). Subsequent research concluded that the ethanol–water aerosol was responsible for the respiratory problems. This incident clearly illustrates, although the nanoparticles were not the causative agent, that thorough research is needed before a product containing nanoparticles or nanomaterials is commercialized (Wolinsky, 2006). A study reporting that inhaled carbon nanotubes cause lesions similar to asbestos, a known carcinogen, has received a lot of media attention in the science community (Poland *et al.*, 2008).

Nanoparticles can be divided into two groups, namely engineered and natural nanoparticles. Engineered nanoparticles are manufactured according to specific properties for specific applications. Natural nanoparticles are also referred to as incidental nanoparticles. Volcanic eruptions and forest fires are responsible for the generation of nanoparticles in nature while man is responsible for the nanoparticles present in fumes generated during automobile exhaust, welding and other industrial processes. Scientific data on the impact of natural nanoparticles on human and environmental health is substantial especially when compared to toxicology data available on engineered nanoparticles. Drawing parallels between the toxicology effects of natural and engineered nanoparticles on human health is however not straightforward since they differ in important aspects. Engineered nanoparticles consists out of homogenous, monodisperse particles with known chemical characteristics compared to natural nanoparticles that consists out of heterogeneous, polydisperse particles with complex chemical characteristics. The unrelated morphology between natural and engineered nanoparticles (branched structures versus spherical or unbranched forms) also makes direct comparisons unreliable (Casals *et al.*, 2008).

This chapter will focus on the hazards of engineered nanoparticles on human health.

Potential hazards of nanomaterials on human health based on *in vitro* results

Genotoxicity

Numerous studies have investigated the genotoxicity of nanoparticles and nanomaterials. The genotoxicity endpoints assessed *in vitro* include the detection of DNA adducts/damage, gene mutations, DNA strand breakage (Comet test), chromosome aberrations, cytokinesis block and centrosome aberrations (Gonzalez *et al.*, 2008). The Comet test, a single-cell agarose gel electrophoresis assay that determines DNA damage, is the most frequent test used to detect for nanoparticle or nanomaterial genotoxicity. The Comet assay detects single-strand breaks, alkaline-labile sites and cross-linking. This is followed by the micronucleus assay where micronuclei are indicative of structural chromosomal aberrations. The Ames test in bacteria detects for gene mutations (Landsiedel *et al.*, 2009).

Fullerenes (Dhawan *et al.*, 2006), SWCNT (Kisin *et al.*, 2007), MWCNT (Karlsson *et al.*, 2008), V_2O_3 and V_2O_5 (Landsiedel *et al.*, 2009), cobalt chrome alloy (Papageorgiou *et al.*, 2007), gold nanoparticles (Zheng *et al.*, 2008), zinc oxide nanoparticles (Karlsson *et al.*, 2008; Sharma *et al.*, 2009; Yang *et al.*, 2009a), copper oxide nanoparticles (Karlsson *et al.*, 2008) and nickel nanopowder (Zhang *et al.*, 1998) have all been shown to cause DNA damage. TiO_2 nanoparticles in the presence (Dunford *et al.*, 1997; Nakagawa *et al.*, 1997) and absence of photoactivation (Gurr *et al.*, 2005; Wang *et al.*, 2007b; Karlsson *et al.*, 2008) also induced DNA damage. DNA damage was determined by a Comet test or other agarose gel electrophoresis tests that was performed on a cell line that the respective researcher selected. Fullerenes (Niwa and Iwai, 2006), MWCNT (Muller *et al.*, 2008), TiO_2 nanoparticles (Rahman *et al.*, 2002; Gurr *et al.*, 2005; Wang *et al.*, 2007b), SiO_2 nanoparticles (Wang *et al.*, 2007c), CoCr nanoparticles (Papageorgiou *et al.*, 2007), nano-cerium-element-doped TiO_2 (Wang *et al.*, 2007d), ZnO nanoparticles (Dufour *et al.*, 2006) and photoactivated TiO_2 nanoparticles (Nakagawa *et al.*, 1997) tested positive in micronucleus assays. ZnO nanoparticles (Dufour *et al.*, 2006) and photoactivated TiO_2 nanoparticles (Nakagawa *et al.*, 1997) tested positive in chromosome aberration studies. Few gene mutation tests in bacteria have resulted in positive outcomes with the Ames assays performed on SiO_2 nanoparticles (Wang *et al.*, 2007c), TiO_2 nanoparticles (Wang *et al.*, 2007b) and MWCNT (Zhu *et al.*, 2007) as examples.

Not all the nanoparticles were positive for DNA damage with tests performed on maghemite (nano-γFe_2O_3) coated with DMSA (Auffan *et al.*, 2006), TiO_2 nanoparticles (Nakagawa *et al.*, 1997; Rehn *et al.*, 2003), ZnO nanoparticles (Yoshida *et al.*, 2009) and SiO_2 nanoparticles (Wang *et al.*, 2007c) revealing negative results in agarose gel electrophoresis tests.

TiO_2 nanoparticles (Nakagawa *et al.*, 1997; Warheit *et al.*, 2007a), ZnO nanoparticles (Dufour *et al.*, 2006), SWCNT (Kisin *et al.*, 2007) and silica-overcoated magnetic nanoparticles labelled with rhodamine B isothiocyanate (Kim *et al.*, 2006a) tested negative for gene mutations in bacteria in the Ames assays.

In contrast to gene mutation tests, were the majority of chromosome mutations tests with nanoparticles positive, with only two micronuclease assays and three chromosome aberration tests giving negative results. The two negative micronucleus assays were performed on TiO_2, V_2O_3 and

V_2O_5 nanoparticles (Landsiedel *et al.*, 2009) and the three chromosome aberration tests on silica-overcoated magnetic nanoparticles labelled with rhodamine B isothiocyanate (Kim *et al.*, 2006a) and TiO_2 nanoparticles (Theogaraj *et al.*, 2007; Warheit *et al.*, 2007a).

Genotoxicity data in literature have given mixed results, depending on the test used, making direct conclusions about the genotoxicty of a nanoparticle difficult. The following must also be taken into consideration when discussing and referring to the genotoxicity of nanoparticles and nanomaterials; the physiochemical characteristics and the preparation of the nanoparticle tested must be known, standardized genotoxicity tests with international agreed reference materials must agreed upon, *in vivo* studies must be used to validate *in vitro* data, knowledge about the mechanism by which nanomaterials exert the genotoxicity effects must be gained and the distribution and uptake of a nanomaterial must be considered. Landsiedel *et al.* (2009) and Gonzalez *et al.* (2008) discuss the genotoxicity of engineered and natural nanomaterials in comprehensive reviews and discuss the outcome of genotoxicity based respectively on the test and the nanoparticle used.

Cytotoxicity

Assessing the safety of nanoparticles and nanomaterials in mammalian cell lines is a simple technique to collect a large amount of data. Numerous *in vitro* studies have revealed that nanoparticles and nanomaterials are cytotoxic towards human and animal cell lines. *In vitro* studies that have shown that nanoparticles and nanomaterials can be cytotoxic are discussed in Table 8.1 with the negative results listed briefly in the text. There is however in nanotoxicology a lack of standardized protocols and a database that is reproducible and credible. This makes comparing the toxicity of different nanoparticles and nanomaterials problematic (Nel *et al.*, 2006; Suh *et al.*, 2009).

Cytotoxicity tests did not always reveal positive results with certain nanoparticles and nanomaterials exhibiting very low or no toxicity. Cytotoxicity of C_{60} was not significant when tested on guinea pig alveolar macrophages (Jia *et al.*, 2005), monocyte-derived macrophages (Fiorito *et al.*, 2006) and human monocyte macrophages (Porter *et al.*, 2006). No cytotoxicity was observed when the highly soluble fullerene derivative, $C_{60}(OH)_{24}$, was tested on human dermal fibroblasts and human liver carcinoma cells (Sayes *et al.*, 2004). SWCNT had no cytotoxic effects when tested on mouse peritoneal macrophage-like cells (Cherukuri *et al.*, 2004), monocyte-derived macrophages (Fiorito *et al.*, 2006) and only very low cytotoxicity against human lung cells (Davoren *et al.*, 2007). Metal traces in commercial SWCNTs was believed to be responsible for the cytotoxicity towards human alveolar epithelial and rat alveolar macrophage cells (Pulskamp *et al.*, 2007). The only cytotoxicity test with MWCNTs that revealed a negative result was on human umbilical vein endothelial cells (Flahaut *et al.*, 2006). Carbon nanofibres were also non-toxic when tested on mouse keratinocytes (Grabinski *et al.*, 2007).

Gold nanomaterials (particles, shells and onions) were non-toxic when tested on human breast carcinoma xenograft cells (Shenoy *et al.*, 2006), macrophage cells (Shukla *et al.*, 2005), leukaemia cell line (Connor *et al.*, 2005), human breast carcinoma cells (Hirsch *et al.*, 2003) and breast adenocarcinoma cells (Loo *et al.*, 2004; 2005). Gold nanorods prepared in cetyltrimethylammonium bromide (CTAB) was cytotoxic to human hepatocarcinoma cells (Niidome *et al.*, 2006; Takahashi *et al.*, 2006; Huff *et al.*, 2007) but toxicity was drastically reduced when prepared in poly(ethylene) glycol (PEG) chains (Huff *et al.*, 2007). Iron oxide nanoparticles coated in PEG were non-toxic towards human dermal fibroblasts (Gupta and Wells, 2004). Polyvinyl-coated and dextran-coated iron oxide nanoparticles were non-toxic towards rat brain-derived endothelial cells, murine microglial cells (Cengelli *et al.*, 2006) and human monocyte-macrophages (Müller *et al.*, 2007). Pullulan-coated and polyvinyl alcohol-coated iron oxide nanoparticles were however cytotoxic towards respectively human dermal fibroblasts (Gupta and Gupta, 2005) and human melanoma cells (Petri-Fink *et al.*, 2005). Uncoated iron oxide nanoparticles were non-toxic towards monkey kidney COS-7 cells (Cheng *et al.*, 2005), and rat liver cells (Hussain *et al.*, 2005). The cytotoxicity of nanoparticles is discussed in an in depth in a review article by Lewinski *et al.* (2008).

Table 8.1 The cytotoxicity of nanomaterials and nanoparticles

Nano-material	Cell line	Effects	References
C_{60}	Human dermal fibroblasts	Cytoxicity was dose dependent with LD50 of 20 ppb	Sayes *et al.* (2004)
	Human liver carcinoma cells	Cytoxicity was dose dependent with LD50 of 20 ppb	Sayes *et al.* (2004)
	Human neuronal astrocytes	Cytotoxic effects observed at 20 ppb with membrane leakage and lipid oxidation observed after 30 h	Sayes *et al.* (2005)
	Human epidermal keratinocytes	Cytotoxic effects observed with amino acid derived C_{60} at 0.04 and 0.4 mg/ml with an increase in proinflammatory cytokines	Rouse *et al.* (2006)
	Human umbilical vein endothelial cells	Cell growth inhibited at 100 μg/ml and cell attachment inhibited at 10 μg/ml	Yamawaki *et al.* (2006)
SWCNT	Human epidermal keratinocytes	Decreases in cell viability were observed in all the studies	Shvedova *et al.* (2003); Manna *et al.* (2005); Tian *et al.* (2006); Zhang *et al.* (2007)
	Human embryonic kidney cells	Decreases in cell proliferation and cell adhesion ability observed with increases in the induction of apoptosis, dose- and time dependent	Cui *et al.* (2005)
	Human dermal fibroblasts	Pristine SWCNTs had higher cytotoxicity than the SWCNTs containing functionalized groups	Sayes *et al.* (2006a)
	Human aortic endothelial cells	Membrane leakage was observed (4.5 μg/ml)	Walker *et al.* (2009)
	Human alveolar epithelial	Decrease in mithondrial membrane potential and increase in intracellular reactive oxygen species, dose and time dependent	Pulskamp *et al.* (2007)
	Lung carcinoma cells	Oxidative stress and decreases in cell proliferation dose- and time dependent	Manna *et al.* (2005)
	Mesothelioma cells	Cytoxicity dose-dependent and increases with increased aggregation	Wick *et al.* (2007)
	Guinea pig alveolar macrophages	Growth inhibition of 20% at 141 μg/cm^2 with necrosis at 3.06 μg/cm^2	Jia *et al.* (2005)
	Rat alveolar macrophage	Cell viability decreased 60–80% at 100 μg/ml, cytotoxicity dose-dependent	Pulskamp *et al.* (2007)
MWCNT	Human skin fibroblasts	Immune and inflammatory gene overexpression at high dosages	Ding *et al.* (2005)
	Human lung fibroblasts	Immune and inflammatory gene overexpression at high dosages	Ding *et al.* (2005)
	Human aortic endothelial cells	Membrane leakage was observed (4.5 μg/ml)	Walker *et al.* (2009)
	Human acute monocytic leukemia cells	Increased TNF-α production (increased inflammation) dose-dependent.	Sato *et al.* (2005)

Table 8.1 Continued

Nano-material	Cell line	Effects	References
	Human T lymphocytes	Cell viability decreased 80% with oxidized CNTs and less than half with pristine CNTs, both at 40 μg/ml	Bottini *et al.* (2006a)
	Human lung tumour cells	Cell viability decreased 33% at 0.2 μg/ml	Magrez *et al.* (2006)
	Human epidermal keratinocytes	Cell viability decreased 73% at 0.4 mg/ml, increase in pro-inflammatory cytokine IL-8 dose dependent	Monteiro-Riviere *et al.* (2005)
	Human lung epithelial cell line A549	Cell viability decreased by 22% at 40 μg/ml	Karlsson *et al.* (2008)
	Guinea pig alveolar macrophages	Decrease in cell viability dose dependent with necrosis at 3.06 μg/cm^2	Jia *et al.* (2005)
	Rat peritoneal macrophages	Increased TNF-α production (increased inflammation) and cytotoxicity dose dependent	Muller *et al.* (2005)
	Murine alveolar macrophages	Cells' membrane integrity compromised	Murr *et al.* (2005); Hirano *et al.* (2008)
Au	Human cervical epithelium cells	Cell viability decreased by 20% at 150 pM	Tkachenko *et al.* (2004)
	Human dermal fibroblasts	Cell area and number decrease, dose and time dependent	Pernodet *et al.* (2006)
	Murine fibroblastoma cells	Cell viability decreased by 5% at 150 pM	Tkachenko *et al.* (2004)
	Monkey kidney COS-7 cells	Cell viability decreased by 70–80% after transfection	Thomas and Klibanov (2003)
	Murine macrophage cells	Cell proliferation significantly decreased ≥10 ppm	Yen *et al.* (2009)
	Mouse vero cells	Cell viability decreased by 15% at 200 μg/ml	Su *et al.* (2007)
Ag	Human lung fibroblasts	Cell viability decreased and apoptosis increased in a dose- and time-dependent manner	Asharani *et al.* (2009)
	Human glioblastoma cells	Cell viability decreased in a dose- and time dependent manner	Asharani *et al.* (2009)
	Murine macrophage cells	Cell proliferation significantly decreased at 10 ppm	Yen *et al.* (2009)
	Rat alveolar macrophages	Cell viability decreased by 88.7% at 50 μg/ml	Carlson *et al.* (2008)
	Mouse fibroblast cells	Cell viability decreased by 50% at 50 μg/ml	Hsin *et al.* (2008)
	Mouse spermatogonial stem cells	Necrosis and apoptosis observed at >10 μg/ml	Braydich-Stolle *et al.* (2005)

Table 8.1 Continued

Nano-material	Cell line	Effects	References
TiO_2	Human bronchial epithelial cells	Cell viability decreased in a dose- and time-dependent manner	Chen *et al.* (2008a); Park *et al.* (2008b)
	Human monoblastoid cells	Cell membrane permeability and necrosis observed at almost all the concentrations tested (0.005–4 mg/ml)	Vamanu *et al.* (2008)
	Human fibroblasts cells	Cell viability decreased by 70% at 100 μg/ml	Lai *et al.* (2008)
	Human astrocytoma U87 cells	Cell viability decreased by >70% at 100 μg/ml, dose-dependent	Lai *et al.* (2008)
	Murine alveolar macrophages	Cell viability decreased in a dose-dependent manner	Kim *et al.* (2009)
Fe_2O_3	Human dermal fibroblasts	Cell viability decreased by 60% at 2 mg/ml	Gupta and Gupta (2005)
	Rodent fibroblast cells	Cell viability decreases after 3.75 ppb	Brunner *et al.* (2006)
	Rat phaeochromocytoma cells	Cell viability decreased by 60% at 15 mM	Pisanic *et al.* (2007)
ZnO	Human T cells	Cell viability decreased by 57% at 10 mM	Reddy *et al.* (2007)
	Human astrocytoma U87 cells	Cell viability decreased by 95% at 20 μg/ml, dose dependent	Lai *et al.* (2008)
	Human lung epithelial cells	Cell viability decreased by 75–85% between 18 and 25 μg/ml	Lin *et al.* (2009)
	Mouse neural stem cells	Cell viability decreased in a dose-dependent manner	Deng *et al.* (2009)
	Mouse embryo fibroblast cells	Cell viability decreased by 26.5% at 20 μg/ml	Yang *et al.* (2009a)
	Human lung epithelial cell line A549	Cell viability decreased by 38% at 40 μg/ml	Karlsson *et al.* (2008)
CuO	Human lung epithelial cell line A549	Cell viability decreased by 96% at 40 μg/ml	Karlsson *et al.* (2008)
Cerium oxide	Human bronchial cells	Oxidative stress resulting in apoptosis	Eom and Choi (2009)
	Human lung epithelial cells	Cell viability decreased by 40–50% at 200 μg/ml	Park *et al.* (2008a)
	Human lung cancer cells	Cell viability decreased in a dose- and time dependent manner with lipid peroxidation and membrane damage	Lin *et al.* (2006)
Dendrimers	Caco-2 cells	Cationic dendrimers cytotoxic > 700 μM	Jevprasesphant *et al.* (2003)

Potential hazards on human health based on *in vivo* results

Nanoparticles and nanomaterials can enter the human body through the following six routes; intravenous, dermal, subcutaneous, inhalation, intraperitoneal and oral. They can be absorbed to various biological components, depending on the physical properties of the nanomaterial and the route of administration, before being translocated to secondary organs. The exposure route will not be the same for all nanoparticles and nanomaterials, inhalation can occur at the workplace, ingestion incidentally through food and drinking water and dermally through sunscreens and cosmetic products (Stern and McNeil, 2008).

Pulmonary translocation and toxicity studies

Nanoparticles and nanomaterials that have entered the lungs can be cleared by mucociliary movement, phagocytosis by alveolar macrophages, interstitial translocation, lymphatic drainage, blood circulation translocation, neuron translocation and epithelial endocytosis (Oberdörster *et al.,* 1992a,b, 2000). Phagocytosis mediated through alveolar macrophages is believed to be a prevalent clearance mechanism. Particles are engulfed in alveolar macrophages during phagocytosis 6–12 h after being deposited into the alveolar region of the lung followed by subsequent slow clearance (Semmler *et al.,* 2004). Nanoparticles are hypothesized, according to this mechanism, to retain in the lungs for 70 days in rats and 700 days in humans. Interstitial translocation is observed to be another important route of clearance with nanoparticles entering the bloodstream of the lymphatic system after crossing the epithelia of the respiratory tract (Oberdörster *et al.,* 2005).

In vivo studies investigating the translocation of nanoparticles from the lungs to secondary organs have given mix results. Kreyling *et al.* (2002) demonstrated that in rats less than 1% of intratracheally instilled iridium nanoparticles translocate from the lungs to the liver, spleen, heart and brain. The majority of iridium nanoparticles were cleared into the gastrointestinal tract and faeces via the airways. Oberdörster *et al.* (2002) demonstrated that inhaled graphite ^{13}C nanoparticles (20–29 nm) translocated to the liver of rats. It was postulated that the particles are deposited directly into the bloodstream but it was difficult to determine if a portion was not ingested and hence no conclusions could be made. Nemmar *et al.* (2002a) detected technetium-99m-labelled carbon nanoparticles in the blood of humans 1 min after inhalation. Contradicting the above studies, Brown *et al.* (2002) determined that inhaled technetium-99m-labelled carbon nanoparticles in humans did not translocate to the bloodstream and secondary organs.

The neuronal translocation of nanoparticles from the respiratory, through the nasal epithelium to the olfactory bulb, has also been investigated. Inhaled manganese oxide and ^{13}C nanoparticles in rats were detected in the olfactory bulb. The drawback in these studies were that the molecular constituents of the nanoparticles were measured and not the nanoparticles themselves, excluding the possibility of measuring nanoparticles dissociating in the central nervous system (Oberdörster *et al.,* 2004; Elder *et al.,* 2006). TiO_2 nanoparticles administered intranasally were detected in the brain, especially in the hippocampus, after passage through the olfactory bulb (Wang *et al.,* 2008a,b). A non-human primate study, more relevant to human nasal exposure, also detected metal nanoparticles, namely gold nanoparticles, in the olfactory bulb after intranasal instillation (De Lorenzo, 1970).

Toxicity studies investigating the effect of nanoparticles and nanomaterials in the respiratory tract are summarized in Table 8.2. Unless otherwise specified were the effects listed in the table observed at all the dosages listed. Mortality was only observed in animals treated with SWCNTs and MWCNTs and in two out of the three cases mortality was attributed to mechanical obstruction of the airways (Warheit *et al.,* 2005; Carrero-Sanchez *et al.,* 2006).

Systematic translocation and toxicity studies following intraperitoneally and intravenous administration

The main target organs of nanoparticles and nanomaterials that enter the bloodstream are the liver and spleen, members of the reticuloendothelial system (Table 8.3). The reticuloendothelial system,

Table 8.2 *In vivo* pulmonary toxicity studies

Nano-material	Species	Dosage and exposure route	Effects	References
C_{60} and $C_{60}(OH)_{24}$	Rats	0.2, 0.4, 1.5 or 3 mg/kg; intratracheal instillation	Low toxicity observed in all the dosage groups with only transient inflammation.	Sayes *et al.* (2007a)
C_{60}	Rats	0.12 mg/m^3; 6h/day for 4 weeks, 5 days a week; whole body inhalation	Low toxicity; Only some genes involved in inflammation up-regulation	Fujita *et al.* (2009)
SWCNT	Rats	1 and 5 µg/animal; intratracheal instillation; 24 h, 1 week, 1 and 3 months	Mortality in high dosage group (5 µg/animal) due to obstruction; Transient inflammation, cell injury effects and multifocal granulomas observed in both groups	Warheit *et al.* (2004)
	Mice	0, 10, 20 or 40 µg/animal; pharyngeal aspiration; 1, 3, 7, 28 and 90 days	Acute inflammation with fibrosis and granulomas observed in all the groups in a dose-dependent manner. Decreased bacterial clearance (40 µg/animal)	Shvedova *et al.* (2005)
	Mice	0, 10 or 40 mµg aspiration; 3 days before infection with *Listeria monocytogenes*	Enhanced acute inflammation and pulmonary injury with delayed bacterial clearance	Shvedova *et al.* (2008a)
	Mice	5 mg/m^3; 5 h/day; 4 days	Inflammation, fibrosis and oxidative stress	Shvedova *et al.* (2008b)
	Mice	Exposed the mesothelial lining of the body cavity	Inflammation with granulomas	Poland *et al.* (2008)
	Mice	0.1 or 0.5 mg/animal; intratracheal instillation; 7 and 90 days	High dosage resulted in 56% mortality (0.5 mg). Weight loss, granulomas, necrosis, interstitial and peribronchial inflammation	Lam *et al.* (2004)
SWCNT, dispersed	Mice	10 µg; pharyngeal aspiration	Incorporated into alveolar interstitium and produced increased collagen deposition	Mercer *et al.* (2008)
MWCNT	Mice	1, 2.5 and 5 mg/kg; intratracheal instillation; 1, 2, 3, 7 and 30 days	Mortality due to obstruction. Inflammation with granulomas observed (5mg/kg)	Carrero-Sanchez *et al.* (2006)
	Guinea pigs	25 mg; intratracheal instillation; 4 weeks	No toxicity	Huczko *et al.* (2001)
	Guinea pigs	15 mg; intratracheal instillation; 90 days	Non-specific desquamative interstitial pneumonia-like reaction	Huczko *et al.* (2005)
	Mice	0.3, 1. 5 mg/m^3; 6 h/day for 7 and 14 days	No inflammation or tissue damage, immunosuppression after 14 days	Mitchell *et al.* (2007)
MWCNT, intact or grounded	Rats	0.5, 2 or 5 mg; intratracheal instillation; 1 and 2 months	Inflammatory and fibrotic reactions, TNF-alpha production were stimulated. Intact CNT induced collagen-rich granulomas and surrounding alveolitis	Muller *et al.* (2005)
MWCNT, dispersed in albumin	Rats	0, 1, 10, 100 µg; intratracheal instillation; 1, 7, 30, 90 and 180 days	Only apoptosis of alveolar macrophages	Elgrabli *et al.* (2008)

Table 8.2 Continued

Nano-material	Species	Dosage and exposure route	Effects	References
TiO_2	Rats	50 mg/kg; 30 days; intranasal instillation	Pathology in kidney and translocation to brain	Wang *et al.* (2008c)
	Rats	1 or 5 mg/kg; intratracheal instillation	Transient inflammation and cell injury effects.	Warheit *et al.* (2006, 2007c)
	Rats	0.1, 1 and 5 mg/kg/day; for 5 days; oropharyngeal aspiration	Pulmonary injury and neutrophil infiltration	Kim *et al.* (2009)
	Rats	500 μg; 2, 10, 20 and 30 days; nasal instillation	Oxidative stress and inflammatory responses in central nervous system	Wang *et al.* (2008a)
	Rats	2, 10, 50 and 110 mg/m³; 6h/day for 5 days; whole body inhalation	Mild inflammation and activation of macrophages	Van Ravenzwaay *et al.* (2008); Ma-Lock *et al.* (2009)
	Rats	0.5, 5 and 50 mk/kg; intratracheal instilled	Inflammatory lesions with infiltration of inflammatory cells and interstitium thickening	Liu *et al.* (2009)
	Mice, rats, hamsters	0.5, 2 or 10 mg/ml; aerosol inhalation	Rats and mice exposed to 10 mg/ml had pulmonary inflammation with increased numbers of macrophages and neutrophils. Progressive fibrosis in rats	Bermudez *et al.* (2004)
	Mice	0.5 mg every 2nd day for 30 days; intranasally instilled	Oxidative stress in hippocampus of brain	Wang *et al.* (2008a)
SiO_2	Rats	0.1, 1 and 5 mg/kg/day; for 5 days; oropharyngeal aspiration	Pulmonary injury and neutrophil infiltration	Kim *et al.* (2009)

also referred to as the mononuclear phagocyte system, consists out of monocytes and macrophages and is an important part of the immune system. Therefore the conclusion can be made that nanoparticles and nanomaterials are taken up by phagocytic cells (Saba, 1970). The kidney is also a common target site and has been shown to be a vital clearance organ of nanoparticles and nanomaterials in the body (Chen *et al.*, 1998; Wang *et al.*, 2004). The kidney was shown in a study by Chen *et al.* (1998) to be the target of clearance and toxicity, with pathology manifesting itself as lysosome-overload nephrosis. Nanoparticles influence lysosomal localization (Shukla *et al.*, 2005a; Bottini *et al.*, 2006b) and enzyme-inhibiting ability (Ueng *et al.*, 1997; Shcharbin *et al.*, 2006) with lysosomal disorders a possible side-effect. Systematic toxicity studies that investigated the effect of intravenous (i.v.) and intraperitoneal (i.p.) administered nanoparticles and nanomaterials are listed in Table 8.3. Unless otherwise specified were the effects listed in the table observed at all the dosages listed.

Gastrointestinal translocation and toxicity studies

The fate and toxicity of ingested nanoparticles and nanomaterials are listed in Table 8.4. Unless otherwise specified were the effects listed in the table observed at all the dosages listed. The liver, spleen and kidney are the main targets of ingested nanomaterials. These organs participate in the clearance of ingested nanoparticles in the body, similar to intravenously and intraperitoneally injected

Table 8.3 In *vivo* systematic toxicity studies

Nano-material	Species	Dosage and exposure route	Effects	References
C_{60}	Mice	2.5 g/kg, i.p.	Absorbed in liver and spleen with hypertrophy of the perisinusoidal cells in the liver	Moussa *et al.* (1996)
C_{60} (polysulphonated)	Rats	0.6 g/kg, i.p.	Lysosomal overload in the kidney	Chen *et al.* (1998)
	Mice	1 g/kg, i.p.	Inhibited mitochondrial function	Ueng *et al.* (1997)
MWCNT	Mice	1, 2.5 and 5 mg/kg	No toxicity	Carrero-Sanchez *et al.* (2006)
TiO_2	Rats	5 mg/kg, i.v.; 1, 14 and 28 days	Low toxicity, TiO_2 detected in the liver, spleen, lung and kidney	Fabian *et al.* (2008)
Fe_3O_4	Mice	50 mg/kg, i.v. 6 mg/kg, i.p.	Low toxicity	Xia *et al.* (2005)
Silica	Mice	50 mg/kg, i.p.	Inflammatory response	Park and Park (2009)
Aluminium oxide	Rats	1 mg/kg and 5 mg/kg; every 2nd day for 30 and 60 days, i.p.	Brain microglia and astrocytes activated in cortex and hippocampus	Li *et al.* (2009)
Carboxyl- and amine quantum dots	Mice	1.44–3600 pmol, i.v.	Pulmonary vascular thrombosis (720 and 3600 pmol), high mortality in 3600 pmol dosage group	Geys *et al.* (2008)

nanoparticles. Mortality was observed in only one study and this was due to intestinal obstruction rather than the toxicity of the ingested nanoparticles (Wang *et al.*, 2006).

Cutaneous translocation and toxicity studies

The number of *in vivo* toxicity studies investigating the effect of cutaneous nanoparticle and nanomaterial exposure is limited especially considering that contact with external surfaces of the body is a likely route of entry. No dermal irritation or allergies were observed when the skin of humans, mice and rabbits were exposed to nanomaterials (Table 8.5). These results do not coincide with *in vitro* studies where nanomaterials were cytotoxic and proinflammatory to cell lines (Shvedova *et al.*, 2003; Gupta and Gupta, 2005; Manna *et al.*, 2005; Pernodet *et al.*, 2006; Rouse *et al.*, 2006; Sayes *et al.*, 2005, 2006b; Tian *et al.*, 2006; Witzmann *et al.*, 2006). This was probably due to limited dermal penetration during *in vivo* studies. Researchers supported this postulation by implanting carbon nanotubes subcutaneously, which resulted in more severe pathology, i.e. foreign body granuloma-like response (Sato *et al.*, 2005; Koyama *et al.*, 2006).

Certain skin penetration studies verify that nanoparticles do not penetrate past the stratum corneum. Zvyagin *et al.* (2008) investigated human skin penetration by ZnO nanoparticles *in vivo* through multiphoton microscopy, scanning electron microscopy and energy-dispersive X-ray. ZnO nanoparticles did not penetrate past the stratum corneum, which has a high turnover rate, and accumulated in the skin folds and hair follicles of human skin. Larese *et al.* (2009) investigated the penetration of silver nanoparticles through human skin by using atomic absorption spectroscopy and transmission electron microscopy. Silver nanoparticles penetrated to the stratum corneum and

Table 8.4 In *vivo* gastrointestinal toxicity studies

Nano-material	Species	Dosage and exposure route	Effects	References
C_{60}	Rats	Received water-soluble C_{60} orally	Low toxicity and majority of compound excreted	Yamago *et al.* (1995)
	Rats	2 g/kg; single dose	Low toxicity	Mori *et al.* (2006)
TiO_2	Rats	Diets contained 0%, 1%, 2% and 5% TiO_2-coated mica, for 130 weeks	No toxicological or carcinogenic effects	Bernard *et al.* (1990)
	Rats	175, 550, 1750 and 5000 mg/kg; Single dose	Low toxicity	Warheit *et al.* (2007a)
	Mice	5g/kg; Single dose	Glomerular swelling in kidney and hydropic degeneration in liver	Wang *et al.* (2007a)
Fe_3O_4	Mice	10mg/kg; single dose	Low toxicity	Xia *et al.* (2005)
Zn	Mice	5 g/kg; 2 weeks	Mortality due to intestinal obstruction. Lesions in liver, renal and heart	Wang *et al.* (2006)
Cu	Mice	413 mg/kg; Single dose	Necrosis in kidney, steatosis in liver and atrophy in spleen	Chen *et al.* (2006)
	Rats	50, 100 and 200 mg/kg/day for 5 days; orally	Hepatotoxicity and nephrotoxicity with mitochondrial failure, enhanced ketogenesis, fatty acid β-oxidation and glycolysis observed (200 mg/kg/day)	Lei *et al.* (2008)
Ag	Rats	30, 300 and 1000 mg/kg for 28 days, orally	Dose-dependent accumulation in all tissue examined. Did not induce genetic toxicity	Kim *et al.* (2008)

the outermost surface of the epidermis of healthy skin. Quantum dots tested on porcine skin also penetrated to the stratum corneum (Zhang *et al.*, 2008b). Only one study reported the penetration of quantum dots past the stratum corneum in healthy skin. Quantum dots with different physicochemical properties were shown to penetrate to the dermal layers of porcine skin (Rymen-Rasmussen *et al.*, 2006).

Flexed skin was shown to be more permeable to nanoparticles which were taken up into the lymphatic system and regional lymph nodes (Tinkle *et al.*, 2003). The localization of nanoparticles in lymph nodes after crossing the dermis was confirmed by a study by Kim *et al.* (2004). Intradermally injected near-infrared quantum dots were detected in the lymph nodes of mice and pigs. Skin macrophages and dendritic cells are responsible for transportation (Sato *et al.*, 1998).

Disease and clinical disorders

Research has shown that exposure to nanoparticles may be associated with neurological conditions such as Alzheimer's and Parkinson's diseases through two separate mechanisms. Firstly, exposure to carbon- and metal-based nanoparticles is associated with elevated levels of autophagic vacuoles in cells (Stern *et al.*, 2008). This is probably as a result of a cellular response trying to clear the cells of nanoparticles. Impaired autophagy can result in morphological changes and subsequent cell death and may therefore play a role in neurological and other diseases (Stern and Johnson, 2008). Secondly,

Table 8.5 In *vivo* dermal toxicity studies

Nanomaterial	Species	Dosage and exposure route	Effects	References
C_{60} soot	Human	Unknown dosage; Patch test; 96 h	No dermal irritation or allergy	Huczko *et al.* (1999)
	Rabbit	Unknown dosage; Ocular irritation; 24, 46 and 72 h	No dermal irritation or allergy	Huczko *et al.* (1999)
SWCNT soot	Human	Unknown dosage; Patch test; 96 h	No dermal irritation or allergy	Hutczko and Lange (2001)
	Rabbit	Unknown dosage; Ocular irritation; 24, 46, and 72 h	No dermal irritation or allergy	Hutczko and Lange (2001)
SWCNT	Mice	2 mg; Subcutaneous implantation	Histocompatibility complex in $CD4^+/CD8^+$ T-cells activated	Koyama *et al.* (2006)
		40, 80, and 160 mμg; 5 days	Dermal toxicity, free radical generation, oxidative stress and inflammation	Murray *et al.* (2009)
	Rat	0.1 mg; subcutaneous implantation; 1 and 4 weeks	Foreign body granuloma with no necrosis or severe inflammation	Sato *et al.* (2005)
TiO_2	Mice	Ears treated with 0%, 5%, 25%, 50% or 100% TiO_2 for 3 days	No dermal sensitization	Warheit *et al.* (2007a)
	Rabbits	0.5 g dermal dose to shaved skin	No dermal irritation	Warheit *et al.* (2007a)

metal nanoparticles have been shown to initiate β-amyloid peptide fibrillation and the formation of senile plagues. This is a key feature of Alzheimer's disease pathology (Wu *et al.*, 2008; Zhang *et al.*, 2008a). Epidemiology studies suggest a possible link between exposure to welding dust (Polizzi *et al.*, 2002; Racette *et al.*, 2005) (essentially incidental nanoparticles) and Alzheimer's and Parkinson's disease-like pathologies. Workers exposed to welding dust containing submicron aluminium particles underwent neuropsychological testing and their aluminium content were determined and compared to that of a control group. There was a strong correlation between the levels of exposure to aluminium-containing welding dust, aluminium content in the blood and preclinical mild cognitive disorder (Polizzi *et al.*, 2002). The prevalence of Parkinson's disease was found in another study to be higher among workers exposed to welding dust than the age-standarsized data for the general population (Racette *et al.*, 2005).

β-Amyloid peptide is an amyloidogenic protein, a group of proteins that form fibril aggregates under specific conditions that are highly insoluble and precipitates to form plagues. They are part not only characteristic in the pathology of Alzheimer's and Parkinson's disease but also Huntington's, Creutzfeld–Jacob and prion disease, to name a few (Hammer *et al.*, 2008). Nanoparticles are also involved in the fibrillation of other amyloidogenic proteins, such as β_2-microglobulin and myoglobin. Copolymer nanoparticles, cerium oxide nanoparticles, quantum dots and carbon nanotubes shorten the lag phase in the nucleation step of human β_2-microglobulin, enhancing the formation of fibril aggregates. β_2-microglobulin also forms multiple layers on the nanoparticle's surface by exchange of the protein between the solution and nanoparticle (Linse *et al.*, 2007). Myoblin interacting with phosphate-grafted zirconia nanoparticles leads to the formation of fibril aggregates (Bellezza *et al.*, 2007). Dendrimers interacting with polypeptide conformations also induced protein fibrillation

(Heegard *et al.*, 2007). In contrast, C_{60} fullerenes inhibit protein fibrillation and can potentially be applied in the development of amyloidogenic therapies (Podolski *et al.*, 2007).

Epidemiology studies have linked natural nanoparticles with cardiovascular diseases (Chen *et al.*, 2008b; Mills *et al.*, 2009). The question has arisen whether engineered nanoparticles will also be linked to cardiovascular diseases. The role engineered nanoparticles plays in the inflammation of the endothelial have been investigated *in vitro* since it is critical for the development of cardiovascular diseases. Atherosclerosis for example is an inflammatory disease associated with cardiovascular complications such as stroke, ischemia and myocardial infarctions (Ross, 1999; Libby, 2002). Inflammation in human aortic endothelial cells was observed after exposure to engineered iron oxide and zinc oxide nanoparticles (Gojova *et al.*, 2007). Carbon nanotubes (both SWCNT and MWCNT) were shown to be cytotoxic towards human aortic endothelial cells (Walker *et al.*, 2009). Alumina particles also elicited a proinflammatory response in porcine pulmonary artery endothelial cells and human umbilical vein endothelial cells (Oesterling *et al.*, 2008).

Metal-derived nanoparticles have shown to induce brain dysfunction and aggravate brain pathology under stress conditions. Mild cognitive impairment and cellular alterations were detected in the brains of normal animals treated with metal nanoparticles. Additional stress, i.e. whole-body hyperthermia, and exposure to metal nanoparticles resulted in severe cognitive and motor deficits, oedema formation, exacerbation of blood–brain barrier disruption and brain pathology (Sharma and Sharma, 2007).

Mechanisms of nanotoxicity

Physiochemical characteristics

The size of nanoparticles influences the deposition within the body. The International Commission on Radiological Protection predicted that nasopharyngeal deposition will increase as particle size decrease and can be influenced by other factors such as aggregation state, integrity of the epithelial lining and particle solubility (International Commission on Radiological Protection, 1994). The absorption of ^{125}I-radiolabelled polysterene nanoparticles in rats to the Peyer's patches of the gut-associated lymphoid tissue was also shown to be size dependent (Jani *et al.*, 1990).

The influence of size on the translocation and clearance of particles in the lungs were also investigated in several studies. The lungs of rats exposed to micron- and nano-sized particles (spherical) were lavaged after 24 h exposure to retrieve the macrophages. Alveolar macrophages remove foreign materials in the lungs by phagocytosis. Macrophages retrieved 80% of the microparticles and only 20% of the nanoparticles. This indicates that 20% of the microparticles remained in the lungs and 80% of the nanoparticles. The majority of inhaled nanoparticles was either in the epithelial cells or translocated to the interstitium (Oberdörster *et al.*, 1992a,b; 2000; Kreyling *et al.*, 2002; Semmler *et al.*, 2004). The biodistribution of nanoparticles injected intravenously is also size dependent. Gold nanoparticles 10 nm in size (spherical) were detected in the blood, liver, spleen, kidney, testis, heart, lung and brain of rats after intravenous administration. Larger gold nanoparticles (50, 100 and 250 nm, spherical) on the other hand were only detected in the blood, liver and spleen (De Jong *et al.*, 2008).

The length of the fibre is a key predictor of a nanomaterial's biopersistence. The longer the fibre the more difficult is it for macrophages to envelop and clear the nanomaterial. Muller *et al.* (2005) have shown that longer fibres of intratracheally instilled MWCNTs had a prolonged residence in the lungs (6 μm versus 1 μm). These fibres are more toxic since it would result in increased fibre–epithelial interactions and thus higher inflammatory responses. Poland *et al.* (2008) also reported that MWCNTs had no toxic effect *in vivo* but longer MWCNTs (>20 μm in length) were responsible for asbestos-like lesions. *In vitro* experiments on human pneumocytes contradicted *in vivo* results, with the length of the carbon nanotubes having no influence on the toxicity (Simon-Deckers *et al.*, 2008). Similarly, the length of SWCNTs (220 nm and 825 nm) had no cytotoxic effect on human acute monocytic leukaemia cells *in vitro*, but the story was different when compared to *in vivo* results. SWCNTs were implanted subcutaneously in rats and it was shown that the longer fibre resulted in a higher inflammatory response (Sato *et al.*, 2005).

The size of nanoparticles has an impact on its genotoxicity. Nanosized TiO_2 (10 and 20 nm, spherical) in the absence of light caused DNA damage in human bronchial epithelial cells. However, no DNA damage were observed when cells were treated with micronsized TiO_2 (200 nm and >200 nm, spherical) (Gurr *et al.*, 2005). Similarly, Syrian hamster embryo fibroblasts exposed to nanosized TiO_2 (<20 nm, spherical) particles had an increase in micronuclei compared to no significant increase after exposure to micronsized TiO_2 particles (>200 nm, spherical) (Rahman *et al.*, 2002). Micron- and nanosized cobalt chrome alloy (2.9 and 29.5 nm respectively, spherical) were both genotoxic towards human dermal fibroblasts but the degree of genotoxicity was significantly higher after exposure to nanosized material (Papageorgiou *et al.*, 2007). Size is though not the only important determinant in genotoxicity since bulk V_2O_5 was positive in the Comet assay and nanosized V_2O_5 negative (Landsiedel *et al.*, 2009).

The surface area of nanoparticles plays a key role in its toxicity. Inhaled nanosized TiO_2 particles (20 nm, spherical) resulted in higher inflammatory responses compared to inhaled submicron-sized TiO_2 particles (250 nm, spherical). The dose-response curves of the different sized particles were similar when the curves were normalized to surface areas. This suggests that the pulmonary inflammation was probably as a result of the surface effects of this particle (Oberdörster *et al.*, 1994). Also, the intratracheal instillation of TiO_2 in the form of 1 or 5 mg/kg 300 nm pigment-grade particles, 200 nm × 35 nm rods and 10 nm dots in rats resulted in the same level of inflammation and tissue injury. It must just be taken into consideration that the crystalline structure of material was not identical, the pigment-grade particles were of rutile-type structure, and the rods and dots were anatase (Warheit *et al.*, 2006). *In vitro* studies had similar findings with smaller TiO_2 nanoparticles effecting human dermal fibroblasts and human lung epithelial cells to the same extent as that of larger TiO_2 nanoparticles. The crystalline structure of the particles and the ability to generate reactive oxygen species was factors that played a role in cytotoxicity (Warheit *et al.*, 2006).

Surface properties of nanoparticles have recently been investigated as a more important determinant in nanotoxicity than surface area (Warheit *et al.*, 2007b, c). The effect of four different quartz particles (12, 50, 300 and 500 nm, spherical) were intratracheally instilled and the inflammation evaluated by measuring the degree of lung cell proliferation, lung weight and lung histopathology. Cytotoxicity and pulmonary inflammation was found not to be related to particle size or surface area but rather surface effect and in particular haemolytic activity (Warheit *et al.*, 2007b). Clouter *et al.* (2001) also found a correlation between the haemolytic activity of micronsized quartz particles and *in vivo* inflammation. A study with intratracheally instilled TiO_2 nanoparticles concluded that their composition and crystal structure (anatase vs. rutile) were the causative factors in lung inflammation, cytotoxicity, cell proliferation and histopathology rather than particle size (Warheit *et al.*, 2007c). Surface charge also has an influence on a nanoparticle's cytotoxicity with cationic surfaces being more toxic than anionic and neutral surfaces. Cationic particles may bind more efficiently to negatively charged cell membranes (Goodman *et al.*, 2004).

Macromolecular perturbation

Nanoparticles and nanomaterials can interact with biological macromolecules because of their small size and unique physiochemical characteristics. Nanoparticles and nanomaterials interacting with macromolecules such as DNA, cytoskeleton elements, collagen, membrane structure and proteins can adversely affect the macromolecules original function in the body.

In vitro genotoxicity studies have shown that some nanomaterials can damage DNA and cause gene and chromosomal mutations (see section on Genotoxicity). Nanomaterials have also been postulated to hinder DNA repair and impair the structure and stability of DNA molecules. Zhao *et al.* (2005) showed through computer stimulation that fullerenes bind strongly to double-stranded and single-stranded DNA. The association between C_{60} and DNA in water was shown to be stronger than the association between two C_{60} molecules. Fullerenes also occupied damaged sites in DNA molecules raising questions about fullerenes ability to impact DNA repair negatively. These results are validated by Shukla and Leszczynski (2009) that reported C_{60} to bind with the nucleic acid base guanine through stacking interaction, forming a stable complex. Yang *et al.* (2009b) determined with atomic force microscopy that silver nanoparticles bind with genomic DNA. The replication capacity of the *rpsL* gene was also compromised after exposure to three kinds of silver nanoparticles.

Lynch *et al.* (2006) hypothesized that the shape of proteins can change after binding to the highly curved surface of nanoparticles. This conformational change will result in core amino acids being exposed to the surface. The immune system would not recognize the proteins and launch an unwanted immune response against these 'cryptic epitopes'. Conformational changes in proteins will also disrupt their original function in the body. These changes can lead to the development of disease and clinical disorders as in the case of amyloidogenic protein fillibration after contact with nanoparticles as is the case in Alzheimer's and Parkinson's disease (see section on Disease and clinical disorders). The interaction between nanoparticles and proteins are discussed in a review by Lynch and Dawson (2008). They propose that nanoparticles are organized into a 'protein corona' after contact with biological fluids and this forms the biological identity of the nanoparticle. The size and surface properties of nanoparticles play a role in the formation of 'protein corona' complexes (Lundqvist *et al.*, 2008). Ehrenberg *et al.* (2009) showed in a recent study that nanoparticles are coated with proteins in a matter of minutes after being suspended in cultured human endothelial. Proteins absorbed to particle surfaces mediated cell association and was shown not to be specific to the type of protein. Gold nanoparticles incubated with human plasma for example were coated with 69 different proteins (Dobrovolskaia *et al.*, 2009).

Some nanoparticles are not haemocompatible and cause blood clotting, thrombocyte activation, plasmatic coagulation, haemolysis, inflammation and immunogenic effects (Jevprasesphant *et al.*, 2003; Li *et al.*, 2008b; Mayer *et al.*, 2008). Li *et al.* (2008b) showed that TiO_2 nanoparticles attach to erythrocytes changing their native surface properties leading to haemagglutination. This triggers oxidative stress which causes the content of the erythrocytes to leak out. Research conducted by Geys *et al.* (2008) detected pulmonary vascular thrombosis after intravenously administering carboxyl- and amine quantum dots. The negative surface charge was responsible for activating the coagulation cascade leading to fibrin formation and simultaneous platelet activation by thrombin. In the case of platelet aggregation enhanced positively charged polysterene nanoparticles aggregated more strongly than negatively charged polysterene nanoparticles. The surface charge had no influence on coagulation aggregation. The zeta potentials and surface chemistry between the quantum dots and polysterene nanoparticles differs drastically making it difficult to make direct comparisons (Nemmar *et al.*, 2002b). This problem was solved by Mayer *et al.* (2009) that investigated the haemocompatibility of nanoparticles using identical polysterene particles as a model. Haemolysis and inflammation was size dependent whereas complement activation was dependent on the particle charge. Hydrophobicity and material composition were just as important as particle charge and size on plasmatic coagulation and thrombocyte activation.

Certain nanoparticles exhibited good haemocompatibility and has been investigated as potential antitrombotic agents. Porous SiO_2/HEMA biocomposites consisting out of silica nanoparticles (Liu *et al.*, 2008b), pHEMA matrixes coupled with CuO nanoparticles (Liu *et al.*, 2008a), poly (L-lactic acid) membrane, layered Au nanoparticle assembly (Yu *et al.*, 2007), and nanomaterial diamond-like carbon-metal composite thin films (Andara *et al.*, 2006) are a few examples. Gold nanoparticles incubated with human blood plasma also did not induce platelet aggregation, complement activation and plasma coagulation (Dobrovolskaia *et al.*, 2008).

The Nanotechnology Characterization Lab has established and optimized the following *in vitro* assays to determine the compatibility of nanomaterials with the immune system; haemolysis, platelet aggregation, plasma coagulation, complement activation, plasma protein binding and phagocytosis (Dobrovolskaia *et al.*, 2008).

Chemical composition

The coating of nanoparticles has an influence on the nanoparticle's toxicity. Genotoxicity tests, using mouse embryonic stem cells and mouse embryonic fibroblasts, indicated that polysaccharide (acacia gum) coated silver nanoparticles exhibited more severe damage than uncoated silver nanoparticles. Researchers speculate that the uncoated nanoparticle agglomerate, limiting the surface area availability and access to cells and tissues (Ahamed *et al.*, 2008). Coating quantum dots (Zhang *et al.*, 2006; Ryman-Rasmussen *et al.*, 2007; Wang *et al.*, 2008d) and gold nanorods (Huff *et al.*, 2007; Alkilany *et al.*, 2009) with PEG have been shown *in vitro* to reduce cytotoxicity. Many nanomaterials, such as MWCNT and QD, water solubility were increased by coating their surfaces with hydrophilic molecules. This modification resulted in lower cell viability in cytotoxicity tests due to the functional groups toxicity (Kirchner *et al.*, 2005; Bottini *et al.*, 2006b; Magrez *et al.*, 2006). The haemocompatibility of nanoparticles and nanomaterials have also been shown to be improved by PEG coating. PEG coating reduced the haemolytic activity of poly (lactide-co-glycolide) acid (Kim *et al.*, 2005) and chitosan (Kim *et al.*, 2006b) nanoparticles.

The coating of nanoparticles will also influence the stabilization and biodistribution of the particle within the body. Gold nanoparticles coated with gum arabic were predominantly recovered form liver tissues after intravenously injection in juvenile swines. In contrast were the majority of maltose-coated gold nanoparticles recovered from lung tissues (Fent *et al.*, 2009).

Generation of reactive oxygen species (ROS)

Nanoparticle's and nanomaterial's ability to produce reactive oxygen species (ROS) disrupts a cell's redox equilibrium and so causes oxidative stress. Nanoparticles and nanomaterials can generate ROS either spontaneously or by targeting subcellular organelles and biological systems (Li *et al.*, 2008a). The final result of oxidative stress is excess cell proliferation, apoptosis, lipid peroxidation or mutagenesis (Kappus, 1987).

The following intrinsic properties of nanoparticles are responsible for spontaneous ROS generation. TiO_2 nanoparticles can form electron hole pairs after UV activation. These hole pairs interacting with the particle surface can participate with donor or capture interactions generating superoxide or hydroxyl radicals. A semiconductor nanoparticle could in excited state cause electrons to jump from the conduction band to O_2 to generate superoxide. Fullerenes, TiO_2 and other metal oxides have been linked to this mechanism. Nanoparticles can also during dissolution release metal ions that catalyse ROS generation. Lastly, superoxides can be generated through transition metals on a nanomaterial's surface (Li *et al.*, 2008a).

ROS can also be generated by nanoparticles through interaction with a cell's subcellular organelles and biological systems after being internalized. In the cells the nanoparticles are enveloped in endosomes and can form ROS through the formation of NADPH oxidase. Endosomes also fuses with lysosomes (Li *et al.*, 2008a). Nanoparticles breaking from lysosomes can disrupt one-electron transfer in the mitochondria inner membrane leading to further ROS production and subsequent interruption of ATP synthesis and DNA damage. The cytotoxicity and genotoxicity of silver nanoparticles to human cells have been attributed to this mechanism (AshaRani *et al.*, 2009). The generation of ROS by silver nanoparticles has also been shown to be size-dependent (Carlson *et al.*, 2008).

The generation of ROS influences the toxicity of nanoparticles through different mechanisms. Iron nanoparticles for example increase the permeability human microvascular endothelial cells through the production of ROS which plays a role in the regulation of the Akt/GSK-3-β-signalling pathway (Apopa *et al.*, 2009). ROS generation elicits pro-inflammatory responses *in vivo* and *in vitro* (Park and Park, 2009). ROS generation and oxidative stress can also result in injury to the mitochondria and endoplasmic reticulum because of increased calcium uptake and structural cellular damage

(Xia *et al.*, 2006). Neurodegeneration by iron-based nanoparticles are attributed to ROS generation and a decrease in glutamate uptake (Alekseenko *et al.*, 2008).

Aggregation

Nanoparticles at high concentrations can aggregate, changing their particle size and structure. This in turn can influence their toxicity. Agglomeration of nanoparticles has been investigated in water and air but can differ drastically in bodily fluids and even within different tissues and subcellular and cellular compartments. Nanoparticles are for example expected to aggregate in blood plasma because of its salinity. This and the question if de-aggregation can occur within the body are important points that need to be investigated (Balbus *et al.*, 2007; Handy *et al.*, 2008).

Carbon nanotubes instilled intratracheally induces pulmonary fibrosis, granulomas or inflammation (Lam *et al.*, 2004; Muller *et al.*, 2005; Shvedova *et al.*, 2005; Warheit *et al.*, 2005; Carrero-Sanchez *et al.*, 2006). Recent studies however revealed that MWCNT dispersed in albumin (Elgrabli *et al.*, 2008) or aerosol (Mitchell *et al.*, 2007) resulted in very low toxicity. This illustrates that the toxicity of nanomaterials can be significantly lowered if the agglomeration can be inhibited *in vivo*.

Limitations in nanotoxicology hazard assessment and future research focus points

Limitations in *in vitro* hazard assessment

There are a number of limitations in *in vitro* assays that are currently used in nanotoxicology such as the ability of nanoparticles to interact with dyes and reagents used the cell viability assays and the agglomeration of nanoparticles with proteins in test medium. Traditional *in vitro* toxicology tests used must therefore first be validated when applied in nanotoxicology. Nanoparticle's and nanomaterial's unique properties may interfere with an assay and lead to false-positives or false-negatives. Cell viability is monitored by dyes absorbing to living cells. Carbon black (Monteiro-Riviere *et al.*, 2009) and SWCNT's (Casey *et al.*, 2007) have been shown to adsorb to various dyes. The MTT assay for example contradicted other cytotoxicity tests, which determined that no loss in cell viability occurred after treatment with carbon nanotubes (Pulskamp *et al.*, 2007). Carbon nanotubes were found to absorb to reduced formazan dye used in MTT cytotoxicity assays resulting in an overestimation of cytotoxicity (Wörle-Knirsch *et al.*, 2006; Casey *et al.*, 2007). Monteiro-Riviere *et al.* (2009) tested the reliability of six cell viability tests, namely the neutral red (NR), calcein AM (CAM), MTT, CellTiter 96® AQueous One (96 AQ), alamar Blue® (aB), CellTiter-Blue® (CTB) and CytoTox OneTM Homogenous Membrane Integrity (CTO) assay. The reliability of all the assays scrutinized was compromised by fullerenes, SWCNTs and MWCNTs. They concluded that the 96 AQ assay was the most accurate assay being affected only by carbon black. Propidium iodide is used during the Annexxin V/propidium iodide stain to detect necrotic cells (Aubry *et al.*, 1999). Gold nanoparticles have been shown to bind to propidium iodide which may lead to false positive results (Shukla *et al.*, 2005b).

The agglomeration of nanoparticles and nanomaterials in cell culture medium can lead to the misinterpretation of cytotoxicity and genotoxicity assays. Carbon nanotubes were shown to agglomerate in KGM-2 cell culture medium (Monteiro-Riviere *et al.*, 2009) and metal nanoparticles in Dulbecco's MEM (Schulze *et al.*, 2008). Agglomeration of nanoparticles and nanomaterials *in vitro* hinders attempts to compare the dosages with *in vivo* studies. Larger agglomerated nanoparticles and nanomaterials will be removed by the RES system in the body (Monteiro-Riviere *et al.*, 2009).

Limitations in *in vivo* hazard assessment

Experimental conditions in *in vivo* toxicity tests needs to be evaluated to determine if there are not factors that can influence the outcome of the test. Dispersive agents, i.e. tetradrohyfuran (THF), are used to hinder the aggregation of nanoparticles or nanomaterials during a toxicity test. In one often-quoted study was fullerenes thought to be the causative agent of oxidative stress in the brain of largemouth bass (Oberdörster, 2004). Further research however concluded that residual THF was responsible for the majority of oxidative stress (Zhu *et al.*, 2006). Impurities such as catalytic metals and amorphous carbon can be produced during the manufacturing of nanoparticles and nanomaterials. These impurities can be responsible for the toxicity and therefore should purified nanoparticles be assessed during toxicity tests (Kaiser *et al.*, 2008).

Research groups use different protocols and experimental set-ups to test the toxicity of a nanoparticle or nanomaterial *in vivo*. Take the pulmonary toxicity studies of SWCNTs as an example. Different laboratory animals (rats versus mice), different measuring units (mg/m^3 versus mg/animal) and different exposure methods (instillation versus aspiration versus inhalation) were used (Table 8.2). The implementation of standardized assays in *in vivo* nanotoxicology studies would make it possible to compare data.

Future research focus points

New technologies employing marker-free and live measurements in *in vitro* assays are currently under evaluation and validation. These *in vitro* toxicity assays are unaffected by the unique physiochemical characteristics of nanoparticles (Ceriotti *et al.*, 2007; Schekenburger and Kemper, 2008). New cell cultures have also been developed that better emulates *in vivo* toxicity (Lilienblum *et al.*, 2008). Toxicogenomics is new a subdiscipline of toxicology that can provide valuable information about the complex nature of nanoparticle toxicity (Aardema and MacGregor, 2002; Robosky *et al.*, 2002). The number of studies involving metabolomics and proteomomics, that falls under this discipline, to date is limited. Toxicogenomics using molecular biomarkers are highly sensitive and can be developed in high-throughput automated systems in nanotechnology (Fowler *et al.*, 2008).

The International Alliance for NanoEHS Harmonization is addressing above mentioned problems by striving to establish reproducible protocols for the hazard assessment of nanoparticles and nanomaterials. This peer group co-organization consists out of scientific experts from around the world. Kenneth Dawson, Centre for BioNano Interactions at University College Dublin, chairs the alliance and other partners are Vince Castranova, Centres of Disease Control and Prevention (CDC); Vicki Colvin, Rice University; Alison Elder, University of Rochester; Vince Hackley, National Institute of Standards and Technology; Mark Hoover, CDC; Gaku Ichihara, Nagoya University; Wolfgang Kreyling, Helmholtz Centre Munich; Harald Krug, EMPA; Kun'ichi Miyazawa, National Institute of Material Science; Vicki Stone, Napier University; Andre Nel, University of California, Los Angeles; Günter Oberdörster, University of Rochester; Dale Porter, CDC; Alexander Stefaniak, CDC and Mark Wiesner, Duke University. The public will have access to the protocols once they have been developed and validated by the alliance (http://www.nanoehsalliance.org). Other groups also currently standardizing and validating hazard assessment protocols include the OECD (Organization for Economic Cooperation and Development) and MEIC (Multicenter Evaluation of *In Vitro* Toxicity) (Kroll *et al.*, 2009).

Conclusions

The physiochemical properties of a nanoparticle or nanomaterial must be known during toxicity assessment. Researchers propose that the following information must be documented; the method by which the nanoparticles are synthesized and prepared, particle size and distribution, the composition and relevant coatings, specific surface area, crystal structure, purity of the sample, absorption capacity, catalytic activity and the aggregation status in the relevant test media (Gonzalez *et al.*, 2008; Kroll *et al.*, 2009).

There is a clear need for standardized assays in nanotoxicology (both for *in vivo* and *in vitro* studies). Currently, standardization efforts in *in vitro* assays for nanoparticles and nanomaterials are under way. The American Society for Testing and Materials International has already approved two methods, one determining the haemolytic properties of nanoparticles and another evaluating the effect of nanoparticles on mouse granulocyte-macrophage colonies (Dobrovolskaia *et al.*, 2008). The development of a database with internationally agreed reference materials is also a necessity that needs special attention.

The majority of research in the emerging field of nanotoxicology focuses on *in vitro* testing, not taking into account the importance of *in vivo* testing (Ostrowski *et al.*, 2009). The impact of foreign materials on the immune system in its entirety can only be evaluated *in vivo*. Macrophage and dendritic cell recruitment are intercellular effects of the immune system that cannot be monitored *in vitro*. Furthermore, the biodistribution and translocation of nanoparticles and nanomaterials to secondary organs can only be assessed during *in vivo* studies (Fischer and Chan, 2007). Data collected from *in vitro* and *in vivo* assays have also been shown not to correlate (Sayes *et al.*, 2007b). *In vitro* assays must thus be validated with *in vivo* results to be considered of importance.

In the future will nanoproducts that have undergone intensive risk assessment have a competitive edge in the market.

Web resources

- http://www.nanoehsalliance.org.

References

Aardema, M.J., and MacGregor, J.T. (2002). Toxicology and genetic toxicology in the new era of 'toxicogenomics': impact of '-omics' technologies. Mutat. Res. *499*, 13–25.

Ahamed, M., Karns, M., Goodson, M., Rowe, J., Hussain, S.M., Schlager, J.J., and Hong, Y. (2008). DNA damage response to different surface chemistry of silver nanoparticles in mammalian cells. Toxicol. Appl. Pharmacol. *233*, 404–410.

Alekseenko, A.V., Waseem, T.V., and Fedorovich, S.V. (2008). Ferritin, a protein containing iron nanoparticles, induces reactive oxygen species formation and inhibits glutamate uptake in rat brain synaptosomes. Brain Res. *1241*, 193–200.

Alkilany, A.M., Nagaria, P.K., Hexel, C.R., Shaw, T.J., Murphy, C.J., and Wyatt, M.D. (2009). Cellular uptake and cytotoxicity of gold nanorods: molecular origin of cytotoxicity and surface effects. Small *5*, 701–708.

Andara, M., Agarwal, A., Scholvin, D., Gerhardt, R.A., Doraiswamy, A., Jin, C.M., Narayan, R.J., Shih, C.C., Shih, C.M., Lin, S.J., and Su, Y.Y. (2006). Hemocompatibility of diamondlike carbon-metal composite thin films. Diamond Rel. Mater. *15*, 1941–1948.

Anonymous. (1994). Human respiratory tract model for radiological protection. A report of a Task Group of the International Commission on Radiological Protection. Ann. ICRP *24*, 1–482.

Apopa, P.L., Qian, Y., Shao, R., Guo, N.L., Schwegler-Berry, D., Pacurari, M., Porter, D., Shi, X., Vallyathan, V., Castranova, V., and Flynn, D.C. (2009). Iron oxide nanoparticles induce human microvascular endothelial cell permeability through reactive oxygen species production and microtubule remodeling. Part Fibre Toxicol. *6*, 1.

AshaRani, P.V., Low Kah Mun, G., Hande, M.P., and Valiyaveettil, S. (2009). Cytotoxicity and genotoxicity of silver nanoparticles in human cells. ACS Nano 3, 279–290.

Aubry, J.P., Blaecke, A., Lecoanet-Henchoz, S., Jeannin, P., Herbault, N., Caron, G., Moine, V., and Bonnefoy, J.Y. (1999). Annexin V used for measuring apoptosis in the early events of cellular cytotoxicity. Cytometry *37*, 197–204.

Auffan, M., Decome, L., Rose, J., Orsiere, T., De Meo, M., Briois, V., Chaneac, C., Olivi, L., Berge-Lefranc, J.L., Botta, A., Wiesner, M.R., and Bottero, J.Y. (2006). *In vitro* interactions between DMSA-coated maghemite nanoparticles and human fibroblasts: A physicochemical and cyto-genotoxical study. Environ. Sci. Technol. *40*, 4367–4373.

Balbus, J.M., Maynard, A.D., Colvin, V.L., Castranova, V., Daston, G.P., Denison, R.A., Dreher, K.L., Goering, P.L., Goldberg, A.M., Kulinowski, K.M., Monteiro-Riviere, N.A., Oberdörster, G., Omenn, G.S., Pinkerton, K.E., Ramos, K.S., Rest, K.M., Sass, J.B., Silbergeld, E.K., and Wong, B.A. (2007). Meeting report: hazard assessment for nanoparticles--report from an interdisciplinary workshop. Environ. Health Perspect. *115*, 1654–1659.

Bellezza, F., Cipiciani, A., Quotadamo, M.A., Cinelli, S., Onori, G., and Tacchi, S. (2007). Structure, stability, and activity of myoglobin adsorbed onto phosphate-grafted zirconia nanoparticles. Langmuir *23*, 13007–13012.

Bermudez, E., Mangum, J.B., Wong, B.A., Asgharian, B., Hext, P.M., Warheit, D.B., and Everitt, J.I. (2004). Pulmonary responses of mice, rats, and hamsters to subchronic inhalation of ultrafine titanium dioxide particles. Toxicol. Sci. *77*, 347–357.

Bernard, B.K., Osheroff, M.R., Hofmann, A., and Mennear, J.H. (1990). Toxicology and carcinogenesis studies of dietary titanium dioxide-coated mica in male and female Fischer 344 rats. J. Toxicol. Environ. Health *29*, 417–429.

Bottini, M., Bruckner, S., Nika, K., Bottini, N., Bellucci, S., Magrini, A., Bergamaschi, A., and Mustelin, T. (2006). Multi-walled carbon nanotubes induce T lymphocyte apoptosis. Toxicol. Lett. *160*, 121–126.

Bottini, M., Cerignoli, F., Dawson, M.I., Magrini, A., Rosato, N., and Mustelin, T. (2006). Full-length single-walled carbon nanotubes decorated with streptavidin-conjugated quantum dots as multivalent intracellular fluorescent nanoprobes. Biomacromolecules 7, 2259–2263.

Braydich-Stolle, L., Hussain, S., Schlager, J.J., and Hofmann, M.C. (2005). *In vitro* cytotoxicity of nanoparticles in mammalian germline stem cells. Toxicol. Sci. *88*, 412–419.

Brown, J.S., Zeman, K.L., and Bennett, W.D. (2002). Ultrafine particle deposition and clearance in the healthy and obstructed lung. Am. J. Respir. Crit. Care Med. *166*, 1240–1247.

Brunner, T.J., Wick, P., Manser, P., Spohn, P., Grass, R.N., Limbach, L.K., Bruinink, A., and Stark, W.J. (2006). *In vitro* cytotoxicity of oxide nanoparticles: comparison to asbestos, silica, and the effect of particle solubility. Environ. Sci. Technol. *40*, 4374–4381.

Carlson, C., Hussain, S.M., Schrand, A.M., Braydich-Stolle, L.K., Hess, K.L., Jones, R.L., and Schlager, J.J. (2008). Unique cellular interaction of silver nanoparticles: size-dependent generation of reactive oxygen species. J. Phys. Chem. B *112*, 13608–13619.

Carrero-Sanchez, J.C., Elias, A.L., Mancilla, R., Arrellin, G., Terrones, H., Laclette, J.P., and Terrones, M. (2006). Biocompatibility and toxicological studies of carbon nanotubes doped with nitrogen. Nano Lett. *6*, 1609–1616.

Casals, E., Vazquez-Campos, S., Bastus, N.G., and Puntes, V. (2008). Distribution and potential toxicity of engineered inorganic nanoparticles and carbon nanostructures in biological systems. Trac-Trend Anal. Chem. *27*, 672–683.

Casey, A., Herzog, E., Davoren, M., Lyng, F.M., Byrne, H.J., and Chambers, G. (2007). Spectroscopic analysis confirms the interactions between single walled carbon nanotubes and various dyes commonly used to assess cytotoxicity. Carbon *45*, 1425–1432.

Cengelli, F., Maysinger, D., Tschudi-Monnet, F., Montet, X., Corot, C., Petri-Fink, A., Hofmann, H., and Juillerat-Jeanneret, L. (2006). Interaction of functionalized superparamagnetic iron oxide nanoparticles with brain structures. J. Pharmacol. Exp. Ther. *318*, 108–116.

Ceriotti, L., Ponti, J., Colpo, P., Sabbioni, E., and Rossi, F. (2007). Assessment of cytotoxicity by impedance spectroscopy. Biosens. Bioelectron. *22*, 3057–3063.

Chen, E., Ruvalcaba, M., Araujo, L., Chapman, R., and Chin, W.C. (2008a). Ultrafine titanium dioxide nanoparticles induce cell death in human bronchial epithelial cells. J. Exp. Nanosci. *3*, 171–183.

Chen, H.H., Yu, C., Ueng, T.H., Chen, S., Chen, B.J., Huang, K.J., and Chiang, L.Y. (1998). Acute and subacute toxicity study of water-soluble polyalkylsulfonated C_{60} in rats. Toxicol. Pathol. *26*, 143–151.

Chen, Z., Meng, H., Xing, G., Chen, C., Zhao, Y., Jia, G., Wang, T., Yuan, H., Ye, C., Zhao, F., Chai, Z., Zhu, C., Fang, X., Ma, B., and Wan, L. (2006). Acute toxicological effects of copper nanoparticles *in vivo*. Toxicol. Lett. *163*, 109–120.

Chen, Z., Meng, H., Xing, G., Yuan, H., Zhao, F., Liu, R., Chang, X., Gao, X., Wang, T., Jia, G., Ye, C., Chai, Z., and Zhao, Y. (2008b). Age-related differences in pulmonary and cardiovascular responses to SiO_2 nanoparticle inhalation: nanotoxicity has susceptible population. Environ. Sci. Technol. *42*, 8985–8992.

Cheng, F.Y., Su, C.H., Yang, Y.S., Yeh, C.S., Tsai, C.Y., Wu, C.L., Wu, M.T., and Shieh, D.B. (2005). Characterization of aqueous dispersions of Fe(3)O(4) nanoparticles and their biomedical applications. Biomaterials *26*, 729–738.

Cherukuri, P., Bachilo, S.M., Litovsky, S.H., and Weisman, R.B. (2004). Near-infrared fluorescence microscopy of single-walled carbon nanotubes in phagocytic cells. J. Am. Chem. Soc. *126*, 15638–15639.

Clouter, A., Brown, D., Hohr, D., Borm, P., and Donaldson, K. (2001). Inflammatory effects of respirable quartz collected in workplaces versus standard DQ12 quartz: particle surface correlates. Toxicol. Sci. *63*, 90–98.

Connor, E.E., Mwamuka, J., Gole, A., Murphy, C.J., and Wyatt, M.D. (2005). Gold nanoparticles are taken up by human cells but do not cause acute cytotoxicity. Small *1*, 325–327.

Cui, D., Tian, F., Ozkan, C.S., Wang, M., and Gao, H. (2005). Effect of single wall carbon nanotubes on human HEK293 cells. Toxicol. Lett. *155*, 73–85.

Davoren, M., Herzog, E., Casey, A., Cottineau, B., Chambers, G., Byrne, H.J., and Lyng, F.M. (2007). *In vitro* toxicity evaluation of single walled carbon nanotubes on human A549 lung cells. Toxicol. In. Vitro. *21*, 438–448.

De Jong, W.H., Hagens, W.I., Krystek, P., Burger, M.C., Sips, A.J., and Geertsma, R.E. (2008). Particle size-dependent organ distribution of gold nanoparticles after intravenous administration. Biomaterials *29*, 1912–1919.

De Lorenzo, A.J. (1970). The olfactory neuron and the blood–brain barrier. In Taste and smell in vertebrates, Wolstenholme, G., and Knight, J., eds. (J&A Churchill, London, UK), pp. 151–176.

Deng, X., Luan, Q., Chen, W., Wang, Y., Wu, M., Zhang, H., and Jiao, Z. (2009). Nanosized zinc oxide particles induce neural stem cell apoptosis. Nanotechnology *20*, 115101.

Dhawan, A., Taurozzi, J.S., Pandey, A.K., Shan, W., Miller, S.M., Hashsham, S.A., and Tarabara, V.V. (2006). Stable colloidal dispersions of C_{60} fullerenes in water: evidence for genotoxicity. Environ. Sci. Technol. *40*, 7394–7401.

Ding, L., Stilwell, J., Zhang, T., Elboudwarej, O., Jiang, H., Selegue, J.P., Cooke, P.A., Gray, J.W., and Chen, F.F. (2005). Molecular characterization of the cytotoxic mechanism of multiwall carbon nanotubes and nano-onions on human skin fibroblast. Nano Lett. *5*, 2448–2464.

Dobrovolskaia, M.A., Aggarwal, P., Hall, J.B., and McNeil, S.E. (2008). Preclinical studies to understand nanoparticle interaction with the immune system and its potential effects on nanoparticle biodistribution. Mol. Pharm. *5*, 487–495.

Dobrovolskaia, M.A., Patri, A.K., Zheng, J., Clogston, J.D., Ayub, N., Aggarwal, P., Neun, B.W., Hall, J.B., and McNeil, S.E. (2009). Interaction of colloidal gold nanoparticles with human blood: effects on particle size and analysis of plasma protein binding profiles. Nanomedicine *5*, 106–117.

Dufour, E.K., Kumaravel, T., Nohynek, G.J., Kirkland, D., and Toutain, H. (2006). Clastogenicity, photo-clastogenicity or pseudo-photo-clastogenicity: Genotoxic effects of zinc oxide in the dark, in pre-irradiated or simultaneously irradiated Chinese hamster ovary cells. Mutat. Res. *607*, 215–224.

Dunford, R., Salinaro, A., Cai, L., Serpone, N., Horikoshi, S., Hidaka, H., and Knowland, J. (1997). Chemical oxidation and DNA damage catalysed by inorganic sunscreen ingredients. FEBS Lett. *418*, 87–90.

Ehrenberg, M.S., Friedman, A.E., Finkelstein, J.N., Oberdorster, G., and McGrath, J.L. (2009). The influence of protein adsorption on nanoparticle association with cultured endothelial cells. Biomaterials *30*, 603–610.

Elder, A., Gelein, R., Silva, V., Feikert, T., Opanashuk, L., Carter, J., Potter, R., Maynard, A., Ito, Y., Finkelstein, J., and Oberdorster, G. (2006). Translocation of inhaled ultrafine manganese oxide particles to the central nervous system. Environ. Health Perspect. *114*, 1172–1178.

Elgrabli, D., Abella-Gallart, S., Robidel, F., Rogerieux, F., Boczkowski, J., and Lacroix, G. (2008). Induction of apoptosis and absence of inflammation in rat lung after intratracheal instillation of multiwalled carbon nanotubes. Toxicology *253*, 131–136.

Eom, H.J., and Choi, J. (2009). Oxidative stress of CeO_2 nanoparticles via p38-Nrf-2 signaling pathway in human bronchial epithelial cell, Beas-2B. Toxicol. Lett. *187*, 77–83.

Fabian, E., Landsiedel, R., Ma-Hock, L., Wiench, K., Wohlleben, W., and van Ravenzwaay, B. (2008). Tissue distribution and toxicity of intravenously administered titanium dioxide nanoparticles in rats. Arch. Toxicol. *82*, 151–157.

Fent, G.M., Casteel, S.W., Kim, D.Y., Kannan, R., Katti, K., Chanda, N., and Katti, K. (2009). Biodistribution of maltose and gum arabic hybrid gold nanoparticles after intravenous injection in juveline swine. Nanomedicine *5*, 128–135.

Fiorito, S., Serafino, A., Andreola, F., Togna, A., and Togna, G. (2006). Toxicity and biocompatibility of carbon nanoparticles. J. Nanosci. Nanotechnol. *6*, 591–599.

Fischer, H.C., and Chan, W.C. (2007). Nanotoxicity: the growing need for *in vivo* study. Curr. Opin. Biotechnol. *18*, 565–571.

Flahaut, E., Durrieu, M.C., Remy-Zolghadri, M., Bareille, R., and Baquey, C. (2006). Investigation of the cytotoxicity of CCVD carbon nanotubes towards human umbilical vein endothelial cells. Carbon *44*, 1093–1099.

Fowler, B.A., Conner, E.A., and Yamauchi, H. (2008). Proteomic and metabolomic biomarkers for III-V semiconductors: and prospects for application to nano-materials. Toxicol. Appl. Pharmacol. *233*, 110–115.

Fujita, K., Morimoto, Y., Ogami, A., Myojyo, T., Tanaka, I., Shimada, M., Wang, W.N., Endoh, S., Uchida, K., Nakazato, Yamamoto, T., K., Fukui, H., Horie, M., Yoshida, Y., Iwahashi, H., and Nakanishi, J. (2009). Gene expression profiles in rat lung after inhalation exposure to C_{60} fullerene particles. Toxicology *258*, 47–55.

Geys, J., Nemmar, A., Verbeken, E., Smolders, E., Ratoi, M., Hoylaerts, M.F., Nemery, B., and Hoet, P.H. (2008). Acute toxicity and prothrombotic effects of quantum dots: impact of surface charge. Environ. Health Perspect. *116*, 1607–1613.

Gojova, A., Guo, B., Kota, R.S., Rutledge, J.C., Kennedy, I.M., and Barakat, A.I. (2007). Induction of inflammation in vascular endothelial cells by metal oxide nanoparticles: effect of particle composition. Environ. Health Perspect. *115*, 403–409.

Gonzalez, L., Lison, D., and Kirsch-Volders, M. (2008). Genotoxicity of engineered nanomaterials: a critical review. Nanotoxicology *2*, 252–273.

Goodman, C.M., McCusker, C.D., Yilmaz, T., and Rotello, V.M. (2004). Toxicity of gold nanoparticles functionalized with cationic and anionic side chains. Bioconjug. Chem. *15*, 897–900.

Grabinski, C., Hussain, S., Lafdi, K., Braydich-Stolle, L., and Schlager, J. (2007). Effect of particle dimension on biocompatibility of carbon nanomaterials. Carbon *45*, 2828–2835.

Gupta, A.K., and Gupta, M. (2005). Cytotoxicity suppression and cellular uptake enhancement of surface modified magnetic nanoparticles. Biomaterials *26*, 1565–1573.

Gupta, A.K., and Wells, S. (2004). Surface-modified superparamagnetic nanoparticles for drug delivery: preparation, characterization, and cytotoxicity studies. i.e.EE Trans. Nanobioscience *3*, 66–73.

Gurr, J.R., Wang, A.S., Chen, C.H., and Jan, K.Y. (2005). Ultrafine titanium dioxide particles in the absence of photoactivation can induce oxidative damage to human bronchial epithelial cells. Toxicology *213*, 66–73.

Hammer, N.D., Wang, X., McGuffie, B.A., and Chapman, M.R. (2008). Amyloids: friend or foe? J. Alzheimers Dis. *13*, 407–419.

Handy, R.D., Owen, R., and Valsami-Jones, E. (2008). The ecotoxicology of nanoparticles and nanomaterials: current status, knowledge gaps, challenges, and future needs. Ecotoxicology *17*, 315–325.

Heegaard, P.M., Boas, U., and Otzen, D.E. (2007). Dendrimer effects on peptide and protein fibrillation. Macromol. Biosci. *7*, 1047–1059.

Hirano, S., Kanno, S., and Furuyama, A. (2008). Multi-walled carbon nanotubes injure the plasma membrane of macrophages. Toxicol. Appl. Pharmacol. *232*, 244–251.

Hirsch, L.R., Stafford, R.J., Bankson, J.A., Sershen, S.R., Rivera, B., Price, R.E., Hazle, J.D., Halas, N.J., and West, J.L. (2003). Nanoshell-mediated near-infrared thermal therapy of tumors under magnetic resonance guidance. Proc. Natl. Acad. Sci. U. S. A. *100*, 13549–13554.

Hsin, Y.H., Chen, C.F., Huang, S., Shih, T.S., Lai, P.S., and Chueh, P.J. (2008). The apoptotic effect of nanosilver is mediated by a ROS- and JNK-dependent mechanism involving the mitochondrial pathway in NIH3T3 cells. Toxicol. Lett. *179*, 130–139.

Huczko, A., and Lange, H. (2001). Carbon nanotubes: experimental evidence for a null risk of skin irritation and allergy. Fuller. Sci. Tech. *9*, 247–250.

Huczko, A., Lange, H., and Calko, E. (1999). Fullerenes: Experimental evidence for a null risk of skin irritation and allergy. Fullerene Sci Technol. *7*, 935–939.

Huczko, A., Lange, H., Bystrzejewski, M., Baranowski, P., Grubek-Jaworska, H., Nejman, P., Przybylowski, T., Czuminska, K., Glapinski, J., Walton, D.R.M., and Kroto, H.W. (2005). Pulmonary toxicity of 1-D nanocarbon materials. Fullerenes Nanotubes and Carbon Nanostructures *13*, 141–145.

Huczko, A., Lange, H., Calko, E., Grubek-Jaworska, H., and Droszcz, P. (2001). Physiological testing of carbon nanotubes: Are they asbestos-like? Fullerene Sci. Technol. *9*, 251–254.

Huff, T.B., Hansen, M.N., Zhao, Y., Cheng, J.X., and Wei, A. (2007). Controlling the cellular uptake of gold nanorods. Langmuir *23*, 1596–1599.

Hussain, S.M., Hess, K.L., Gearhart, J.M., Geiss, K.T., and Schlager, J.J. (2005). *In vitro* toxicity of nanoparticles in BRL 3A rat liver cells. Toxicol. In. Vitro. *19*, 975–983.

Hutchison, J.E. (2008). Greener nanoscience: a proactive approach to advancing applications and reducing implications of nanotechnology. ACS Nano *2*, 395–402.

Jani, P., Halbert, G.W., Landridge, J., and Florence, A.T. (1990). Nanoparticles uptake by the rat gastrointestinal mucosa: quantitation and particle size dependency. J. Pharm. Pharmacol. *42*, 821–826.

Jevprasesphant, R., Penny, J., Jalal, R., Attwood, D., McKeown, N.B., and D'Emanuele, A. (2003). The influence of surface modification on the cytotoxicity of PAMAM dendrimers. Int. J. Pharm. *252*, 263–266.

Jia, G., Wang, H., Yan, L., Wang, X., Pei, R., Yan, T., Zhao, Y., and Guo, X. (2005). Cytotoxicity of carbon nanomaterials: single-wall nanotube, multi-wall nanotube, and fullerene. Environ. Sci. Technol. *39*, 1378–1383.

Kaiser, J.P., Wick, P., Manser, P., Spohn, P., and Bruinink, A. (2008). Single walled carbon nanotubes (SWCNT) affect cell physiology and cell architecture. J. Mater. Sci. Mater. Med. *19*, 1523–1527.

Kappus, H. (1987). Oxidative stress in chemical toxicity. Arch. Toxicol. *60*, 144–149.

Karlsson, H.L., Cronholm, P., Gustafsson, J., and Moller, L. (2008). Copper oxide nanoparticles are highly toxic: a comparison between metal oxide nanoparticles and carbon nanotubes. Chem. Res. Toxicol. *21*, 1726–1732.

Kim, D., El-Shall, H., Dennis, D., and Morey, T. (2005). Interaction of PLGA nanoparticles with human blood constituents. Colloids Surf. B Biointerfaces *40*, 83–91.

Kim, H.W., Ahn, E.K., Jee, B.K., Yoon, H.K., Lee, K.H., and Lim, Y. (2009). Nanoparticle-induced toxicity and related mechanisms *in vitro* and *in vivo*. J. Nanopart. Res. *11*, 55–65.

Kim, J.S., Yoon, T.J., Yu, K.N., Kim, B.G., Park, S.J., Kim, H.W., Lee, K.H., Park, S.B., Lee, J.K., and Cho, M.H. (2006a). Toxicity and tissue distribution of magnetic nanoparticles in mice. Toxicol. Sci. *89*, 338–347.

Kim, S., Lim, Y.T., Soltesz, e.g. De Grand, A.M., Lee, J., Nakayama, A., Parker, J.A., Mihaljevic, T., Laurence, R.G., Dor, D.M., Cohn, L.H., Bawendi, M.G., and Frangioni, J.V. (2004). Near-infrared fluorescent type II quantum dots for sentinel lymph node mapping. Nat. Biotechnol. *22*, 93–97.

Kim, T.H., Nah, J.W., Cho, M.H., Park, T.G., and Cho, C.S. (2006b). Receptor-mediated gene delivery into antigen presenting cells using mannosylated chitosan/DNA nanoparticles. J. Nanosci. Nanotechnol. *6*, 2796–2803.

Kim, Y.S., Kim, J.S., Cho, H.S., Rha, D.S., Kim, J.M., Park, J.D., Choi, B.S., Lim, R., Chang, H.K., Chung, Y.H., Kwon, I.H., Jeong, J., Han, B.S., and Yu, I.J. (2008). Twenty-eight-day oral toxicity, genotoxicity, and gender-related tissue distribution of silver nanoparticles in Sprague-Dawley rats. Inhal. Toxicol. *20*, 575–583.

Kirchner, C., Liedl, T., Kudera, S., Pellegrino, T., Munoz Javier, A., Gaub, H.E., Stolzle, S., Fertig, N., and Parak, W.J. (2005). Cytotoxicity of colloidal CdSe and CdSe/ZnS nanoparticles. Nano Lett. *5*, 331–338.

Kisin, E.R., Murray, A.R., Keane, M.J., Shi, X.C., Schwegler-Berry, D., Gorelik, O., Arepalli, S., Castranova, V., Wallace, W.E., Kagan, V.E., and Shvedova, A.A. (2007). Single-walled carbon nanotubes: geno- and cytotoxic effects in lung fibroblast V79 cells. J. Toxicol. Environ. Health A *70*, 2071–2079.

Koyama, S., Endo, M., Kim, Y.A., Hayashi, T., Yanagisawa, T., Osaka, K., Koyama, H., Haniu, H., and Kuroiwa, N. (2006). Role of systemic T-cells and histopathological aspects after subcutaneous implantation of various carbon nanotubes in mice. Carbon *44*, 1079–1092.

Kreyling, W.G., Semmler, M., Erbe, F., Mayer, P., Takenaka, S., Schulz, H., Oberdorster, G., and Ziesenis, A. (2002). Translocation of ultrafine insoluble iridium particles from lung epithelium to extrapulmonary organs is size dependent but very low. J. Toxicol. Environ. Health A *65*, 1513–1530.

Kroll, A., Pillukat, M.H., Hahn, D., and Schnekenburger, J. (2009). Current *in vitro* methods in nanoparticle risk assessment: Limitations and challenges. Eur. J. Pharm. Biopharm. *72*, 370–377.

Lai, J.C., Lai, M.B., Jandhyam, S., Dukhande, V.V., Bhushan, A., Daniels, C.K., and Leung, S.W. (2008). Exposure to titanium dioxide and other metallic oxide nanoparticles induces cytotoxicity on human neural cells and fibroblasts. Int. J. Nanomedicine *3*, 533–545.

Lam, C.W., James, J.T., McCluskey, R., and Hunter, R.L. (2004). Pulmonary toxicity of single-wall carbon nanotubes in mice 7 and 90 days after intratracheal instillation. Toxicol. Sci. *77*, 126–134.

Landsiedel, R., Kapp, M.D., Schulz, M., Wiench, K., and Oesch, F. (2009). Genotoxicity investigations on nanomaterials: methods, preparation and characterization of test material, potential artifacts and limitations--many questions, some answers. Mutat. Res. *681*, 241–258.

Larese, F.F., D'Agostin, F., Crosera, M., Adami, G., Renzi, N., Bovenzi, M., and Maina, G. (2009). Human skin penetration of silver nanoparticles through intact and damaged skin. Toxicology *255*, 33–37.

Lei, R., Wu, C., Yang, B., Ma, H., Shi, C., Wang, Q., Wang, Q., Yuan, Y., and Liao, M. (2008). Integrated metabolomic analysis of the nano-sized copper particle-induced hepatotoxicity and nephrotoxicity in rats: a rapid *in vivo* screening method for nanotoxicity. Toxicol. Appl. Pharmacol. *232*, 292–301.

Lewinski, N., Colvin, V., and Drezek, R. (2008). Cytotoxicity of nanoparticles. Small *4*, 26–49.

Li, N., Xia, T., and Nel, A.E. (2008a). The role of oxidative stress in ambient particulate matter-induced lung diseases and its implications in the toxicity of engineered nanoparticles. Free Radic. Biol. Med. *44*, 1689–1699.

Li, S.Q., Zhu, R.R., Zhu, H., Xue, M., Sun, X.Y., Yao, S.D., and Wang, S.L. (2008b). Nanotoxicity of TiO_2 nanoparticles to erythrocyte *in vitro*. Food Chem. Toxicol. *46*, 3626–3631.

Li, X., Zheng, H., Zhang, Z., Li, M., Huang, Z., Schluesener, H.J., Li, Y., and Xu, S. (2009). Glia activation induced by peripheral administration of alunium oxide nanoparticles in rat brains. Nanomedicine *5*, 473–479.

Libby, P. (2002). Inflammation in atherosclerosis. Nature *420*, 868–874.

Lilienblum, W., Dekant, W., Foth, H., Gebel, T., Hengstler, J.G., Kahl, R., Kramer, P.J., Schweinfurth, H., and Wollin, K.M. (2008). Alternative methods to safety studies in experimental animals: role in the risk assessment of chemicals under the new European Chemicals Legislation (REACH). Arch. Toxicol. *82*, 211–236.

Lin, W., Huang, Y.W., Zhou, X.D., and Ma, Y. (2006). Toxicity of cerium oxide nanoparticles in human lung cancer cells. Int. J. Toxicol. *25*, 451–457.

Lin, W.S., Xu, Y., Huang, C.C., Ma, Y.F., Shannon, K.B., Chen, D.R., and Huang, Y.W. (2009). Toxicity of nano- and micro-sized ZnO particles in human lung epithelial cells. Journal of Nanoparticle Research *11*, 25–39.

Linse, S., Cabaleiro-Lago, C., Xue, W.F., Lynch, I., Lindman, S., Thulin, E., Radford, S.E., and Dawson, K.A. (2007). Nucleation of protein fibrillation by nanoparticles. Proc. Natl. Acad. Sci. U. S. A. *104*, 8691–8696.

Liu, R., Yin, L.H., Pu, Y.P., Liang, G.Y., Zhang, J., Su, Y.Y., Xiao, Z.P., and Ye, B. (2009). Pulmonary toxicity induced by three forms of titanium dioxide nanoparticles via intra-tracheal instillation in rats. Prog. Nat. Sci. *19*, 573–579.

Liu, Y.Y., Liu, D.M., Chen, S.Y., Tung, T.H., and Liu, T.Y. (2008a). In situ synthesis of hybrid nanocomposite with highly order arranged amorphous metallic copper nanoparticle in poly(2-hydroxyethyl methacrylate) and its potential for blood-contact uses. Acta Biomater. *4*, 2052–2058.

Liu, Y.Y., Liu, T.Y., Chen, S.Y., and Liu, D.M. (2008b). Synthesis and characterization of nanoporous SiO_2/pHEMA biocomposites. J. Mater. Sci. Mater. Med. *19*, 2903–2911.

Loo, C., Lin, A., Hirsch, L., Lee, M.H., Barton, J., Halas, N., West, J., and Drezek, R. (2004). Nanoshell-enabled photonics-based imaging and therapy of cancer. Technol. Cancer. Res. Treat. *3*, 33–40.

Loo, C., Lowery, A., Halas, N., West, J., and Drezek, R. (2005). Immunotargeted nanoshells for integrated cancer imaging and therapy. Nano Lett. *5*, 709–711.

Lundqvist, M., Stigler, J., Elia, G., Lynch, I., Cedervall, T., and Dawson, K.A. (2008). Nanoparticle size and surface properties determine the protein corona with possible implications for biological impacts. Proc. Natl. Acad. Sci. U. S. A. *105*, 14265–14270.

Lynch, I., and Dawson, K.A. (2008). Protein–nanoparticle interactions. Nano Today *3*, 40–47.

Lynch, I., Dawson, K.A., and Linse, S. (2006). Detecting cryptic epitopes created by nanoparticles. Sci. STKE *2006*, pe14.

Magrez, A., Kasas, S., Salicio, V., Pasquier, N., Seo, J.W., Celio, M., Catsicas, S., Schwaller, B., and Forro, L. (2006). Cellular toxicity of carbon-based nanomaterials. Nano Lett. *6*, 1121–1125.

Ma-Hock, L., Burkhardt, S., Strauss, V., Gamer, A.O., Wiench, K., van Ravenzwaay, B., and Landsiedel, R. (2009). Development of a short-term inhalation test in the rat using nano-titanium dioxide as a model substance. Inhal. Toxicol. *21*, 102–118.

Manna, S.K., Sarkar, S., Barr, J., Wise, K., Barrera, E.V., Jejelowo, O., Rice-Ficht, A.C., and Ramesh, G.T. (2005). Single-walled carbon nanotube induces oxidative stress and activates nuclear transcription factor-kappaB in human keratinocytes. Nano Lett. *5*, 1676–1684.

Mayer, A., Muhry, N., Novak, A., and Fröhlich, E. (2008). Hemocompatibility of various nanoparticles in human blood. Toxicol. Lett. *180*, 223–224.

Mayer, A., Vadon, M., Rinner, B., Novak, A., Wintersteiger, R., and Frohlich, E. (2009). The role of nanoparticle size in hemocompatibility. Toxicology *258*, 139–147.

Mercer, R.R., Scabilloni, J., Wang, L., Kisin, E., Murray, A.R., Schwegler-Berry, D., Shvedova, A.A., and Castranova, V. (2008). Alteration of deposition pattern and pulmonary response as a result of improved dispersion of aspirated single-walled carbon nanotubes in a mouse model. Am. J. Physiol. Lung Cell. Mol. Physiol. *294*, L87–97.

Mills, N.L., Donaldson, K., Hadoke, P.W., Boon, N.A., MacNee, W., Cassee, F.R., Sandstrom, T., Blomberg, A., and Newby, D.E. (2009). Adverse cardiovascular effects of air pollution. Nat. Clin. Pract. Cardiovasc. Med. *6*, 36–44.

Mitchell, L.A., Gao, J., Wal, R.V., Gigliotti, A., Burchiel, S.W., and McDonald, J.D. (2007). Pulmonary and systemic immune response to inhaled multiwalled carbon nanotubes. Toxicol. Sci. *100*, 203–214.

Monteiro-Riviere, N.A., Inman, A.O., and Zhang, L.W. (2009). Limitations and relative utility of screening assays to assess engineered nanoparticle toxicity in a human cell line. Toxicol. Appl. Pharmacol. *234*, 222–235.

Monteiro-Riviere, N.A., Nemanich, R.J., Inman, A.O., Wang, Y.Y., and Riviere, J.E. (2005). Multi-walled carbon nanotube interactions with human epidermal keratinocytes. Toxicol. Lett. *155*, 377–384.

Mori, T., Takada, H., Ito, S., Matsubayashi, K., Miwa, N., and Sawaguchi, T. (2006). Preclinical studies on safety of fullerene upon acute oral administration and evaluation for no mutagenesis. Toxicology *225*, 48–54.

Moussa, F., Trivin, F., Ceolin, R., Hadchouel, M., Sizaret, P.Y., Greugny, V., Fabre, C., Rassat, A., and Szwarc, H. (1996). Early effects of C-60 administration in Swiss mice: A preliminary account for *in vivo* C-60 toxicity. Fullerene Sci. Technol. *4*, 21–29.

Muller, J., Decordier, I., Hoet, P.H., Lombaert, N., Thomassen, L., Huaux, F., Lison, D., and Kirsch-Volders, M. (2008). Clastogenic and aneugenic effects of multi-wall carbon nanotubes in epithelial cells. Carcinogenesis *29*, 427–433.

Muller, J., Huaux, F., Moreau, N., Misson, P., Heilier, J.F., Delos, M., Arras, M., Fonseca, A., Nagy, J.B., and Lison, D. (2005). Respiratory toxicity of multi-wall carbon nanotubes. Toxicol. Appl. Pharmacol. *207*, 221–231.

Muller, K., Skepper, J.N., Posfai, M., Trivedi, R., Howarth, S., Corot, C., Lancelot, E., Thompson, P.W., Brown, A.P., and Gillard, J.H. (2007). Effect of ultrasmall superparamagnetic iron oxide nanoparticles (Ferumoxtran-10) on human monocyte-macrophages *in vitro*. Biomaterials *28*, 1629–1642.

Murr, L.E., Garza, K.M., Soto, K.F., Carrasco, A., Powell, T.G., Ramirez, D.A., Guerrero, P.A., Lopez, D.A., and Venzor, J., 3rd. (2005). Cytotoxicity assessment of some carbon nanotubes and related carbon nanoparticle aggregates and the implications for anthropogenic carbon nanotube aggregates in the environment. Int. J. Environ. Res. Public. Health. *2*, 31–42.

Murray, A.R., Kisin, E., Leonard, S.S., Young, S.H., Kommineni, C., Kagan, V.E., Castranova, V., and Shvedova, A.A. (2009). Oxidative stress and inflammatory response in dermal toxicity of single-walled carbon nanotubes. Toxicology *257*, 161–171.

Nakagawa, Y., Wakuri, S., Sakamoto, K., and Tanaka, N. (1997). The photogenotoxicity of titanium dioxide particles. Mutat. Res. *394*, 125–132.

Nel, A., Xia, T., Madler, L., and Li, N. (2006). Toxic potential of materials at the nanolevel. Science *311*, 622–627.

Nemmar, A., Hoet, P.H.M., Vanquickenborne, B., Dinsdale, D., Thomeer, M., Hoylaerts, M.F., Vanbilloen, H., Mortelmans, L., and Nemery, B. (2002a). Passage of inhaled particles into the blood circulation in humans. Circulation *105*, 411–414.

Nemmar, A., Hoylaerts, M.F., Hoet, P.H., Dinsdale, D., Smith, T., Xu, H., Vermylen, J., and Nemery, B. (2002b). Ultrafine particles affect experimental thrombosis in an *in vivo* hamster model. Am. J. Respir. Crit. Care Med. *166*, 998–1004.

Niidome, T., Yamagata, M., Okamoto, Y., Akiyama, Y., Takahashi, H., Kawano, T., Katayama, Y., and Niidome, Y. (2006). PEG-modified gold nanorods with a stealth character for *in vivo* applications. J. Control. Release *114*, 343–347.

Niwa, Y., and Iwai, N. (2006). Genotoxicity in cell lines induced by chronic exposure to water-soluble fullerenes using micronucleus test. Environmental Health and Preventive Medicine *11*, 292–297.

Oberdörster, E. (2004). Manufactured nanomaterials (fullerenes, C_{60}) induce oxidative stress in the brain of juvenile largemouth bass. Environ. Health Perspect. *112*, 1058–1062.

Oberdörster, G. (2000). Toxicology of ultrafine particles: *in vivo* studies. Philosophical Transactions of the Royal Society of London Series A-Mathematical Physical and Engineering Sciences *358*, 2719–2739.

Oberdörster, G., Ferin, J., and Lehnert, B.E. (1994). Correlation between particle size, *in vivo* particle persistence, and lung injury. Environ. Health Perspect. *102 Suppl 5*, 173–179.

Oberdörster, G., Ferin, J., Gelein, R., Soderholm, S.C., and Finkelstein, J. (1992a). Role of the alveolar macrophage in lung injury: studies with ultrafine particles. Environ. Health Perspect. *97*, 193–199.

Oberdörster, G., Ferin, J., and Morrow, P.E. (1992b). Volumetric loading of alveolar macrophages (AM): a possible basis for diminished AM-mediated particle clearance. Exp. Lung Res. *18*, 87–104.

Oberdörster, G., Finkelstein, J.N., Johnston, C., Gelein, R., Cox, C., Baggs, R., and Elder, A.C. (2000). Acute pulmonary effects of ultrafine particles in rats and mice. Res. Rep. Health Eff. Inst. *96*, 5–74; disc. 75–86.

Oberdörster, G., Oberdörster, E., and Oberdörster, J. (2005). Nanotoxicology: an emerging discipline evolving from studies of ultrafine particles. Environ. Health Perspect. *113*, 823–839.

Oberdörster, G., Sharp, Z., Atudorei, V., Elder, A., Gelein, R., Kreyling, W., and Cox, C. (2004). Translocation of inhaled ultrafine particles to the brain. Inhal. Toxicol. *16*, 437–445.

Oberdörster, G., Sharp, Z., Atudorei, V., Elder, A., Gelein, R., Lunts, A., Kreyling, W., and Cox, C. (2002). Extrapulmonary translocation of ultrafine carbon particles following whole-body inhalation exposure of rats. J. Toxicol. Environ. Health A *65*, 1531–1543.

Oesterling, E., Chopra, N., Gavalas, V., Arzuaga, X., Lim, E.J., Sultana, R., Butterfield, D.A., Bachas, L., and Hennig, B. (2008). Alumina nanoparticles induce expression of endothelial cell adhesion molecules. Toxicol. Lett. *178*, 160–166.

Ostrowski, A.D., Martin, T., Conti, J., Hurt, I., and Harthorn, B.H. (2009). Nanotoxicology: characterizing the scientific literature, 2000–2007. J. Nanopart. Res. *11*, 251–257.

Papageorgiou, I., Brown, C., Schins, R., Singh, S., Newson, R., Davis, S., Fisher, J., Ingham, E., and Case, C.P. (2007). The effect of nano- and micron-sized particles of cobalt-chromium alloy on human fibroblasts *in vitro*. Biomaterials *28*, 2946–2958.

Park, E.J., and Park, K. (2009). Oxidative stress and pro-inflammatory responses induced by silica nanoparticles *in vivo* and *in vitro*. Toxicol. Lett. *184*, 18–25.

Park, E.J., Choi, J., Park, Y.K., and Park, K. (2008a). Oxidative stress induced by cerium oxide nanoparticles in cultured BEAS-2B cells. Toxicology *245*, 90–100.

Park, E.J., Yi, J., Chung, K.H., Ryu, D.Y., Choi, J., and Park, K. (2008b). Oxidative stress and apoptosis induced by titanium dioxide nanoparticles in cultured BEAS-2B cells. Toxicol. Lett. *180*, 222–229.

Pernodet, N., Fang, X., Sun, Y., Bakhtina, A., Ramakrishnan, A., Sokolov, J., Ulman, A., and Rafailovich, M. (2006). Adverse effects of citrate/gold nanoparticles on human dermal fibroblasts. Small *2*, 766–773.

Petri-Fink, A., Chastellain, M., Juillerat-Jeanneret, L., Ferrari, A., and Hofmann, H. (2005). Development of functionalized superparamagnetic iron oxide nanoparticles for interaction with human cancer cells. Biomaterials *26*, 2685–2694.

Pisanic, T.R., 2nd, Blackwell, J.D., Shubayev, V.I., Finones, R.R., and Jin, S. (2007). Nanotoxicity of iron oxide nanoparticle internalization in growing neurons. Biomaterials *28*, 2572–2581.

Podolski, I.Y., Podlubnaya, Z.A., Kosenko, E.A., Mugantseva, E.A., Makarova, e.g. Marsagishvili, L.G., Shpagina, M.D., Kaminsky, Y.G., Andrievsky, G.V., and Klochkov, V.K. (2007). Effects of hydrated forms of C60 fullerene on amyloid 1-peptide fibrillization *in vitro* and performance of the cognitive task. J. Nanosci. Nanotechnol. *7*, 1479–1485.

Poland, C.A., Duffin, R., Kinloch, I., Maynard, A., Wallace, W.A., Seaton, A., Stone, V., Brown, S., Macnee, W., and Donaldson, K. (2008). Carbon nanotubes introduced into the abdominal cavity of mice show asbestos-like pathogenicity in a pilot study. Nat. Nanotechnol. *3*, 423–428.

Polizzi, S., Pira, E., Ferrara, M., Bugiani, M., Papaleo, A., Albera, R., and Palmi, S. (2002). Neurotoxic effects of aluminium among foundry workers and Alzheimer's disease. Neurotoxicology *23*, 761–774.

Porter, A.E., Muller, K., Skepper, J., Midgley, P., and Welland, M. (2006). Uptake of C_{60} by human monocyte macrophages, its localization and implications for toxicity: studied by high resolution electron microscopy and electron tomography. Acta Biomater. *2*, 409–419.

Pulskamp, K., Diabate, S., and Krug, H.F. (2007). Carbon nanotubes show no sign of acute toxicity but induce intracellular reactive oxygen species in dependence on contaminants. Toxicol. Lett. *168*, 58–74.

Racette, B.A., Tabbal, S.D., Jennings, D., Good, L., Perlmutter, J.S., and Evanoff, B. (2005). Prevalence of parkinsonism and relationship to exposure in a large sample of Alabama welders. Neurology *64*, 230–235.

Rahman, Q., Lohani, M., Dopp, E., Pemsel, H., Jonas, L., Weiss, D.G., and Schiffmann, D. (2002). Evidence that ultrafine titanium dioxide induces micronuclei and apoptosis in Syrian hamster embryo fibroblasts. Environ. Health Perspect. *110*, 797–800.

Reddy, K.M., Feris, K., Bell, J., Wingett, D.G., Hanley, C., and Punnoose, A. (2007). Selective toxicity of zinc oxide nanoparticles to prokaryotic and eukaryotic systems. Appl. Phys. Lett. *90*, 2139021–2139023.

Rehn, B., Seiler, F., Rehn, S., Bruch, J., and Maier, M. (2003). Investigations on the inflammatory and genotoxic lung effects of two types of titanium dioxide: untreated and surface treated. Toxicol. Appl. Pharmacol. *189*, 84–95.

Robosky, L.C., Robertson, D.G., Baker, J.D., Rane, S., and Reily, M.D. (2002). *In vivo* toxicity screening programs using metabonomics. Comb. Chem. High Throughput Screen. *5*, 651–662.

Ross, R. (1999). Atherosclerosis is an inflammatory disease. Am. Heart J. *138*, S419–20.

Rouse, J.G., Yang, J., Barron, A.R., and Monteiro-Riviere, N.A. (2006). Fullerene-based amino acid nanoparticle interactions with human epidermal keratinocytes. Toxicol. In. Vitro. *20*, 1313–1320.

Ryman-Rasmussen, J.P., Riviere, J.E., and Monteiro-Riviere, N.A. (2007). Surface coatings determine cytotoxicity and irritation potential of quantum dot nanoparticles in epidermal keratinocytes. J. Invest. Dermatol. *127*, 143–153.

Ryman-Rasmussen, J.P., Riviere, J.E., and Monteiro-Riviere, N.A. (2006). Penetration of intact skin by quantum dots with diverse physicochemical properties. Toxicol. Sci. *91*, 159–165.

Saba, T.M. (1970). Physiology and physiopathology of the reticuloendothelial system. Arch. Intern. Med. *126*, 1031–1052.

Sato, K., Imai, Y., and Irimura, T. (1998). Contribution of dermal macrophage trafficking in the sensitization phase of contact hypersensitivity. J. Immunol. *161*, 6835–6844.

Sato, Y., Yokoyama, A., Shibata, K., Akimoto, Y., Ogino, S., Nodasaka, Y., Kohgo, T., Tamura, K., Akasaka, T., Uo, M. *et al.* (2005). Influence of length on cytotoxicity of multi-walled carbon nanotubes against human acute monocytic leukemia cell line THP-1 *in vitro* and subcutaneous tissue of rats *in vivo*. Mol. Biosyst. *1*, 176–182.

Sayes, C.M., Fortner, J.D., Guo, W., Lyon, D., Boyd, A.M., Ausman, K.D., Tao, Y.J., Sitharaman, B., Wilson, L.J., Hughes, J.B., West, J.L., and Colvin, V.L. (2004). The differential cytotoxicity of water-soluble fullerenes. Nano Letters *4*, 1881–1887.

Sayes, C.M., Gobin, A.M., Ausman, K.D., Mendez, J., West, J.L., and Colvin, V.L. (2005). Nano-C_{60} cytotoxicity is due to lipid peroxidation. Biomaterials *26*, 7587–7595.

Sayes, C.M., Liang, F., Hudson, J.L., Mendez, J., Guo, W., Beach, J.M., Moore, V.C., Doyle, C.D., West, J.L., Billups, W.E., Ausman, K.D., and Colvin, V.L. (2006a). Functionalization density dependence of single-walled carbon nanotubes cytotoxicity *in vitro*. Toxicol. Lett. *161*, 135–142.

Sayes, C.M., Marchione, A.A., Reed, K.L., and Warheit, D.B. (2007a). Comparative pulmonary toxicity assessments of C60 water suspensions in rats: few differences in fullerene toxicity *in vivo* in contrast to *in vitro* profiles. Nano Lett. *7*, 2399–2406.

Sayes, C.M., Reed, K.L., and Warheit, D.B. (2007b). Assessing toxicity of fine and nanoparticles: comparing *in vitro* measurements to *in vivo* pulmonary toxicity profiles. Toxicol. Sci. *97*, 163–180.

Sayes, C.M., Wahi, R., Kurian, P.A., Liu, Y., West, J.L., Ausman, K.D., Warheit, D.B., and Colvin, V.L. (2006b). Correlating nanoscale titania structure with toxicity: a cytotoxicity and inflammatory response study with human dermal fibroblasts and human lung epithelial cells. Toxicol. Sci. *92*, 174–185.

Schnekenbuger, J., and Kemper, B. (2008). Digital holography imaging – a 3D view on drug induced apoptosis. Screen. Trends Drug. Discov. *9*, 24–26.

Schulze, C., Kroll, A., Lehr, C.M., Schafer, U.F., Becker, K., Schnekenburger, J., Isfort, C.S., Landsiedel, R., and Wohlleben, W. (2008). Not ready to use – overcoming pitfalls when dispersing nanoparticles in physiological media. Nanotoxicology *2*, 51-U17.

Semmler, M., Seitz, J., Erbe, F., Mayer, P., Heyder, J., Oberdorster, G., and Kreyling, W.G. (2004). Long-term clearance kinetics of inhaled ultrafine insoluble iridium particles from the rat lung, including transient translocation into secondary organs. Inhal. Toxicol. *16*, 453–459.

Sharma, H.S., and Sharma, A. (2007). Nanoparticles aggravate heat stress induced cognitive deficits, blood–brain barrier disruption, edema formation and brain pathology. Prog. Brain Res. *162*, 245–273.

Sharma, V., Shukla, R.K., Saxena, N., Parmar, D., Das, M., and Dhawan, A. (2009). DNA damaging potential of zinc oxide nanoparticles in human epidermal cells. Toxicol. Lett. *185*, 211–218.

Shcharbin, D., Jokiel, M., Klajnert, B., and Bryszewska, M. (2006). Effect of dendrimers on pure acetylcholinesterase activity and structure. Bioelectrochemistry *68*, 56–59.

Shenoy, D., Fu, W., Li, J., Crasto, C., Jones, G., DiMarzio, C., Sridhar, S., and Amiji, M. (2006). Surface functionalization of gold nanoparticles using hetero-bifunctional poly(ethylene glycol) spacer for intracellular tracking and delivery. Int. J. Nanomedicine *1*, 51–57.

Shukla, M.K., and Leszczynski, J. (2009). Fullerene (C-60) forms stable complex with nucleic acid base guanine. Chemical Physics Letters *469*, 207–209.

Shukla, R., Bansal, V., Chaudhary, M., Basu, A., Bhonde, R.R., and Sastry, M. (2005a). Biocompatibility of gold nanoparticles and their endocytotic fate inside the cellular compartment: a microscopic overview. Langmuir *21*, 10644–10654.

Shukla, S., Priscilla, A., Banerjee, M., Bhonde, R.R., Ghatak, J., Satyam, P.V., and Sastry, M. (2005b). Porous gold nanospheres by controlled transmetalation reaction: A novel material for application in cell imaging. Chem. Mater. *17*, 5000–5005.

Shvedova, A.A., Castranova, V., Kisin, E.R., Schwegler-Berry, D., Murray, A.R., Gandelsman, V.Z., Maynard, A., and Baron, P. (2003). Exposure to carbon nanotube material: assessment of nanotube cytotoxicity using human keratinocyte cells. J. Toxicol. Environ. Health A *66*, 1909–1926.

Shvedova, A.A., Fabisiak, J.P., Kisin, E.R., Murray, A.R., Roberts, J.R., Tyurina, Y.Y., Antonini, J.M., Feng, W.H., Kommineni, C., Reynolds, J., Barchowsky, A., Castranova, V., and Kagan, V.E. (2008a). Sequential exposure to carbon nanotubes and bacteria enhances pulmonary inflammation and infectivity. Am. J. Respir. Cell Mol. Biol. *38*, 579–590.

Shvedova, A.A., Kisin, E., Murray, A.R., Johnson, V.J., Gorelik, O., Arepalli, S., Hubbs, A.F., Mercer, R.R., Keohavong, P., Sussman, N., Jin, J., Yin, J.., Stone, S., Chen, B.T., Deye, G., Maynard, A., Castranova, V., Baron, P.A., and Kagan, V.E. (2008b). Inhalation vs. aspiration of single-walled carbon nanotubes in C57BL/6 mice: inflammation, fibrosis, oxidative stress, and mutagenesis. Am. J. Physiol. Lung Cell. Mol. Physiol. *295*, L552–65.

Shvedova, A.A., Kisin, E.R., Mercer, R., Murray, A.R., Johnson, V.J., Potapovich, A.I., Tyurina, Y.Y., Gorelik, O., Arepalli, S., Schwegler-Berry, D., Hubbs, A.F., Antonini, J., Evans, D.E., Ku, B.K., Ramsey, D., Maynard, A., Kagan, V.E., Castranova, V., and Baron, P. (2005). Unusual inflammatory and fibrogenic pulmonary responses to single-walled carbon nanotubes in mice. Am. J. Physiol. Lung Cell. Mol. Physiol. *289*, L698–708.

Simon-Deckers, A., Gouget, B., Mayne-L'hermite, M., Herlin-Boime, N., Reynaud, C., and Carriere, M. (2008). *In vitro* investigation of oxide nanoparticle and carbon nanotube toxicity and intracellular accumulation in A549 human pneumocytes. Toxicology *253*, 137–146.

Stern, S.T., and Johnson, D.N. (2008). Role for nanomaterial–autophagy interaction in neurodegenerative disease. Autophagy *4*, 1097–1100.

Stern, S.T., and McNeil, S.E. (2008). Nanotechnology safety concerns revisited. Toxicol. Sci. *101*, 4–21.

Stern, S.T., Zolnik, B.S., McLeland, C.B., Clogston, J., Zheng, J., and McNeil, S.E. (2008). Induction of autophagy in porcine kidney cells by quantum dots: a common cellular response to nanomaterials? Toxicol. Sci. *106*, 140–152.

Su, C.H., Sheu, H.S., Lin, C.Y., Huang, C.C., Lo, Y.W., Pu, Y.C., Weng, J.C., Shieh, D.B., Chen, J.H., and Yeh, C.S. (2007). Nanoshell magnetic resonance imaging contrast agents. J. Am. Chem. Soc. *129*, 2139–2146.

Suh, W.H., Suslick, K.S., Stucky, G.D., and Suh, Y.H. (2009). Nanotechnology, nanotoxicology, and neuroscience. Prog. Neurobiol. *87*, 133–170.

Takahashi, H., Niidome, Y., Niidome, T., Kaneko, K., Kawasaki, H., and Yamada, S. (2006). Modification of gold nanorods using phosphatidylcholine to reduce cytotoxicity. Langmuir *22*, 2–5.

Theogaraj, E., Riley, S., Hughes, L., Maier, M., and Kirkland, D. (2007). An investigation of the photo-clastogenic potential of ultrafine titanium dioxide particles. Mutat. Res. *634*, 205–219.

Thomas, M., and Klibanov, A.M. (2003). Conjugation to gold nanoparticles enhances polyethylenimine's transfer of plasmid DNA into mammalian cells. Proc. Natl. Acad. Sci. U. S. A. *100*, 9138–9143.

Tian, F., Cui, D., Schwarz, H., Estrada, G.G., and Kobayashi, H. (2006). Cytotoxicity of single-wall carbon nanotubes on human fibroblasts. Toxicol. In. Vitro. *20*, 1202–1212.

Tinkle, S.S., Antonini, J.M., Rich, B.A., Roberts, J.R., Salmen, R., DePree, K., and Adkins, E.J. (2003). Skin as a route of exposure and sensitization in chronic beryllium disease. Environ. Health Perspect. *111*, 1202–1208.

Tkachenko, A.G., Xie, H., Liu, Y., Coleman, D., Ryan, J., Glomm, W.R., Shipton, M.K., Franzen, S., and Feldheim, D.L. (2004). Cellular trajectories of peptide-modified gold particle complexes: comparison of nuclear localization signals and peptide transduction domains. Bioconjug. Chem. *15*, 482–490.

Ueng, T.H., Kang, J.J., Wang, H.W., Cheng, Y.W., and Chiang, L.Y. (1997). Suppression of microsomal cytochrome P450-dependent monooxygenases and mitochondrial oxidative phosphorylation by fullerenol, a polyhydroxylated fullerene C60. Toxicol. Lett. *93*, 29–37.

Vamanu, C.I., Cimpan, M.R., Hol, P.J., Sornes, S., Lie, S.A., and Gjerdet, N.R. (2008). Induction of cell death by TiO_2 nanoparticles: studies on a human monoblastoid cell line. Toxicol. In. Vitro. *22*, 1689–1696.

van Ravenzwaay, B., Landsiedel, R., Fabian, E., Burkhardt, S., Strauss, V., and Ma-Hock, L. (2009). Comparing fate and effects of three particles of different surface properties: Nano-TiO_2, pigmentary TiO_2 and quartz. Toxicol. Lett. *186*, 152–159.

Walker, V.G., Li, Z., Hulderman, T., Schwegler-Berry, D., Kashon, M.L., and Simeonova, P.P. (2009). Potential *in vitro* effects of carbon nanotubes on human aortic endothelial cells. Toxicol. Appl. Pharmacol. *236*, 319–328.

Wang, B., Feng, W.Y., Wang, T.C., Jia, G., Wang, M., Shi, J.W., Zhang, F., Zhao, Y.L., and Chai, Z.F. (2006). Acute toxicity of nano- and micro-scale zinc powder in healthy adult mice. Toxicol. Lett. *161*, 115–123.

Wang, H., Wang, J., Deng, X., Sun, H., Shi, Z., Gu, Z., Liu, Y., and Zhao, Y. (2004). Biodistribution of carbon single-wall carbon nanotubes in mice. J. Nanosci. Nanotechnol. *4*, 1019–1024.

Wang, J., Chen, C., Liu, Y., Jiao, F., Li, W., Lao, F., Li, Y., Li, B., Ge, C., Zhou, G., Gao, Y., Zhao, Y., and Chai, Z. (2008a). Potential neurological lesion after nasal instillation of TiO_2 nanoparticles in the anatase and rutile crystal phases. Toxicol. Lett. *183*, 72–80.

Wang, J., Liu, Y., Jiao, F., Lao, F., Li, W., Gu, Y., Li, Y., Ge, C., Zhou, G., Li, B., Zhao. Y., Chai, Z., and Chen, C. (2008b). Time-dependent translocation and potential impairment on central nervous system by intranasally instilled TiO_2 nanoparticles. Toxicology *254*, 82–90.

Wang, J., Zhou, G., Chen, C., Yu, H., Wang, T., Ma, Y., Jia, G., Gao, Y., Li, B., Sun, J., Li, Y., Jiao, F., Zhao, Y., and Chai, Z. (2007a). Acute toxicity and biodistribution of different sized titanium dioxide particles in mice after oral administration. Toxicol. Lett. *168*, 176–185.

Wang, J.J., Sanderson, B.J., and Wang, H. (2007b). Cyto- and genotoxicity of ultrafine TiO_2 particles in cultured human lymphoblastoid cells. Mutat. Res. *628*, 99–106.

Wang, J.J., Sanderson, B.J., and Wang, H. (2007c). Cytotoxicity and genotoxicity of ultrafine crystalline SiO_2 particulate in cultured human lymphoblastoid cells. Environ. Mol. Mutagen. *48*, 151–157.

Wang, J.X., Li, Y.F., Li, W., Chen, C.Y., Li, B., and Zhao, Y.L. (2008c). Biological effect of intranasally instilled titanium dioxide nanoparticles on female mice. Nano. *3*, 279–285.

Wang, L., Mao, J., Zhang, G.H., and Tu, M.J. (2007d). Nano-cerium-element-doped titanium dioxide induces apoptosis of Bel 7402 human hepatoma cells in the presence of visible light. World J. Gastroenterol. *13*, 4011–4014.

Wang, L., Nagesha, D.K., Selvarasah, S., Dokmeci, M.R., and Carrier, R.L. (2008d). Toxicity of CdSe Nanoparticles in Caco-2 Cell Cultures. J. Nanobiotechnology *6*, 11.

Warheit, D.B., Brock, W.J., Lee, K.P., Webb, T.R., and Reed, K.L. (2005). Comparative pulmonary toxicity inhalation and instillation studies with different TiO_2 particle formulations: impact of surface treatments on particle toxicity. Toxicol. Sci. *88*, 514–524.

Warheit, D.B., Hoke, R.A., Finlay, C., Donner, E.M., Reed, K.L., and Sayes, C.M. (2007a). Development of a base set of toxicity tests using ultrafine TiO_2 particles as a component of nanoparticle risk management. Toxicol. Lett. *171*, 99–110.

Warheit, D.B., Webb, T.R., Colvin, V.L., Reed, K.L., and Sayes, C.M. (2007b). Pulmonary bioassay studies with nanoscale and fine-quartz particles in rats: toxicity is not dependent upon particle size but on surface characteristics. Toxicol. Sci. *95*, 270–280.

Warheit, D.B., Webb, T.R., Reed, K.L., Frerichs, S., and Sayes, C.M. (2007). Pulmonary toxicity study in rats with three forms of ultrafine-TiO_2 particles: differential responses related to surface properties. Toxicology *230*, 90–104.

Warheit, D.B., Webb, T.R., Sayes, C.M., Colvin, V.L., and Reed, K.L. (2006). Pulmonary instillation studies with nanoscale TiO_2 rods and dots in rats: toxicity is not dependent upon particle size and surface area. Toxicol. Sci. *91*, 227–236.

Weiss R. Nanotech product recalled in Germany. April 6, 2006. Washington Post.

Wick, P., Manser, P., Limbach, L.K., Dettlaff-Weglikowska, U., Krumeich, F., Roth, S., Stark, W.J., and Bruinink, A. (2007). The degree and kind of agglomeration affect carbon nanotube cytotoxicity. Toxicol. Lett. *168*, 121–131.

Witzmann, F.A., and Monteiro-Riviere, N.A. (2006). Multi-walled carbon nanotube exposure alters protein expression in human keratinocytes. Nanomedicine *2*, 158–168.

Wolinsky, H. (2006). Nanoregulation – A recent scare involving nanotech products reveals that the technology is not yet properly regulated. EMBO Rep. *7*, 858–861.

Worle-Knirsch, J.M., Pulskamp, K., and Krug, H.F. (2006). Oops they did it again! Carbon nanotubes hoax scientists in viability assays. Nano Lett. *6*, 1261–1268.

Wu, W.H., Sun, X., Yu, Y.P., Hu, J., Zhao, L., Liu, Q., Zhao, Y.F., and Li, Y.M. (2008). TiO_2 nanoparticles promote beta-amyloid fibrillation *in vitro*. Biochem. Biophys. Res. Commun. *373*, 315–318.

Xia, T., Kovochich, M., Brant, J., Hotze, M., Sempf, J., Oberley, T., Sioutas, C., Yeh, J.I., Wiesner, M.R., and Nel, A.E. (2006). Comparison of the abilities of ambient and manufactured nanoparticles to induce cellular toxicity according to an oxidative stress paradigm. Nano Lett. *6*, 1794–1807.

Xia, Z., Wang, G., Tao, K., Li, J., and Tian, Y. (2005). Preparation and acute toxicology of nano-magnetic ferrofluid. J. Huazhong Univ. Sci. Technolog. Med. Sci. *25*, 59–61.

Yamago, S., Tokuyama, H., Nakamura, E., Kikuchi, K., Kananishi, S., Sueki, K., Nakahara, H., Enomoto, S., and Ambe, F. (1995). *In vivo* biological behavior of a water-miscible fullerene: 14C labeling, absorption, distribution, excretion and acute toxicity. Chem. Biol. *2*, 385–389.

Yamawaki, H., and Iwai, N. (2006). Cytotoxicity of water-soluble fullerene in vascular endothelial cells. Am. J. Physiol. Cell. Physiol. *290*, C1495–502.

Yang, H., Liu, C., Yang, D., Zhang, H., and Xi, Z. (2009a). Comparative study of cytotoxicity, oxidative stress and genotoxicity induced by four typical nanomaterials: the role of particle size, shape and composition. J. Appl. Toxicol. *29*, 69–78.

Yang, W.J., Shen, C.C., Ji, Q.L., An, H.J., Wang, J.J., Liu, Q.D., and Zhang, Z.Z. (2009b). Food storage material silver nanoparticles interfere with DNA replication fidelity and bind with DNA. Nanotechnology *20*, 085102.

Yen, H.J., Hsu, S.H., and Tsai, C.L. (2009). Cytotoxicity and immunological response of gold and silver nanoparticles of different sizes. Small *13*, 1553–1561.

Yoshida, R., Kitamura, D., and Maenosono, S. (2009). Mutagenicity of water-soluble ZnO nanoparticles in Ames test. J. Toxicol. Sci. *34*, 119–122.

Yu, D.G., Lin, W.C., and Yang, M.C. (2007). Surface modification of poly(L-lactic acid) membrane via layer-by-layer assembly of silver nanoparticle-embedded polyelectrolyte multilayer. Bioconjug. Chem. *18*, 1521–1529.

Zhang, L.H., Wang, X., Stoltenberg, M., Danscher, G., Huang, L., and Wang, Z.Y. (2008a). Abundant expression of zinc transporters in the amyloid plaques of Alzheimer's disease brain. Brain Res. Bull. *77*, 55–60.

Zhang, L.W., Yu, W.W., Colvin, V.L., and Monteiro-Riviere, N.A. (2008b). Biological interactions of quantum dot nanoparticles in skin and in human epidermal keratinocytes. Toxicol. Appl. Pharmacol. *228*, 200–211.

Zhang, L.W., Zeng, L., Barron, A.R., and Monteiro-Riviere, N.A. (2007). Biological interactions of functionalized single-wall carbon nanotubes in human epidermal keratinocytes. Int. J. Toxicol. *26*, 103–113.

Zhang, Q., Kusaka, Y., Sato, K., Nakakuki, K., Kohyama, N., and Donaldson, K. (1998). Differences in the extent of inflammation caused by intratracheal exposure to three ultrafine metals: role of free radicals. J. Toxicol. Environ. Health A *53*, 423–438.

Zhang, T., Stilwell, J.L., Gerion, D., Ding, L., Elboudwarej, O., Cooke, P.A., Gray, J.W., Alivisatos, A.P., and Chen, F.F. (2006). Cellular effect of high doses of silica-coated quantum dot profiled with high throughput gene expression analysis and high content cellomics measurements. Nano Lett. *6*, 800–808.

Zhao, X., Striolo, A., and Cummings, P.T. (2005). C_{60} binds to and deforms nucleotides. Biophys. J. *89*, 3856–3862.

Zheng, Y., Hunting, D.J., Ayotte, P., and Sanche, L. (2008). Radiosensitization of DNA by gold nanoparticles irradiated with high-energy electrons. Radiat. Res. *169*, 19–27.

Zhu, L., Chang, D.W., Dai, L., and Hong, Y. (2007). DNA damage induced by multiwalled carbon nanotubes in mouse embryonic stem cells. Nano Lett. *7*, 3592–3597.

Zhu, S., Oberdorster, E., and Haasch, M.L. (2006). Toxicity of an engineered nanoparticle (fullerene, C60) in two aquatic species, Daphnia and fathead minnow. Mar. Environ. Res. *62 Suppl*, S5–9.

Zvyagin, A.V., Zhao, X., Gierden, A., Sanchez, W., Ross, J.A., and Roberts, M.S. (2008). Imaging of zinc oxide nanoparticle penetration in human skin *in vitro* and *in vivo*. J. Biomed. Opt. *13*, 064031.

Index